Scientific Computing

WITH CASE STUDIES

Scientific Computing
WITH CASE STUDIES

Dianne P. O'Leary

University of Maryland
College Park, Maryland

siam.

Society for Industrial and Applied Mathematics
Philadelphia

10 9 8 7 6 5 4 3 2 1

The images in Figure 1.1 were taken from http://nightglow.gsfc.nasa.gov/eric_journal_files/sydney_bridge.jpg and http://www.cpsc.gov/cpscpub/prerel/prhtml07/07267a.jpg

Figure 26.1 (http://www.myrmecos.net/insects/Tribolium1.html) is owned by Alex Wild.

Figures 11.1 and 11.2 were taken by Timothy O'Leary.

Library of Congress Cataloging-in-Publication Data

O'Leary, Dianne P.
Scientific computing with case studies / Dianne P. O'Leary.
p. cm.
Includes bibliographical references and index.
ISBN 978-0-898716-66-5
1. Mathematical models--Data processing--Case studies. I. Title.
QA401.O44 2008
510.285--dc22

2008031493

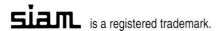

To Gene H. Golub, my first research mentor.
To my parents, Raymond and Anne Prost.
To my husband, Timothy.
To my children, Theresa, Thomas, and Brendan.
With love.

Contents

Preface

A master carpenter does not need to know how her hammer was designed or what Newton's laws say about the force that the hammer applies. But she does need to know how to use the hammer, when to use a ball-peen hammer instead, and what to do when things go wrong, for example, when a nail bends as it is driven.

We take the same viewpoint in this book. Although there are fascinating stories to tell in the details of how basic numerical algorithms are designed and how they operate, we view them as tools in our virtual toolbox, discussing the innards just enough to be able to master their uses. Instead we focus on how to choose the most appropriate algorithm, how to make use of it, how to evaluate the results, and what to do when things go wrong.

This viewpoint frees us to explore many diverse applications of our tools, and through such case studies we practice the analysis and experimentation that are the mainstays of computational science.

The reader should have background knowledge equivalent to a first course in scientific computing or numerical analysis. Excellent textbooks for learning this information include those by Michael Heath [71], Cleve Moler [108], and Charles Van Loan [148].

Examples and illustrations use the MATLAB® programming language. Standard MATLAB functions provide us with our basic numerical algorithms, and the graphics interface is quite useful. For some problems, we make use of some of the MATLAB toolboxes, in particular, the Optimization Toolbox. If you do not have access to MATLAB, the basic numerical algorithms can also be obtained from NETLIB and other sources noted in the text. Sample programs for each case study are available at the website

`www.cs.umd.edu/users/oleary/SCCS/`

No single book can give a computational scientist all of the background needed for a career. In fact, computational science is primarily a collaborative enterprise, since it is rare that a single individual has all of the computational and scientific background necessary to complete a project. My hope is that this particular slice of knowledge will prove useful in your work and will lead you to further study, exciting applications, and productive collaborations.

I'm grateful to my many mentors, collaborators, and students, who through their probing questions forced me to seek deeper understanding and clearer explanations. May you too be blessed with good colleagues.

Notes to Students

This book is written as a textbook for a second course in scientific computing, so it assumes that you have had a semester (or equivalent) of background using a standard textbook such as that by Heath [71], Moler [108], Van Loan [148], or equivalent. The Basics box at the beginning of each unit tells you what part of this material you might want to review in preparation for the unit. The Mastery box is a checklist of points to master in working through the unit.

The basic premise behind this book is that people learn by doing. Therefore, the book is best read with a pencil, paper, and MATLAB window close at hand. Challenges are sprinkled throughout the text, and they are meant be worked as they are encountered, or at least before the end of the chapter. Answers are provided for most challenges at

`www.cs.umd.edu/users/oleary/SCCS/`

There you can see examples of how someone else worked through the challenges. Mastery will be best if the answers are used to verify and refine your own approach to the problem. Merely reading the answer, though tempting, is (unfortunately) no substitute for trying to work the challenge on your own.

Pointers give important information and references to additional literature and software. I hope the content of this book leads you to want to learn more about scientific computing.

Notes to Instructors

The material in this book has been used for a semester and a half in a graduate level course in the applied mathematics program at the University of Maryland.

- I lecture from the introductory material in each unit, with material from the Case Studies used to occasionally provide extra information and motivation. Students can become quite passionate about some of the Case Studies, especially the more visual ones such as the image deblurring problem (Chapter 6), the data clustering problem (Chapter 11), and the epidemiology models (Chapter 19 and 21).

- For quizzes and exams, I derive problems from the Mastery points at the beginning of each unit.

- If possible, I like to allow "laboratory time" in class for students to work on some of the Challenges. The opportunity to see how other people solve problems is helpful even to the best students. This is especially true if, as at the University of Maryland, the students in this course come from backgrounds in mathematics, computer science, and engineering. This provides a remarkably diverse set of viewpoints on the material and enriches the dialog.

- Many of the **Case Studies** were originally homeworks.

- For a term project, I often ask students to develop a Case Study, using the tools presented in the course to solve a problem in their application area. Such projects can then be adapted for use in later terms. My students Nargess Memarsadeghi, David A. Schug, and Yalin E. Sagduyu developed particularly interesting case studies, and adapted versions of them are included here.

- There are not many unsolved exercises in this book. In the age of the Internet, there are very few textbook problems for which solutions cannot be found somewhere, and providing solutions here at least puts all students on equal footing. Some unsolved exercises and Case Studies are available on the book's website, and I would be grateful for your contribution of additional ones to post there.

There is a great deal of flexibility in choice and ordering of units, except that the optimization unit should be covered before nonlinear equations, and dense matrix computations should be discussed before optimization. The first six units form the syllabus for a one semester course at Maryland, while the final one is combined with a textbook in numerical solution of partial differential equations for the second semester.

Acknowledgments

I am grateful for the help of many, including the following:

- *Computing in Science and Engineering*, published by the American Institute of Physics and the IEEE Computer Society, for permission to include chapters derived from the case studies published there: Chapters
  1 (Vol. 8, No. 5, 2006, pp. 86–90),
  3 (Vol. 8, No. 3, 2006, pp. 86–89),
  4 (Vol. 7, No. 6, 2006, pp. 78–80),
  6 (Vol. 5, No. 3, 2003, pp. 82–85),
  7 (Vol. 8, No. 2, 2006, pp. 66–70),
  8 (Vol. 5, No. 6, 2003, pp. 60–63),
  11 (Vol. 5, No. 5, 2003, pp. 54–57),
  12 (Vol. 6, No. 5, 2004; pp. 60–62),
  13 (Vol. 6, No. 3, 2004, pp. 66–69),
  14 (Vol. 7, No. 1, 2005, pp. 56–59),
  15 (Vol. 7, No. 2, 2005, pp. 60–62),
  17 (Vol. 9, No. 1, 2007, pp. 72–76),
  18 (Vol. 6, No. 6, 2004; pp. 58–62),
  19 (Vol. 6, No. 1, 2004, pp. 68–70),
  21 (Vol. 6, No. 2, 2004, pp. 50–53),
  22 (Vol. 5, No. 4, 2003, pp. 68–71),
  23 (Vol. 7, No. 3, 2005, pp. 20–23),
  26 (Vol. 9, No. 2, 2007, pp. 96–99),
  27 (Vol. 7, No. 5, 2005, pp. 62–67),
  28 (Vol. 8, No. 4, 2006, pp. 74–78),
  29 (Vol. 6, No. 4, 2004, pp. 74–76),
  30 (Vol. 7, No. 6, 2005, pp. 74–77),
  31 (Vol. 7, No. 4, 2005, pp. 68–70),
  32 (Vol. 8, No. 5, 2006, pp. 86–90).

- Jennifer Stout, Lead Editor of *Computing in Science and Engineering*, who patiently edited the case studies.

- Mei Huang, for her work on Chapter 18.

- Jin Hyuk Jung, who as a teaching assistant wrote supplementary lecture notes from which some of the figures were taken, particularly those in Chapters 5, 9, and 24.
- Nargess Memarsadeghi, David Schug, and Yalin Sagduyu, whose term projects were so interesting that they led to case studies included here.
- Staff in the Technical Support Department at The MathWorks, for discussions about the sources of overhead in MATLAB interpreted and compiled instructions.
- James G. Nagy, a master teacher, who inspired the case studies and coauthored the first one.
- The National Science Foundation and the National Institute of Standards and Technology, for supporting my research into many of the problems discussed in the case studies.
- Timothy O'Leary for the photo of Charlie in Chapter 11.
- Students in the University of Maryland courses Scientific Computing I and II: (especially Samuel Lamphier) for their patience and debugging as the notes were developed.
- G. W. Stewart, for his example of clearly written textbooks and for the privilege of being his colleague at Maryland.
- Howard Elman, David Gilsinn, Vadim Kavalerov, Tamara Kolda, Samuel Lamphier, K.J.R. Liu, Brendan O'Leary, Bert Rust, Simon P. Schurr, Elisa Sotelino, G. W. Stewart, and Layne T. Watson for helpful comments.

The images in Figure 1.1 were taken from `http://nightglow.gsfc.nasa.gov/eric_journal_files/sydney_bridge.jpg` and `http://www.cpsc.gov/cpscpub/prerel/prhtml07/07267a.jpg`, and that in Figure 26.1 (`http://www.myrmecos.net/insects/Tribolium1.html`) is owned by Alex Wild.

Unit I

Preliminaries: Mathematical Modeling, Errors, Hardware and Software

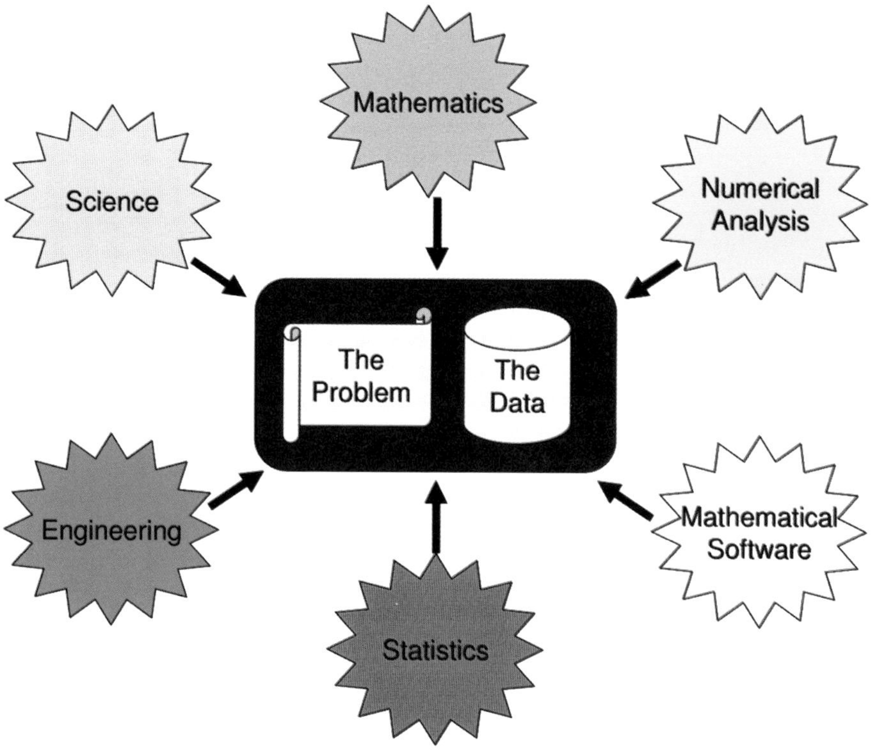

The topic of this book is efficient and accurate computation with mathematical models. In this unit, we discuss the basic facts that we need to know about error, software, and computers.

We begin our study with some basics. First in Chapter 1 we consider how errors are introduced in scientific computing and how to measure them. We apply these principles in Chapter 2, studying how small changes in our data can affect our answers. In Chapter 3, we see how computer memory is organized and how that impacts the efficiency of our algorithms. Then in Chapter 4, we study the principles behind writing and documenting our algorithms.

BASICS: To understand this unit, the following background is helpful:

- MATLAB programming [78].
- Gauss elimination; see a linear algebra textbook or a beginning book on numerical analysis or scientific computing [148].

MASTERY: After you have worked through this unit, you should be able to do the following:

- Identify the sources of error in scientific computing.
- Represent an integer in a fixed-point number system and a real number in a floating-point number system.
- Use the parameter ϵ_m (machine epsilon) to determine the error introduced in floating-point representation.
- Measure relative and absolute errors and determine how they are magnified during computation.
- Write algorithms that compute values such as the sum of a series, avoiding unnecessary inaccuracies.
- Determine ways to avoid catastrophic cancellation in designing algorithms.
- Use forward and backward error analysis to assess the quality of a computed solution to a problem.
- Determine whether a problem is well-conditioned or ill-conditioned.
- Discuss the importance of stability in an algorithm.
- Measure the sensitivity of a problem using derivatives, condition numbers, Monte Carlo experiments, and confidence intervals.
- Distinguish between a row-oriented matrix algorithm and a column-oriented matrix algorithm, and be able to write them for simple tasks.
- Explain how matrices are stored in main memory and moved to cache, and perform counts of page moves.
- Count the number of multiplications in a given MATLAB algorithm.
- Explain what the BLAS are and why they are useful.
- Document computer programs effectively.
- Understand the principles of modular design.
- Write a program to validate a function that you have written.

Chapter 1

Errors and Arithmetic

What better way to start a book than with error? We need to know how errors arise, how they are propagated in calculations, and how to measure and bound errors.

1.1 Sources of Error

Suppose an engineer wants to study the stresses in the bridge shown in Figure 1.1. The study would begin by gathering some data, including the lengths and angles for the girders and cables and the material properties for each component. There is some **measurement error**, though, since no measuring device gives full precision. Therefore, the measurements would typically be recorded as a value plus or minus an uncertainty.

The engineer would then need to model the stresses on the bridge. The bridge might be approximated by a "finite element model" to limit the number of unknowns in the problem, and this is an additional source of error. Simplifying assumptions might be made; for example, we might assume that the material in each girder is homogeneous. **Modeling error** is the result of the difference between the true bridge and our computable model.

Now we have a mathematical model, and we need to compute the stresses. If the model is large or nonlinear, then a numerical analyst might develop an algorithm that computes the solution as

$$\lim_{n \to \infty} G(n),$$

where, for example, $G(n)$ might be the result of n iterations of Newton's method. In general, we can't take this limit on a computer, so we might decide that $G(150)$ is good enough. This introduces **truncation error**.

Finally, we implement the algorithm and run it on our favorite computer. This introduces additional error, since we don't compute with real numbers but with **finite-precision** numbers: a fixed number of digits are carried in the computation. The effect of this is **rounding error**.

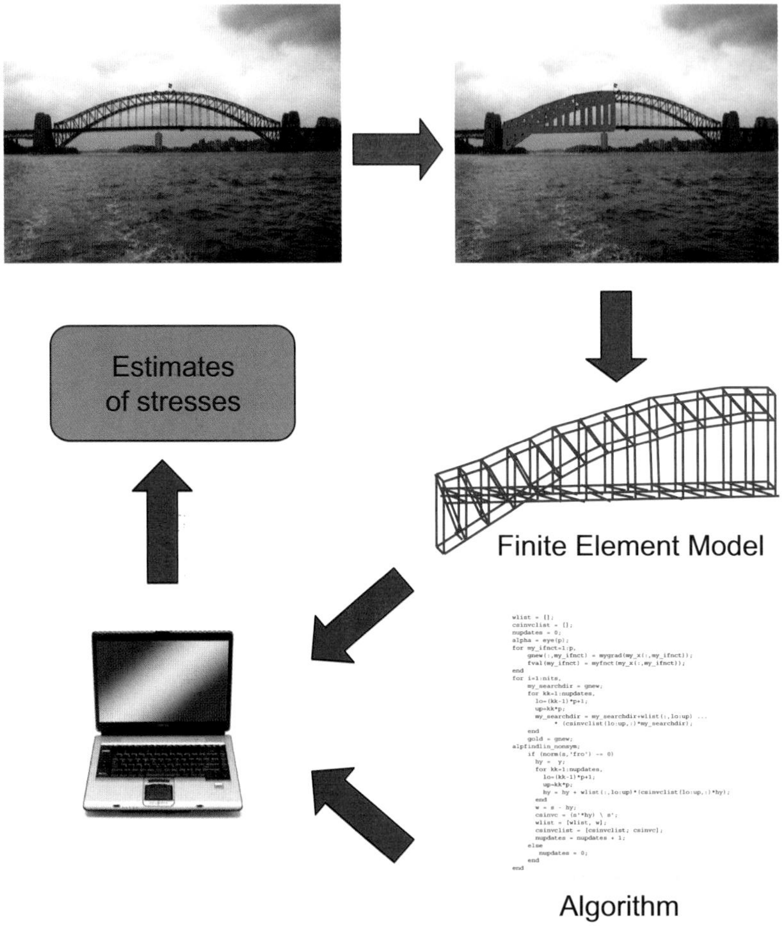

Figure 1.1. *Computing the stresses in a bridge involves measurement error, modeling error, truncation error, and rounding error.*

Therefore, the results obtained for the stresses on the bridge are contaminated by these four types of error: measurement error, modeling error, truncation error, and rounding error. It is important to note that no **mistakes** were made:

- The engineer did not misread the measurement device.
- The model was a good approximation to the true bridge.
- The programmer did not type the value of π incorrectly.
- The computer worked flawlessly.

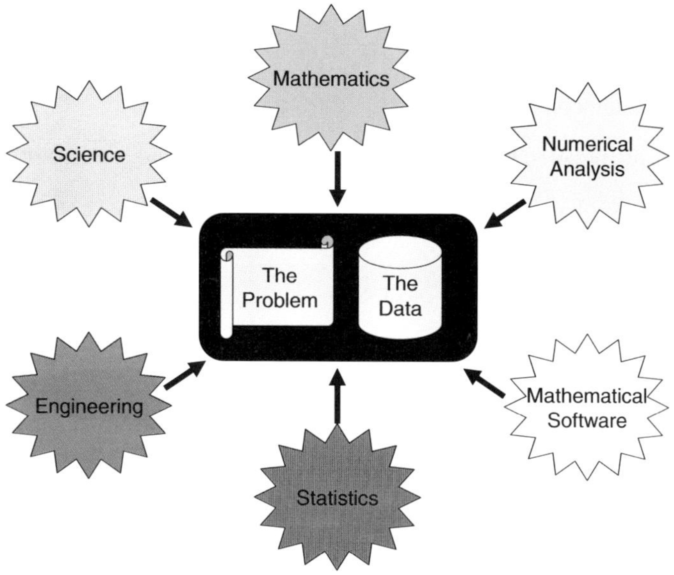

Figure 1.2. *Computational science involves knowledge from many disciplines.*

POINTER 1.1. Modeling the Error.

Developing a realistic understanding of the errors in the data is often the most challenging part of scientific computing. If you are solving a spectroscopy problem, for example, ideally you would first want to take a sample for which the composition is known and obtain several sets of sample data from the spectrometer. Using that data, you could develop a model for the error and see how your algorithms are affected by it.

But at the end of the process, the engineer needs to ask what the computed solution has to do with the stresses on the bridge!

1.2 Computational Science and Scientific Computing

In order to answer the question posed at the end of the previous section, we require several types of expertise. We use **science** and **engineering** to formulate the problem and determine what data is needed. We use **mathematics** and **statistics** to design the model. We use **numerical analysis** to design and analyze the algorithms, develop mathematical software, and answer questions about how accurate the final answer is. Therefore, our project could easily involve an interdisciplinary team of four or more experts; see Figure 1.2. Often, though, if the model is more or less routine, one person might do it all.

A **computational scientist** is a team member whose focus is on **scientific computing**: intelligently using mathematical software to analyze mathematical models. To do this requires a basic understanding of how computers do arithmetic.

POINTER 1.2. Matrix and Vector Notation.

Throughout this book we use the following notational conventions:

- All vectors are column vectors.
- Matrices are denoted by boldface upper case letters; vectors are boldface lower case.
- The elements of a matrix or vector are denoted by subscripted values: the element of $\boldsymbol{A}$ in row i and column j is a_{ij} or $A(i,j)$.
- The elements of matrices and vectors can be real or complex numbers.
- $\boldsymbol{I}$ is the identity matrix, and $\boldsymbol{e}_i$ is the ith column of $\boldsymbol{I}$.
- $\boldsymbol{B} = \boldsymbol{A}^T$ means that $\boldsymbol{B}$ is the **transpose** of $\boldsymbol{A}$: $b_{ij} = a_{ji}$. Therefore, $(\boldsymbol{A}^T)^T = \boldsymbol{A}$.
- $\boldsymbol{B} = \boldsymbol{A}^*$ means that $\boldsymbol{B}$ is the complex conjugate transpose of $\boldsymbol{A}$: $b_{ij} = \bar{a}_{ji}$, where the bar denotes complex conjugate. Therefore, $(\boldsymbol{A}^*)^* = \boldsymbol{A}$. If $\boldsymbol{A}$ is real, then $\boldsymbol{A}^* = \boldsymbol{A}^T$.
- We'll use MATLAB notation when convenient. For example, $\boldsymbol{A}(3:5, 1:7)$ denotes the submatrix of $\boldsymbol{A}$ with row entries between 3 and 5 and column entries between 1 and 7 (inclusive), and $\boldsymbol{A}(:,5)$ denotes column 5 of the matrix $\boldsymbol{A}$.

1.3 Computer Arithmetic

Computers use **binary arithmetic**, representing each number as a **binary number**, a finite sum of integer powers of 2. Some numbers can be represented exactly, but others, such as $1/10, 1/100, 1/1000, \ldots$, cannot. For example,

$$2.125 = 2^1 + 2^{-3}$$

has an exact representation in binary (base 2), but

$$3.1 \approx 2^1 + 2^0 + 2^{-4} + 2^{-5} + 2^{-8} + \cdots$$

does not. And, of course, there are numbers like π that have no finite representation in either our usual decimal number system or in binary.

Computers use two formats for numbers. **Fixed-point** numbers are used to store integers. Typically, each number is stored in a **computer word** consisting of 32 binary digits (**bits**) with values 0 and 1. Therefore, at most 2^{32} different numbers can be stored. If we allow for negative numbers, then we can, for example, represent integers in the range $-2^{31} \le x \le 2^{31} - 1$, since there are 2^{32} such numbers. Since $2^{31} \approx 2.1 \times 10^9$, the range for fixed-point numbers is too limited for scientific computing. Therefore, they are used mostly for indices and counters.

As an alternative to fixed-point numbers, **floating-point** numbers approximate real numbers. We'll discuss features of the most common floating-point number system, the IEEE Standard Floating Point Arithmetic.

The format for a floating-point number is

$$x = \pm z \times 2^p,$$

where z is called the **mantissa** or **significand**. This representation is not unique; for example,

$$1 \times 2^2 = 4 \times 2^0 = 8 \times 2^{-1}.$$

Therefore we make the rule that if $x \neq 0$, we **normalize** so that $1 \leq z < 2$, choosing the first of the three alternatives in the example.

To fit a floating-point number in a single word, we need to limit the number of digits in the mantissa and the exponent. For these **single-precision** numbers, 24 digits are used to represent the mantissa, and the exponent is restricted to the range $-126 \leq p \leq 127$. This allows us to represent numbers as close to zero as $2^{-126} \approx 1.18 \times 10^{-38}$ and as far as almost $2^{128} \approx 3.40 \times 10^{38}$, a considerably larger range than for fixed-point.

If this range is not large enough, or if 24 digits of precision are not enough, we turn to **double-precision** numbers, stored in two words, using 53 digits for the mantissa, with an exponent $-1022 \leq p \leq 1023$. This allows us to represent numbers as close to zero as $2^{-1022} \approx 2.23 \times 10^{-308}$ and as large as almost $2^{1024} \approx 1.80 \times 10^{308}$.

If we perform a computation in which the exponent of the answer is outside the allowed range, we have a more or less serious error.

- If the exponent is too big, then we cannot store the answer, and our computation has produced an **overflow** error. The answer is set to a special representation called `Inf` or `-Inf` to signal an error.

- If the exponent is too small, then the computation produced an **underflow**. If the number can be stored using the smallest possible exponent and a mantissa that is less than 1 in magnitude, then the IEEE Standard produces this as the answer.[1]

- If we divide zero by zero, then the answer is set to a code indicating **not-a-number**, `NaN`.

In double precision, at most 2^{64} different numbers can be represented (including `NaN` and $\pm$`Inf`) so any other number must be approximated by one of the representable numbers. For example, numbers in the range $1+2^{-53} \leq x < 1+(2^{-52}+2^{-53})$ might be rounded to $x_m = 1+2^{-52}$, which can be represented exactly. This introduces a very small error: the **absolute error** in the representation is

$$|x - x_m| \leq 2^{-53}.$$

Similarly, numbers in the range $1024+2^{-43} \leq x < 1024+(2^{-42}+2^{-43})$ might be rounded to $x_m = 1024+2^{-42}$, with absolute error bound 2^{-43}, which is 1024 times bigger than the bound for numbers near 1. In each case, though, the **relative error**

$$\frac{|x - x_m|}{|x|}$$

[1] Some hardware manufacturers implement this gradual underflow only in software, with the faster default option of setting the answer to zero, thus reducing the reliability of computations and causing difficulties with portability of software [121, Chap. 14].

POINTER 1.3. IEEE Standard Floating-Point Arithmetic.

Up until the mid-1980s, each computer manufacturer had a different representation for floating-point numbers and different rules for rounding the answer to a computation. Therefore, a program written for one machine would not compute the same answers on other machines. The situation improved somewhat with the introduction in 1987 of the **IEEE standard floating-point arithmetic**, now used by virtually all computers.

For more detailed information on floating-point computer arithmetic, see the excellent book by Overton [121]. In particular, a careful reader might note that we seem to be storing 33 bits of floating-point information in a 32 bit word, and the trick that enables us to avoid storing the leading bit in the mantissa is explained in that book.

POINTER 1.4. Internal Representation vs. Printed Numbers.

In interpreting MATLAB results, remember that if a number `x` is displayed as 1.0000, it is not necessarily equal to 1. All you know is that if you round the number to the nearest decimal number with 5 significant (trusted) digits, you get 1. If you want to see whether it equals 1 exactly, then display `x-1`. Alternatively, typing `format hex` changes the display to the internal machine representation.

---

is bounded by 2^{-53} when 53 digits are used for the mantissa.

Let's pause to consider the difference between the fixed-point number system and the floating-point number system.

CHALLENGE 1.1.

For each machine-representable number `r`, define `f(r)` to be the next larger machine-representable number. Consider the following statements:

(a) For fixed-point (integer) arithmetic, the distance between `r` and `f(r)` is constant.
(b) For floating-point arithmetic, the relative distance `|(f(r)-r)/r|` is constant (for `r` $\neq 0$).

Are the statements true or false? Give examples or counterexamples to explain your reasoning.[2]

This brings us to a very important parameter that characterizes machine precision: **machine epsilon** ϵ_m is defined as the gap between 1 and the next bigger number; for double precision, $\epsilon_m = 2^{-52}$. The relative error in rounding a number is bounded by $\epsilon_m/2$. Note that ϵ_m is much larger than the *smallest* positive number that the machine can store exactly!

[2]The solutions to challenges, except those marked "Extra," can be found on the book's website.

POINTER 1.5. Floating-Point Precision.

By default, MATLAB computes using double-precision floating-point numbers, and that is what we use in all of our computations.

There are two features of computer arithmetic that can make predictions of results difficult:

- Although floating-point arithmetic uses 64 bits to store a result, sometimes intermediate values are stored in 80 bits, giving them extra precision. For example, in the statement `z = a + b + c`, the value of `a + b` might be stored in 80 bits; then `c` would be added on, and the rounded result would be stored in 64 bits, specifying the value of `z`.

- In some languages such as C and FORTRAN, the sequence of arithmetic operations that you specify might be modified by the **compiler** (the software that translates your program into machine language) in order to shorten the computation time, making use of mathematical properties such as commutivity of addition and multiplication, or the distributive property of multiplication over addition. Since these properties do not always hold for floating-point arithmetic (See Challenge 1.4), such **optimization** of your program by the compiler can change the computed answer.

The next two challenges provide some practice with floating-point number systems, first in base 10 and then in base 2.

CHALLENGE 1.2.

Assume you have a base 10 computer that stores floating-point numbers using a 5 digit normalized mantissa (x.xxxx), a 4 digit exponent, and a sign for each.

(a) For this machine, what is machine epsilon?

(b) What is the smallest positive normalized number that can be represented exactly in this machine?

CHALLENGE 1.3.

Assume you have a base 2 computer that stores floating-point numbers using a 6 digit (bit) normalized mantissa (x.xxxxx), a 4 digit exponent, and a sign for each.

(a) For this machine, what is machine epsilon?

(b) What is the smallest positive normalized number that can be represented exactly in this machine?

(c) What mantissa and exponent are stored for the value 1/10? Hint:

$$\frac{1}{10} = \frac{1}{16} + \frac{1}{32} + \frac{1}{256} + \frac{1}{512} + \frac{1}{4096} + \frac{1}{8192} + \cdots.$$

We'll experiment a bit with the oddities of floating-point arithmetic.

CHALLENGE 1.4.

(a) Consider the following program:

```
x = 1;
delta = 1 / 2^(53);
for j=1:2^(20),
    x = x + delta;
end
```

Using mathematical reasoning, we expect the final value of x to be $1+2^{-33}$. Use your knowledge of floating-point arithmetic to predict what it actually is. Verify by running the program. Explain the result.

(b) Using mathematical reasoning, we know that for any positive number x, $2x$ is a number greater than x. Is this true of floating-point numbers? Run this program fragment and explain your result:

```
x = 1;
twox = 2 * x;
k = 0;
while (twox > x)
   x = twox;
   twox = 2*x;
   k = k + 1;
end
```

(c) Using mathematical reasoning, we know that addition and multiplication are commutative

$$x+y=y+x, \quad xy=yx,$$

and associative

$$((x+y)+z)=x+(y+z), \quad (xy)z=x(yz)$$

and that multiplication distributes over addition:

$$x(y+z)=xy+xz.$$

Give examples of floating-point numbers x, y, and z for which addition is not associative. Find a similar example for multiplication, and a third example showing that floating-point multiplication does not always distribute over addition. (Avoid expressions that evaluate to ±`Inf` or `NaN`, even though examples can be constructed using these values.)

(d) Write a MATLAB expression that gives an answer of `NaN` and one that gives `-Inf`.

(e) Given a floating-point number x, what is the distance between x and the next larger floating-point number? (Answer this either by analyzing the machine representation scheme or by experimenting in MATLAB.) Approximate your answer as a multiple of ϵ_m.

Our experiments have shown us the following:

- Unlike the fixed-point numbers, the numbers that we can store in floating-point representation are not equally spaced.
- When we do a floating-point operation (addition, subtraction, multiplication, or division), we get either exactly the right answer, or a rounded version of it, or `NaN`, or an indication of overflow.
- The main advantage of floating-point representation is the wide range of values that can be approximated with it.

Because of the errors introduced in floating-point computation, small changes in the way the data is stored can make large changes in the answer, as we see in the next challenge.

CHALLENGE 1.5.

Suppose we solve the linear system

$$Ax \equiv \begin{bmatrix} 2.00 & 1.00 \\ 1.99 & 1.00 \end{bmatrix} x = \begin{bmatrix} 1.00 \\ -1.00 \end{bmatrix} \equiv b.$$

Now suppose that the units for b_1 are centimeters, while the units for b_2 are meters. If we convert the problem to meters we obtain the linear system

$$Cz \equiv \begin{bmatrix} 0.02 & 0.01 \\ 1.99 & 1.00 \end{bmatrix} z = \begin{bmatrix} 0.01 \\ -1.00 \end{bmatrix} \equiv d.$$

Solve both systems in MATLAB using the backslash operator and explain why x is not exactly equal to z.

If all data were exact and if computers did their arithmetic using real numbers, then mathematical analysis would tell us all we need to know. Because of uncertainty in data and use of the floating-point number system, we need to understand how errors propagate through computation.

POINTER 1.6. Numerical Disasters.

Studying error propagation and catastrophic cancellation is not just an academic exercise. These kinds of errors have led to real disasters: explosion of the Ariane 5 rocket due to an overflow error, and many errors due to rounding, including the 1982 miscalculation of the index on the Vancouver stock exchange, the 1991 Patriot Missile failure in Saudi Arabia, and a vote counting error in a 1992 German election. See the websites of Douglas Arnold [3], Kees Vuik [149], and Thomas Huckle [82] for discussion of these and other examples. Walter Gander explains the strange "Heisenberg effects in computer arithmetic" arising from the difference in length between registers and words of memory [53].

1.4 How Errors Propagate

If answers to our calculations were always represented as the floating-point number closest to the true answer, then designing accurate algorithms would be easy. Unfortunately, the computed answer tends to drift away from the true answer due to accumulation of rounding error. This happens whenever the number of digits is limited, so for convenience, we'll look at examples in decimal arithmetic rather than binary.

Suppose we have measured the lengths of two cables (meters):

$$a = 2.003 \pm 0.001,$$
$$b = 2.000 \pm 0.001\,.$$

The absolute error in each measurement is bounded by 0.001, and the relative error in the second is at most $.001/1.999 \approx 0.05\%$ (since the true value is at least 1.999). The relative error in the first is also about 0.05%.

What can we conclude about the difference between the two values? The true difference is at most $2.004 - 1.999 = .005$ and at least $2.002 - 2.001 = .001$. We obtain the same information by taking the difference between the measurements and adding the uncertainties: $a - b = 0.003 \pm 0.002$.

When we subtracted the numbers, our bounds on the absolute errors were added. What happened to our bound on the relative error? If the true answer is 0.001, the relative error would be $(0.003 - 0.001)/0.001 = 200\%$. This enormous magnification of the relative error bound resulted from **catastrophic cancellation** of the significant (trusted) digits in the two measurements: although the measured values have 4 significant digits, the difference has only 1. Any subsequent computation involving this difference propagates the error.

We could generalize this example to prove a theorem: when adding or subtracting, the bounds on absolute error add.

What about multiplication and division?

CHALLENGE 1.6.

Suppose x and y are true (nonzero) values and $\tilde{x}$ and $\tilde{y}$ are our approximations to them. Let's express the errors as

$$\tilde{x} = x(1 - r),$$
$$\tilde{y} = y(1 - s).$$

(a) Show that the relative error in $\tilde{x}$ is $|r|$ and the relative error in $\tilde{y}$ is $|s|$.

(b) Show that we can bound the relative error in $\tilde{x}\tilde{y}$ as an approximation to xy by

$$\left|\frac{\tilde{x}\tilde{y}-xy}{xy}\right| \leq |r|+|s|+|rs|.$$

Since we expect the relative errors r and s to be much less than 1, the quantity $|rs|$ is expected to be very small compared to $|r|$ and $|s|$. Therefore, when multiplying or dividing, the bounds on relative errors (approximately) add.

Notice that these statements about errors after arithmetic operations assume that the computed solution is stored exactly; additional error may result from rounding to the nearest floating-point number.

CHALLENGE 1.7.

Consider the following MATLAB program:

```
x = .1;
sum = 0;
for i=1:100,
    sum = sum + x;
end
```

Is the final value of sum equal to 10? If not, why not?

In computations where error build-up can occur, it is good to rearrange the computation to avoid cancellation whenever possible. We'll consider a familiar example, finding the roots of a quadratic polynomial, next.

1.5 Mini Case Study: Avoiding Catastrophic Cancellation

Suppose we are asked to find the roots of the polynomial

$$x^2 - 56x + 1 = 0.$$

The usual formula, which you may have learned in an algebra class, computes

$$\begin{aligned} x_1 &= 28+\sqrt{783} &\approx 28+27.982 &= 55.982 \quad (\pm 0.0005), \\ x_2 &= 28-\sqrt{783} &\approx 28-27.982 &= 0.018 \quad (\pm 0.0005). \end{aligned}$$

The error arose from approximating $\sqrt{783}$ by its correctly rounded value, 27.982. The absolute error bounds are the same, but the relative error bounds are about 10^{-5} for x_1 and 0.02 for x_2 – vastly different!

The problem, of course, was catastrophic cancellation in the computation of x_2, and it is easy to convince yourself that for any quadratic with real roots, the quadratic formula causes some cancellation during the computation of one of the roots.

We can avoid this cancellation by using other facts about quadratic equations and about square roots. We consider three possibilities.

- Use an alternate formula. The product of the two roots equals the constant term in the polynomial,[3] so $x_1 x_2 = 1$. If we compute

$$x_2 = \frac{1}{x_1},$$

then our relative error is bounded by 10^{-5}, the relative error in our value for x_1, so we obtain

$$x_2 \approx .0178629(\pm 2 \times 10^{-7}),$$

accurate to the same relative error.

- Rewrite the formula. Notice that

$$x_2 = 28 - \sqrt{783} = \sqrt{784} - \sqrt{783}.$$

Let's derive a better formula for the difference of these square roots:

$$\begin{aligned}\sqrt{z+e} - \sqrt{z} &= (\sqrt{z+e} - \sqrt{z})\frac{\sqrt{z+e}+\sqrt{z}}{\sqrt{z+e}+\sqrt{z}}\\ &= \frac{z+e-z}{\sqrt{z+e}+\sqrt{z}}\\ &= \frac{e}{\sqrt{z+e}+\sqrt{z}}.\end{aligned}$$

Therefore, letting $z = 783$ and $e = 1$, we calculate

$$x_2 = \frac{1}{28+\sqrt{783}},$$

giving the same result as above but from a different approach.

- Use Taylor series. Let $f(x) = \sqrt{x}$. Then

$$f(z+e) - f(z) = f'(z)e + \frac{1}{2}f''(z)e^2 + \cdots,$$

so we can approximate the difference by $f'(z)e = 1/(2\sqrt{783})$.

CHALLENGE 1.8. (Extra)

Write a MATLAB function that computes the two roots of a quadratic polynomial with good relative precision.

---

[3] This is true since $x^2 - 56x + 1 = (x - x_1)(x - x_2)$.

POINTER 1.7. Symbolic Computation.

Some people claim that the pitfalls in floating-point arithmetic are best avoided by avoiding floating-point arithmetic altogether, and instead using **symbolic computation** systems such as MAPLE (`http://www.maplesoft.com`) (available with MATLAB) or MATHEMATICA (`www.wolfram.com`). These systems are incredibly useful, but eventually they produce a formula that needs to be evaluated using arithmetic. These systems have pitfalls of their own: the computation can use a tremendous amount of time and storage, and they can produce formulas that lead to unnecessarily high relative and absolute errors.

1.6 How Errors Are Measured

Error analysis determines the cumulative effects of error. We have been using forward error analysis, but there are alternatives, including backward error analysis.

- In **forward error analysis**, we find an estimate for the answer and bounds on the error. Schematically, we see in the top of Figure 1.3 that we have a space of all possible problems and a space of their solutions. We are given a problem whose true solution is unknown. We compute a solution and report that solution along with a bound on the distance between the computed solution and the true solution. For example, we might compute the answer 5.348 and determine that the true answer is $5.348 \pm .001$. Or, for a vector solution, we might report that $\|\boldsymbol{x}_c - \boldsymbol{x}_{true}\| \leq 10^{-5}$, where $\boldsymbol{x}_c$ is the computed solution and $\boldsymbol{x}_{true}$ is the true solution.

- In **backward error analysis**, we again are given a problem whose true solution is unknown. We compute a solution, and report that solution along with a bound on the difference between the problem we solved and the given problem. This is illustrated in the bottom of Figure 1.3.

Let's determine forward and backward error bounds for a simple problem.

CHALLENGE 1.9.

Suppose the sides of a rectangle have lengths $3.2 \pm .005$ and $4.5 \pm .005$. Consider approximating the area of the rectangle by $A = 14$.

(a) Give a forward error bound for A as an approximation to the true area.

(b) Give a backward error bound.

It might be hard to imagine a situation in which backward error analysis provides any useful information, but think back to our bridge. Suppose we compute a solution to a problem for which the measurements differ from our measurements by 10^{-5}. If the error bounds in our measurements are greater than 10^{-5}, then we may have computed the stresses for the true bridge! In any case, the solution we computed is as reasonable as one for any

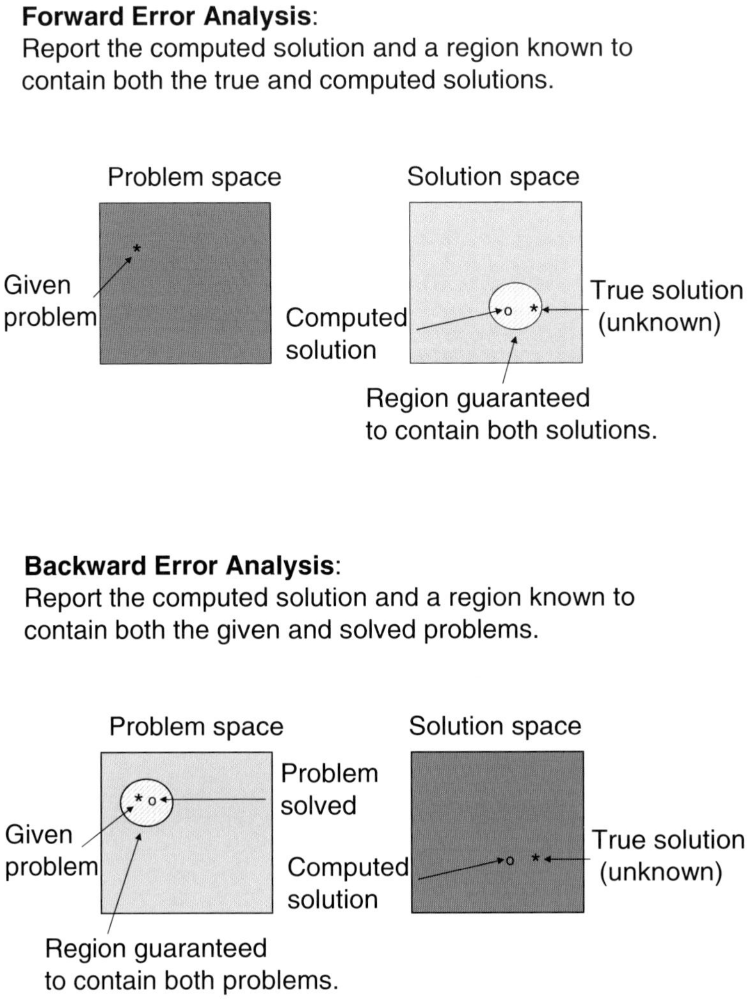

Figure 1.3. *In forward error analysis, we find bounds on the distance between the computed solution and the true solution. In backward error analysis, we find bounds on the distance between the problem we solved and the problem we wanted to solve.*

other problem in the uncertainty intervals, so we can be quite satisfied with the outcome. In general, backward error statements are quite useful when the data has uncertainty.

Backward error estimates also tend to be less pessimistic than forward error estimates, since they don't involve taking a worst-case bound after every computation. Backward error estimates are usually derived at the end of the algorithm. For example, if we compute an approximate solution $\boldsymbol{x}_c$ to a linear system of equations

$$\boldsymbol{A}\boldsymbol{x} = \boldsymbol{b},$$

then we can test how good it is by evaluating the **residual**

$$\boldsymbol{r} = \boldsymbol{b} - \boldsymbol{A}\boldsymbol{x}_c.$$

If $\boldsymbol{x}_c$ equals the true solution, then $\boldsymbol{r} = \boldsymbol{0}$; if it is a good approximation, then we expect $\boldsymbol{r} \approx \boldsymbol{0}$. In any case, we know that our computed solution $\boldsymbol{x}_c$ is the exact solution to the nearby problem

$$\boldsymbol{A}\boldsymbol{x}_c = \boldsymbol{b} - \boldsymbol{r},$$

so $\|\boldsymbol{r}\|$ gives us a backward error bound.

Here are two examples to provide some experience in computing error bounds.

CHALLENGE 1.10.

Bound the backward error in approximating the solution to

$$\begin{bmatrix} 2 & 1 \\ 3 & 6 \end{bmatrix} \begin{bmatrix} x_1 \\ x_2 \end{bmatrix} = \begin{bmatrix} 5.244 \\ 21.357 \end{bmatrix} \text{ by } \boldsymbol{x}_c = \begin{bmatrix} 1 \\ 3 \end{bmatrix}.$$

CHALLENGE 1.11.

Suppose that you have measured the length of the side of a cube as $(3.00 \pm .005)$ meters. Give an estimate of the volume of the cube and a (good) bound on the absolute error in your estimate.

1.7 Conditioning and Stability

It is important to distinguish between difficult problems and bad algorithms.

We say that a problem is **well-conditioned** if small changes in the data always make small changes in the solution; otherwise it is **ill-conditioned**. Similarly, an algorithm is **stable** if it always produces the solution to a nearby problem, and **unstable** otherwise.[4]

To illustrate these ideas, consider the linear system of equations

$$\begin{bmatrix} \delta & 1 \\ 1 & 1 \end{bmatrix} \begin{bmatrix} x_1 \\ x_2 \end{bmatrix} = \begin{bmatrix} 1 \\ 0 \end{bmatrix},$$

where δ is a small positive number. Suppose we solve the system on a computer for which $\delta < \epsilon_m/2$. If use Gauss elimination without pivoting, we compute

$$\begin{bmatrix} \delta & 1 \\ 0 & -1/\delta \end{bmatrix} \begin{bmatrix} x_1 \\ x_2 \end{bmatrix} = \begin{bmatrix} 1 \\ -1/\delta \end{bmatrix}$$

so

$$x_2 = 1, \qquad x_1 = 0.$$

The true solution is

$$\boldsymbol{x}_{true} = \begin{bmatrix} -\frac{1}{1-\delta} \\ \frac{1}{1-\delta} \end{bmatrix},$$

so our answer is very bad. The problem is well-conditioned, though; we can see this graphically on the left in Figure 1.4, since small changes in any of the coefficients of the two lines move the intersection point by just a little. Therefore, Gauss elimination without pivoting must be an unstable algorithm. If we use pivoting, our answer improves: the linear system is rewritten as

$$\begin{bmatrix} 1 & 1 \\ \delta & 1 \end{bmatrix} \begin{bmatrix} x_1 \\ x_2 \end{bmatrix} = \begin{bmatrix} 0 \\ 1 \end{bmatrix},$$

so the elimination gives us

$$\begin{bmatrix} 1 & 1 \\ 0 & 1 \end{bmatrix} \begin{bmatrix} x_1 \\ x_2 \end{bmatrix} = \begin{bmatrix} 0 \\ 1 \end{bmatrix}$$

from which we determine that

$$x_2 = 1,\ x_1 = -1.$$

This is quite close to the true solution.

Consider a second example with 3 digit decimal arithmetic:

$$\begin{bmatrix} 0.661 & 0.991 \\ 0.500 & 0.750 \end{bmatrix} \begin{bmatrix} x_1 \\ x_2 \end{bmatrix} = \begin{bmatrix} 0.330 \\ 0.250 \end{bmatrix}. \tag{1.1}$$

If we compute the solution with pivoting, truncating all intermediate results to 3 digits, we obtain

$$\boldsymbol{x}_c = \begin{bmatrix} -.470 \\ .647 \end{bmatrix}.$$

[4]Actually, for historical reasons, well-conditioned problems are sometimes called stable in some areas of scientific computing, but it is best to use the term well-conditioned to avoid confusion.

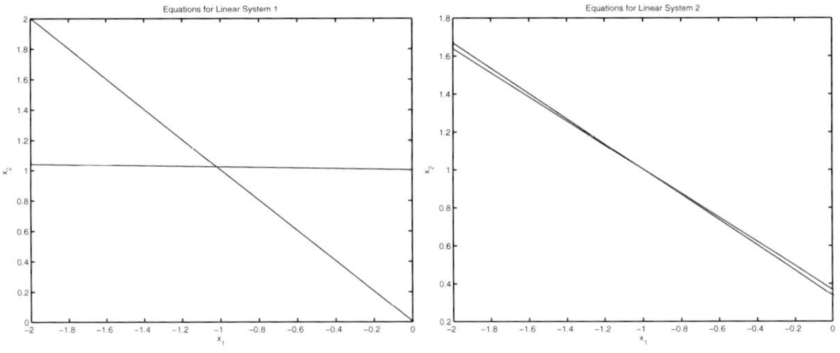

Figure 1.4. *(Left) Plot of a well-conditioned system of linear equations. Points on the red line satisfy* $\delta x_1 + x_2 = 1$*, and points on the blue line satisfy* $x_1 + x_2 = 0$*. Small changes in the data move the intersection of the two lines by a small amount. (Right) Plot of an ill-conditioned system of linear equations. Small changes in the data can move the intersection of the two lines by a large amount. This is the example in equation* (1.1) *except that the* (1,1) *coefficient in the matrix has been changed from* 0.661 *to* 0.630 *so that the two lines are visually distinct.*

The true solution is quite far from this:

$$\boldsymbol{x}_{true} = \begin{bmatrix} -1.000 \\ 1.000 \end{bmatrix}.$$

But when we substitute our computed solution back into (1.1), we see that the residual, or difference between the left and right sides, is

$$\boldsymbol{r} = \begin{bmatrix} -.000507 \\ -.000250 \end{bmatrix}.$$

Gauss elimination with pivoting produced a small residual because it is a stable algorithm, so it is guaranteed to solve a nearby problem. But the $\boldsymbol{x}$-error is not small, since the problem is ill-conditioned. We can see this graphically on the right in Figure 1.4; if we wiggle the coefficients of the two lines, we can make the intersection move quite a bit.

Sometimes we have additional information about the solution to a problem that gives us some guidance about improving a computed solution, as in the next challenge.

CHALLENGE 1.12.

Suppose you solve the nonlinear equation $f(x) = 0$ using a MATLAB routine, and the answers are complex numbers with small imaginary parts. If you know that the true answers are real numbers, what would you do?

POINTER 1.8. Further Reading.

Further information on this material can be found in the book by Overton [121] on IEEE floating-point arithmetic and the book by Higham [79], which focuses on the impact of floating-point arithmetic on algorithms.

Life may toss us some ill-conditioned problems, but there is no good reason to settle for an unstable algorithm.

In the next chapter we illustrate various ways of measuring the sensitivity or conditioning of a problem.

Chapter 2

Sensitivity Analysis: When a Little Means a Lot

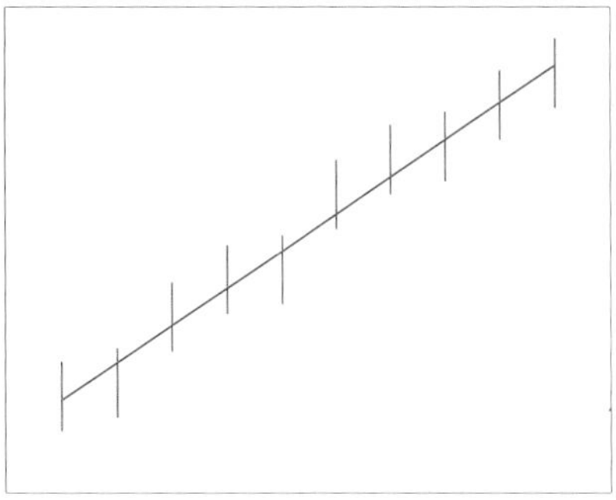

In contrast to classroom exercises, it is rare to be given a problem in which the data is known with absolute certainty. There are some parameters, such as π, that we can define with certainty, and others like Planck's constant $\hbar$ that we know to high precision, but most data is measured and therefore contains significant measurement error.

So what we really solve is not the problem we want, but some nearby problem, and in addition to reporting the computed solution, we really need to report a bound on either

- the difference between the true solution and the computed solution (a **forward error bound**), or
- the difference between the problem we solved and the problem we wanted to solve (a **backward error bound**).

This need occurs throughout computational science. Consider these examples:

- If we compute the resonant frequencies of a building, we want to know how these frequencies change if the load within the building changes a little.
- If we compute the stresses on a bridge, we want to know how sensitive these values are to changes that might occur as the bridge ages.
- If we develop a model for our data and fit the parameters, we want to know how much the parameters change when the data is wiggled within the uncertainty limits.

In this chapter we use some simple problems to investigate the use of several tools for **sensitivity analysis**.

2.1 Sensitivity Is Measured by Derivatives.

The best way to measure the sensitivity of a variable x to a small change in a parameter t is to compute $\mathrm{d}x/\mathrm{d}t$, since, by Taylor series, if δ is a small number, then

$$x(t+\delta) \approx x(t) + \delta \frac{\mathrm{d}x}{\mathrm{d}t}(t).$$

Therefore, the change in x due to a small change in t is approximately proportional to dx/dt (whenever the second derivative d^2x/dt^2 is bounded). Sometimes the derivative does not exist, and sometimes it is too expensive to compute, but whenever possible, this is the method of choice.

As an example, let's use the derivative to get insight into the sensitivity of roots of a quadratic equation to changes in the coefficients.

CHALLENGE 2.1.

Suppose x_1 and x_2 are the roots of the quadratic equation

$$x^2 + bx + c = 0.$$

(a) Use implicit differentiation to compute dx/db.

(b) We know that the roots are

$$x_{1,2} = \frac{1}{2}(-b \pm \sqrt{b^2 - 4c}).$$

Differentiate this expression with respect to b and show that the answer is equivalent to the one you obtained in (a).

(c) Find values of b and c for which the roots are very sensitive to small changes in b, and values for which they are not very sensitive.

If there are several parameters to vary, then the partial derivatives of the variable with respect to the parameters yield the sensitivity information.

Derivatives also give sensitivity information for constrained problems. For example, if we want to minimize a function $f(\boldsymbol{x})$ subject to the constraints $\boldsymbol{h}(\boldsymbol{x}) = \mathbf{0}$, then we learned in calculus to introduce **Lagrange multipliers** $\boldsymbol{\lambda}$, one per constraint, and look for solutions $(\boldsymbol{x}_{sol}, \boldsymbol{\lambda}_{sol})$ for which the Lagrangian function

$$L(\boldsymbol{x}, \boldsymbol{\lambda}) = f(\boldsymbol{x}) - \boldsymbol{\lambda}^T \boldsymbol{h}(\boldsymbol{x})$$

has a zero gradient. The "artificial" variables $\boldsymbol{\lambda}$ actually have a physical meaning: λ_i is the partial derivative of L with respect to the constraint h_i. Therefore, the value of a multiplier at a point where $\boldsymbol{h}(\boldsymbol{x}) = \mathbf{0}$ measures the sensitivity of f to small changes in the corresponding constraint. For this reason, Lagrange multipliers are sometimes called **marginal values** or **reduced costs**. We'll see an example of their use in Challenge 2.3.

2.2 Condition Numbers Give Bounds on Sensitivity.

Although the derivatives of the variable with respect to the parameters provide the "gold standard" for sensitivity analysis, differentiation is not always practical. For example, a linear system of equations with 100 variables has 10^4 coefficients in the matrix, and that is a lot of partial derivatives to compute and assess! Because of that, shortcuts have been developed to give less specific information but summarize what can happen in the worst

POINTER 2.1. Matrix and Vector Norms.

The size of an $n \times 1$ vector $\boldsymbol{x}$ is usually measured using the familiar 2-norm:

$$\|\boldsymbol{x}\|_2 = \sqrt{|x_1|^2 + \cdots + |x_n|^2}.$$

There are alternatives, though, including

$$\|\boldsymbol{x}\|_1 = |x_1| + \cdots + |x_n|,$$
$$\|\boldsymbol{x}\|_\infty = \max_j |x_j|.$$

Similarly, there are many **matrix norms**. In particular, for each vector norm $\|\cdot\|$, we define a matrix norm by

$$\|\boldsymbol{A}\| = \max_{\boldsymbol{x} \neq \boldsymbol{0}} \frac{\|\boldsymbol{A}\boldsymbol{x}\|}{\|\boldsymbol{x}\|}.$$

Computing a matrix norm using this definition is a difficult problem, but fortunately there are shortcuts. It can be shown that for an $m \times n$ matrix $\boldsymbol{A}$

$$\|\boldsymbol{A}\|_1 = \max_j \sum_{i=1}^{m} |a_{ij}|,$$

$$\|\boldsymbol{A}\|_2 = \sqrt{\max_j \lambda_j(\boldsymbol{A}^*\boldsymbol{A})},$$

$$\|\boldsymbol{A}\|_\infty = \max_i \sum_{j=1}^{n} |a_{ij}|,$$

where $\lambda_j(\boldsymbol{A}^*\boldsymbol{A})$ $(j = 1, \ldots, n)$ denotes the eigenvalues of the conjugate transpose of $\boldsymbol{A}$ times $\boldsymbol{A}$. One other norm, the **Frobenius norm**, is also useful. It is the 2-norm of the matrix $\boldsymbol{A}$ after stacking its columns:

$$\|\boldsymbol{A}\|_F = \sqrt{\sum_{j=1}^{n} \sum_{i=1}^{m} |a_{ij}|^2}.$$

In this book, if you see a norm without a subscript, assume that the 2-norm is used.

Two properties of norms are useful to know: if $\boldsymbol{A}$ and $\boldsymbol{B}$ are matrices and $\boldsymbol{x}$ is a vector, then for all of these norms, $\|\boldsymbol{A}\boldsymbol{B}\| \leq \|\boldsymbol{A}\|\,\|\boldsymbol{B}\|$ and $\|\boldsymbol{A}\boldsymbol{x}\| \leq \|\boldsymbol{A}\|\,\|\boldsymbol{x}\|$.

---

case for perturbations of a given size. These shortcuts involve computing a **condition number** for the problem.

Given a condition number, we can make statements such as the following: If the matrix $\boldsymbol{A}$ is changed to $\boldsymbol{A} + \boldsymbol{\Delta A}$, where

$$\frac{\|\boldsymbol{\Delta A}\|}{\|\boldsymbol{A}\|} \leq \delta,$$

POINTER 2.2. Eigenvalues, Determinants, and Condition Numbers.

People often try to measure the sensitivity of the solution to the problem $\boldsymbol{A}\boldsymbol{x}=\boldsymbol{b}$ using eigenvalues or the determinant of $\boldsymbol{A}$, saying that if there is an eigenvalue close to zero, or if the determinant is very small, then the problem is ill-conditioned. We notice, though, that the solution to $\boldsymbol{A}\boldsymbol{x}=\boldsymbol{b}$ is also the solution to $(c\boldsymbol{A})\boldsymbol{x}=(c\boldsymbol{b})$ for any nonzero constant c. Multiplying by c changes the eigenvalues by a factor of c and the determinant by a factor c^n. By changing c, we can make the determinant of $\boldsymbol{A}$ (or some eigenvalue of $\boldsymbol{A}$) arbitrarily big or arbitrarily close to zero. The condition number of $\boldsymbol{A}$ is independent of scale and is the number that correctly characterizes the sensitivity of $\boldsymbol{x}$ to relative changes in the data, as we will see in Pointer 5.3.

---

and δ is a small number, then the difference between the solution $\boldsymbol{x}$ to the linear system

$$\boldsymbol{A}\boldsymbol{x}=\boldsymbol{b}$$

and the solution $\boldsymbol{y}$ to the linear system

$$(\boldsymbol{A}+\boldsymbol{\Delta A})\boldsymbol{y}=\boldsymbol{b}$$

is bounded by

$$\frac{\|\boldsymbol{x}-\boldsymbol{y}\|}{\|\boldsymbol{x}\|} \leq \frac{\kappa(\boldsymbol{A})}{1-\kappa(\boldsymbol{A})\delta}\delta, \tag{2.1}$$

where $\kappa(\boldsymbol{A})$ is the condition number of $\boldsymbol{A}$. This bound is valid when $\delta < 1/\kappa(\boldsymbol{A})$ and holds for the 1, 2, or ∞ norm.

By its very nature, (2.1) is a worst-case statement, since it needs to hold for all perturbations $\boldsymbol{\Delta A}$ that are small enough. For some values of $\boldsymbol{\Delta A}$, the error bound (2.1) can be a serious overestimate, but there always exists some particular matrix $\boldsymbol{\Delta A}$ for which the bound is tight.

Condition numbers enable us to replace the full set of partial derivatives by a single number, but even that one number may be hard to compute. For example, for (2.1), the condition number of $\boldsymbol{A}$ is defined to be $\kappa(\boldsymbol{A})=\|\boldsymbol{A}\|\|\boldsymbol{A}^{-1}\|$. The norm of $\boldsymbol{A}$ is usually easy to compute; see Pointer 2.1. Even so, computing the norm of $\boldsymbol{A}^{-1}$ is quite expensive, since computing the inverse of a matrix is expensive (and generally not a good idea). Therefore, the condition number is usually estimated. In MATLAB we can use `cond(A,normtype)` to compute the condition number (setting `normtype` to 1, 2, or `inf`), or use the cheaper function `condest` to estimate the condition number.

CHALLENGE 2.2.

Consider the linear system $\boldsymbol{A}\boldsymbol{x}=\boldsymbol{b}$ with

$$\boldsymbol{A}=\begin{bmatrix} \delta & 1 \\ 1 & 1 \end{bmatrix}, \quad \boldsymbol{b}=\begin{bmatrix} 1 \\ 0 \end{bmatrix},$$

and $\delta = 0.002$.

(a) Plot the two equations defined by this system and compute the condition number of $\boldsymbol{A}$ (`cond(A)`).

(b) Compute the solution $\boldsymbol{x}_{true}$ to $\boldsymbol{Ax} = \boldsymbol{b}$ and also compute the solution to the nearby systems

$$(\boldsymbol{A} + \boldsymbol{\Delta A}^{(i)})\boldsymbol{x}^{(i)} = \boldsymbol{b}$$

for $i = 1, \ldots, 1000$, where the elements of $\boldsymbol{\Delta A}^{(i)}$ are independent and normally distributed with mean 0 and standard deviation $\tau = .0001$. (You can generate these examples by setting each `DeltaA = tau*randn(2,2)`.) Plot the 1000 points $(e_1^{(i)}, e_2^{(i)})$ with $\boldsymbol{e}^{(i)} = \boldsymbol{x}^{(i)} - \boldsymbol{x}_{true}$. This plot reveals the forward error in using $(\boldsymbol{A} + \boldsymbol{\Delta A}^{(i)})$ as an approximation to $\boldsymbol{A}$. In a separate figure, plot the 1000 residuals $(r_1^{(i)}, r_2^{(i)})$ with $\boldsymbol{r}^{(i)} = \boldsymbol{b} - \boldsymbol{Ax}^{(i)}$ (the backward error) for each of the computed solutions.

(c) Repeat (a) and (b) with the linear system $\boldsymbol{Ax} = \boldsymbol{b}$ with

$$\boldsymbol{A} = \begin{bmatrix} 1+\delta & \delta-1 \\ \delta-1 & 1+\delta \end{bmatrix}, \quad \boldsymbol{b} = \begin{bmatrix} 2 \\ -2 \end{bmatrix}.$$

(d) Discuss your results. Why do the forward error plots for the two problems look so different? How does the condition number relate to what you see in the forward error plots? What do the backward error plots tell you?

2.3 Monte Carlo Experiments Can Estimate Sensitivity.

In Challenge 2.2(b) you did a Monte Carlo experiment, discussed in more detail in Chapter 16. The idea is to randomly sample nearby problems and see how the solution changes. This is a fine way to measure sensitivity if a condition number would not give enough information and if derivatives cannot be obtained, but the process can be expensive.

Let's try two more applications of Monte Carlo to estimate sensitivity, one involving linear programming and one involving a differential equation.

CHALLENGE 2.3.

Investigate the sensitivity of the **linear programming problem**

$$\min_{\boldsymbol{x}} \boldsymbol{c}^T \boldsymbol{x}$$

subject to

$$\boldsymbol{Ax} \le b,$$
$$\boldsymbol{x} \ge \boldsymbol{0}.$$

(a) For Example 1, let $\boldsymbol{A} = [1, 1]$, $b = 1$, and $\boldsymbol{c}^T = [-3, -1]$. Solve the linear program (using, for example, MATLAB's `linprog` from the Optimization Toolbox) and use the

Lagrange multipliers (also called **dual variables**) to evaluate the sensitivity of $\mathbf{c}^T\mathbf{x}$ to small changes in the constraints. Illustrate this sensitivity using a Monte Carlo experiment, solving 100 problems with $\mathbf{A}$ replaced by $\mathbf{A} + \mathbf{\Delta A}^{(i)}$, where the elements of $\mathbf{\Delta A}^{(i)}$ are uniformly distributed on the interval $[-\tau, \tau]$, with $\tau = 0.001$. (You can do this by setting each `deltaA = 2*tau*(rand(1,2)-.5*ones(1,2))`.) Plot all of the solutions in one figure, and all of the function values $\mathbf{c}^T\mathbf{x}$ in another.

(b) Repeat for two more examples:
Example 2: $\mathbf{A} = [1,1]$, $b = 1$, $\mathbf{c}^T = [-1.0005, -0.9995]$.
Example 3: $\mathbf{A} = [0.01,\ 5]$, $b = 0.01$, $\mathbf{c}^T = [-1, 0]$.

(c) Explain how the Lagrange multiplier for the constraint $\mathbf{Ax} \leq b$ gives insight into the sensitivity observed in the corresponding Monte Carlo experiment.

CHALLENGE 2.4.

Consider the very simple differential equation

$$y'(t) = ay(t),$$

where $y(0) = 1$ and $a(t)$ is given. Let's investigate the sensitivity of the equation to perturbations in a.

(a) To make the problem concrete, the population growth rate $a(t)$ of the U.S. can be divided into two pieces: a rate of 0.006 due to births and deaths, and a rate of 0.003 due to migration. Determine how much the population increases over the next 50 years if this rate stays constant, and how much it increases if migration is set to zero.

(b) Then perform Monte Carlo experiments to see how sensitive the solution is to changes in the population growth rate. Assume that the rate can change each year. For each experiment, choose the birth/death rate for each year from a normal distribution with mean 0.006 and standard deviation 0.001, and choose the migration rate from a uniform distribution on the interval $[0, 0.003]$.

(c) Also experiment with what happens if years of high growth rate come early, followed by years of low growth rate, and vice versa.

2.4 Confidence Intervals Give Insight into Sensitivity

Another way to assess sensitivity is to make a statement like the following: If we repeat the data measurement many times, then we expect that 95% of the time the solution lies in the interval $[x^{lo}, x^{up}]$. Such an interval is called a **confidence interval**, and $\alpha = .95$ is the **confidence level**, determined using statistical estimation, assuming that the errors in the measurements are random.

As an example, consider a linear system $\boldsymbol{A}\boldsymbol{x} = \boldsymbol{b}$ and assume that the error $\boldsymbol{b} - \boldsymbol{b}_{true}$ is (multivariate) normally distributed with mean $\boldsymbol{0}$ and variance $\boldsymbol{S}^2$. Suppose that we want to estimate the value of $\boldsymbol{w}^T\boldsymbol{x}$; taking $\boldsymbol{w}$ to be the first column of the identity matrix, for example, gives us an estimate of x_1. We proceed as follows:

- Determine a value κ from the cumulative normal distribution, so that

$$\frac{1}{\sqrt{2\pi}} \int_{-\kappa}^{+\kappa} e^{-z^2/2}\, dz = \alpha. \tag{2.2}$$

  For 95% confidence intervals, $\alpha = 0.95$ and $\kappa \approx 1.960$.

- Given the value κ, let $\widehat{\boldsymbol{x}}$ solve $\boldsymbol{A}\boldsymbol{x} = \boldsymbol{b}$ and compute

$$\phi^{lo} = \boldsymbol{w}^T\widehat{\boldsymbol{x}} - \kappa\sqrt{\boldsymbol{w}^T(\boldsymbol{A}^T\boldsymbol{S}^{-2}\boldsymbol{A})^{-1}\boldsymbol{w}},$$

$$\phi^{up} = \boldsymbol{w}^T\widehat{\boldsymbol{x}} + \kappa\sqrt{\boldsymbol{w}^T(\boldsymbol{A}^T\boldsymbol{S}^{-2}\boldsymbol{A})^{-1}\boldsymbol{w}}.$$

Then $[\phi^{lo}, \phi^{up}]$ is a $100\alpha\%$ confidence interval for $\boldsymbol{w}^T\boldsymbol{x}$.

There are more general forms of the procedure. It is possible to construct (wider) confidence intervals that have joint probability α so that, for example, we can bound all the components of $\boldsymbol{x}$ simultaneously. There is also a nonparametric form of the result that allows us to compute confidence intervals when the error is not normally distributed. See Pointer 2.3 for some references.

Let's apply our procedure to the examples from Challenge 2.2.

CHALLENGE 2.5.

Using the first linear system from Challenge 2.2, perform a Monte Carlo experiment that computes the solution to the nearby systems

$$\boldsymbol{A}\boldsymbol{x}^{(i)} = \boldsymbol{b} + \boldsymbol{e}^{(i)}$$

for $i = 1, \ldots, 1000$, where the elements of $\boldsymbol{e}^{(i)}$ are independent and normally distributed with mean 0 and standard deviation $\tau = 0.0001$ (so that $\boldsymbol{S} = \tau\boldsymbol{I}$). Compute 95% confidence intervals on each component of the solution and see how many of the components of the Monte Carlo samples lie within the confidence limits.

Repeat for the second linear system from Challenge 2.2.

We have experimented with several ways to measure sensitivity in our problems, and Pointer 2.3 gives additional alternatives. A good computational scientist computes not just an answer to a problem but an assessment of how good it is, and sensitivity analysis is an important component in this assessment.

POINTER 2.3. Further Reading.

Lagrange multipliers are discussed in Unit III, in standard advanced calculus textbooks, and in textbooks on optimization, such as that by Nash and Sofer [111].

Condition numbers are commonly used in numerical linear algebra to measure sensitivity of linear systems of equations, eigenvalues, eigenvectors, and other quantities. The book by Higham [79] is a good reference.

See the book by Fishman [49] for more information on Monte Carlo estimation.

Standard statistical textbooks explain the use of confidence intervals. These methods can also be applied to constrained problems, as seen, for example, in [130].

One alternative to the methods we considered is **interval analysis**. In this method, we carry upper and lower bounds on each quantity along through our calculations. The result is a rigorous, although often pessimistic, set of forward error bounds on the answer. The method's most forceful advocate was R. E. Moore [110], and there are many textbooks that apply the method to scientific computing.

A second alternative is the use of **symbolic computation**, in which we carry analytic expressions for each quantity; see Pointer 1.7.

Chapter 3

Computer Memory and Arithmetic: A Look under the Hood

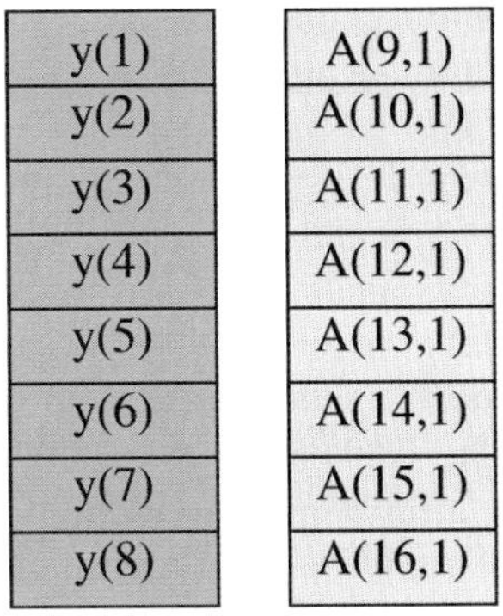

You have many places where you store data. You might keep your identification and credit cards in your wallet, where you can get to them quickly. Space is limited, though, so you can't keep all important information there. You might carry current papers for work or school in a backpack or briefcase. Older papers might be stored in your desk or file cabinet. And papers that you don't think you need but are afraid to throw out might be stored in an attic, basement, or storage locker. Your wallet, backpack, desk, and attic form a hierarchy of storage spaces. The small ones give you fast access to data that you often need, while the larger ones give slower access but more space.

For the same reasons, computers also have a hierarchy of storage units. Memory management systems try to store information that you will soon need in a unit that gives fast access. This means that large vectors and arrays are broken up and moved piece by piece as needed. You can write a correct computer program without ever knowing about memory management, but attention to memory management allows you to consistently write programs that don't have excessive memory delays.

In this chapter, we'll consider a model of computer memory organization. We'll hide some detail but discuss making decisions about how to organize our computations for efficiency. We'll use mathematical modeling to estimate the memory parameters for a typical computer. Then we'll see how important these parameters are relative to the speed of floating-point arithmetic.

Our discussions assume that the computation is being performed by a single processor on a machine that possibly has many processors (**multicore**). When the full power of a multicore system is used to solve a problem, then obviously memory management becomes more complicated!

3.1 A Motivating Example

Suppose we have an $m \times n$ matrix $\boldsymbol{A}$ and an $n \times 1$ vector $\boldsymbol{x}$. To form $\boldsymbol{y} = \boldsymbol{A}\boldsymbol{x}$ in MATLAB, we just write `y = A * x`, but let's consider how this might be implemented. The vector

$\boldsymbol{y}$ can be defined by inner-products (also called dot-products) between $\boldsymbol{x}$ and rows of $\boldsymbol{A}$: for $i = 1,\ldots,m$,

$$y_i = \boldsymbol{A}(i,\ :) * \boldsymbol{x}. \tag{3.1}$$

In contrast to this **row-oriented algorithm**, we can also define $\boldsymbol{Ax}$ using a **column-oriented algorithm**:

$$\boldsymbol{y} = \boldsymbol{Ax} = x(1) * \boldsymbol{A}(:\ ,1) + x(2) * \boldsymbol{A}(:\ ,2) + \cdots + x(n) * \boldsymbol{A}(:\ ,n). \tag{3.2}$$

This scheme is based on an operation called `axpy`, which is an abbreviation for $a\boldsymbol{x} + \boldsymbol{y}$. We work left to right through our expression, taking a scalar times a vector and adding it to a previously accumulated vector. Thus we initialize $\boldsymbol{y}$ to zero and then compute $\boldsymbol{y} = \boldsymbol{y} + x(j) * \boldsymbol{A}(:\ ,j)$ for $j = 1,\ldots,n$.

These algorithms both have the same rounding properties and take almost the same number of numeric operations: mn multiplications and $m(n-1)$ or mn additions. But, surprisingly, the time taken by the two algorithms is quite different.

CHALLENGE 3.1.

Program the two algorithms in MATLAB. Time them for a random matrix $\boldsymbol{A}$ and a random vector $\boldsymbol{x}$ for $m = n = 1024$, and verify that they yield the same product $\boldsymbol{Ax}$ but take a different amount of time (as measured by `tic` and `toc`).

Your results should have indicated that the second algorithm is much faster than the first when m and n are large. The speed difference is due to memory management; MATLAB stores matrices column by column, and if we want a fast implementation, we must use this fact in design of our algorithm.

3.2 Memory Management

The computer memory hierarchy includes **registers**, **cache**, **main memory**, and **disk**. Arithmetic and logical operations are performed on the contents of registers. The other storage units are accessed when data is on its way to or from the registers. It is as if whenever you need to change some data in your attic, you move it first to your desk, then to your backpack, then to your wallet, make the correction, and move it back through wallet, backpack, and desk, finally storing it back in the attic.

A small illustration of memory management is given in Figure 3.1. We'll consider a 1-level cache, although most machines have a hierarchy of cache units. Let's see how information is moved between main memory and cache in forming $\boldsymbol{y} = \boldsymbol{Ax}$ using (3.2). Suppose that $\boldsymbol{A}$ is a matrix of size $m \times n$, with $m = 128$ and $n = 32$, and suppose for ease of counting that the first element in each matrix and vector lies in the first element of some page of main memory. The matrix elements are stored in the order `A(1,1), ..., A(128,1), A(1,2), ..., A(128,2), ..., A(1,32), ..., A(128,32)`. Cache memory is loaded by **block** (also called a **cache line**); in this example, this means 8 elements at a time. So in the `axpy` implementation, where we successively add $x_j * \boldsymbol{A}(:, j)$ to $\boldsymbol{y}$, the computer first loads `A(1,1), ..., A(8,1)` into one block of the cache, `x(1), ..., x(8)` into a second block (since the value of `x(1)` is needed), and lets `y(1), ..., y(8)` occupy a third block (Figure 3.1, top). (Note that `y` must be loaded into cache, since at later

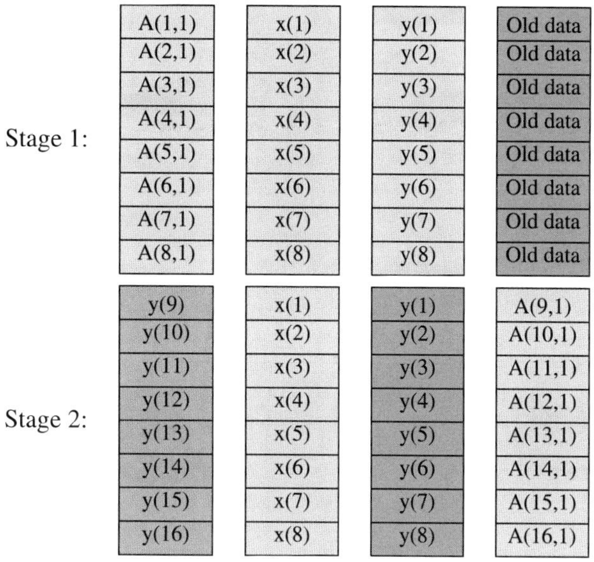

Figure 3.1. *State of a cache memory of 4 blocks, 8 words each, during two stages of the matrix-vector product algorithm. The red blocks are the ones that were least recently used.*

times through the loop it is nonzero.) After `x(1)*A(1:8,1)` is added into `y(1:8)`, we then need `A(9,1),...,A(16,1)` and `y(9),...,y(16)`. In this example we have only 4 blocks of cache, though, so to access the fifth block of data we overwrite some old block, after being sure that any updated values are changed in main memory. In our case, the old `y`-block or `A`-block disappears from cache (Figure 3.1, bottom).

Loading 5 blocks from main memory into the cache has allowed us to do 16 of our $128 * 32$ multiplications. We continue the count in the next challenge.

CHALLENGE 3.2.

(a) Count the number of blocks that are moved from main memory into the cache memory in Figure 3.1 for each of the two matrix-vector multiplication algorithms (3.1) and (3.2) with $m = 128$ and $n = 32$. If a cache block needs to be written over, choose the least-recently-used block.

(b) How does your answer change if matrices are stored row-by-row (as in C, C++, or Java) rather than column-by-column (as in MATLAB or FORTRAN)?

In contrast to row- and column-oriented algorithms, we might choose to store our matrix in block-matrix order. For example, if our matrix is 4×4, we might divide it into four 2×2 submatrices:

$$\begin{bmatrix} a_{11} & a_{12} & a_{13} & a_{14} \\ a_{21} & a_{22} & a_{23} & a_{24} \\ a_{31} & a_{32} & a_{33} & a_{34} \\ a_{41} & a_{42} & a_{43} & a_{44} \end{bmatrix}$$

and store the elements in the order $a_{11}, a_{21}, a_{12}, a_{22}$, $a_{31}, a_{41}, a_{32}, a_{42}$, $a_{13}, a_{23}, a_{14}, a_{24}$, a_{33}, a_{43}, a_{34}, a_{44}. If each submatrix fits in a single memory block, then it can be moved efficiently and used in computation.

CHALLENGE 3.3. (Extra)

Suppose that the $m \times n$ matrix $\boldsymbol{A}$ is stored in $k \times k$ blocks, with $m = 128$, $n = 32$, and $k = 8$. Write an efficient algorithm for forming $\boldsymbol{Ax}$ for a given vector $\boldsymbol{x}$.

3.3 Determining Hardware Parameters

We see that memory management matters in matrix multiplication! It is certainly faster to use the algorithm that has fewer moves of blocks into cache, but how much it matters depends on memory parameters such as the following:

- b = the number of blocks that can be held in cache memory.
- ℓ = the number of double-precision words in a block.
- α = the time to access a double-precision word in cache (nanoseconds).
- μ = the extra time needed for access if the word is not already in cache, also known as the **cache miss penalty** (nanoseconds).

To estimate the memory parameters, we can run a program that constructs a long vector z of length m and then steps through it, incrementing some elements. When we step through every element, we are almost always accessing a value that can be found in cache. If we step through elements $z(1), z(1+s), z(1+2s), \ldots$, where the stride s is bigger than the block size ℓ, then we always get a cache miss penalty (unless the hardware correctly guesses our next request and pre-fetches the block before we ask for it). By varying s we can estimate the block size, and by cycling through the computation several times, we can estimate the size of the cache. Consider the following program fragment, written in C:

```
steps = 0;
i = 1;
do
{  z(i);
   steps = steps + 1;
   i = i + s;
   if (i > m)
      i = 1;
   end }
while (steps < naccess)
```

The loop makes `naccess` accesses to the array `z`, where `naccess` is a suitably large number. If we time the loop, subtract off the loop overhead (estimated by timing a similar loop with the statement `z(i)` omitted), and divide the resulting time by `naccess`, we estimate the average time for one access.

We see how this works in the next challenge.

CHALLENGE 3.4.

Suppose our cache memory has parameters $b = 4$, $\ell = 8$, $\alpha = 1$ ns, and $\mu = 16$ ns. Assume that when necessary we replace the block in cache that was least-recently used, and that we set `naccess`=256 in the program fragment above. Consider the following table of estimated times per access in nanoseconds and show how each entry is derived.

s	$m = 16$	$m = 32$	$m = 64$	$m = 128$
1	1.125	1.250	3.000	3.000
2	1.125	1.250	5.000	5.000
4	1.125	1.250	9.000	9.000
8	1.125	1.250	17.000	17.000
16	1.063	1.125	1.250	17.000

If we work in a high-performance "compiled" language such as FORTRAN or a C-variant, we can use our timings of program fragments to estimate the cache miss penalty. In "interpreted" MATLAB, overhead masks the penalty.

Whenever we time a program, though, there are many sources of uncertainty:

- Other processes are running. Even if you are running on a laptop on which you are the only user, the operating system (Windows, Linux, ...) is still doing many other tasks, e.g., refreshing the screen, updating the clock, tracking the cursor. Most systems have two timers, one that gives the elapsed time (e.g., `tic, toc`) and one that tries to capture the time used by this process alone (e.g., `cputime`).

- There is uncertainty in the timer, so the data you collect are noisy. Most timers give trash unless they are timing intervals that are at least a millisecond, and they are much better at intervals near one second. Therefore, the loop you are timing should do as many operations as possible, but not so many that interruptions by other active processes contaminate elapsed time.

- The time for arithmetic operations often depends on the values of the operands. For instance, dividing by a power of 2 is usually much faster than dividing by other numbers, and adding zero is usually faster than other additions.

- The computer uses pipelining. This occurs on many levels but the fundamental idea is that the execution of each instruction that we give the computer is partially overlapped with other instructions, so it is difficult to assign a cost to a single instruction.

- Compilers optimize the statements that we write. For example, a compiler might recognize that `z(i)` is not changed by our program fragment above and therefore remove that statement from the loop.

Table 3.1. *Average memory access times (nanoseconds) on a Sun workstation for various lengths m of (single precision) arrays and various strides s. The negative entry that occurred when overhead was subtracted off is indicative of the uncertainty in the data.*

	$\log_2 m =$										
$\log_2 s$	10	11	12	13	14	15	16	17	18	19	20
0	2	3	4	5	5	7	7	8	7	8	9
1	4	4	3	4	4	8	10	9	8	10	11
2	7	7	6	7	8	15	18	17	18	20	22
3	6	7	7	7	8	19	21	21	21	21	25
4	13	12	12	10	11	21	22	21	22	23	23
5	14	15	15	14	15	25	26	26	27	27	26
6	10	12	10	10	9	20	19	17	17	18	19
7	11	11	10	9	10	19	20	18	17	19	19
8	1	11	11	10	9	18	18	17	18	17	18
9	1	0	10	11	10	20	18	18	19	17	18
10	3	0	1	10	11	21	18	18	17	17	19
11	3	2	0	11	30	57	58	58	60	59	62
12	1	3	2	0	10	51	60	60	61	58	62
13	3	3	3	3	0	10	48	61	60	60	61
14	2	3	3	2	4	-1	10	50	59	61	60
15	3	2	2	2	3	2	0	11	49	60	59
16	2	3	4	3	4	3	4	0	9	49	61
17	3	2	2	2	2	3	3	4	0	9	51
18	3	3	3	4	3	2	3	4	1	0	10
19	3	3	2	3	2	3	2	4	3	2	0

- Cache blocks might be prefetched. Programs often access data in order, so computers might predict that the next adjacent block of memory should be loaded into cache while you are operating on the current one.

Therefore, real data values are not as clean as those in the previous challenge. Running a program like that considered in Challenge 3.4 on an old Sun workstation gives the data in Table 3.1. Let's try to estimate the workstation's cache parameters.

CHALLENGE 3.5.
Estimate the cache parameters from the data of Table 3.1.

3.4 Speed of Computer Arithmetic

We saw in Section 3.3 that the order in which we access the elements of a matrix affects the time, but is it a significant effect? Let's time some arithmetic operations and memory accesses to see whether it matters.

POINTER 3.1. Fast Algorithms for Matrix Multiplication.

Matrix multiplication algorithms built on the techniques discussed in this chapter require $O(mnp)$ multiplications. Some faster algorithms have been proposed, variants on Strassen's algorithm [143], which makes recursive use of a clever algorithm for multiplying 2×2 matrices using 7 multiplications instead of 8. The best currently known uses $O(n^{2.376})$ operations when $m = n = p$, but the constant factor is very large [30]. Strassen's algorithm requires fewer multiplications when $m = n = p \geq 655$, but the memory management is quite different and the stability properties are somewhat weaker [79, Sec. 23.2.2].

CHALLENGE 3.6.

Determine for your computer the time for floating-point operations (addition, subtraction, multiplication, division, square-root) and the time for integer operations (addition, subtraction, multiplication, division). One way is to look up the peak speed claimed by the manufacturer. An alternative is to write a timing loop in a high-performance language.

CHALLENGE 3.7. (Extra)

Write a program in a high-performance language such as C or FORTRAN to estimate the cache size, the block size for the cache, the time to access a value in cache, and the cache miss penalty. Run it on your favorite computer. Average the time over enough operations to get an accurate estimate, and also estimate the variance in your measurements.

Find the manufacturer's claims for at least some of these parameters and determine whether your estimates agree or disagree, and why.

Knowing our machine's timing parameters allows us to design matrix algorithms that achieve speeds close to the manufacturer's peak performance claims, as illustrated in the next challenge.

CHALLENGE 3.8. (Extra)

Use the information you gathered about the memory access properties of your machine to write the best program that you can for doing matrix-matrix multiplication on your computer. Use a high-performance language. The inputs to the function are two matrices $\boldsymbol{A}$ of dimension $m \times n$ and $\boldsymbol{B}$ of dimension $n \times p$, along with n, m, and p. The output of the function is the $m \times p$ matrix $\boldsymbol{F} = \boldsymbol{AB}$. The program should order the computations in order to minimize the number of cache misses.

A (better) alternative to writing our own fast algorithms for basic matrix operations is to use the ones provided in the BLAS implementation discussed in Chapter 5 and in Pointer 3.2. In MATLAB we access the BLAS for matrix-vector product by typing `A*x`. For other matrix tasks, we now know to keep our MATLAB algorithms column oriented whenever possible.

POINTER 3.2. Further Reading.

A good, detailed description of memory hierarchies is given in the textbook by Hennessy and Patterson [74, Chapter 5].

In MATLAB, the underlying matrix decomposition software is drawn from the LAPACK FORTRAN suite [2]. This is based on a set of Basic Linear Algebra Subroutines (BLAS), provided by hardware manufacturers to optimize operations such as inner product, `axpy`, matrix-matrix multiplication, etc. The LAPACK routines implement stable algorithms and provide high performance on a variety of hardware.

Matrices are not the only kinds of data that need to be carefully organized in memory. Samet [133] gives an encyclopedic treatment of more complicated data structures, especially multidimensional ones.

Floating-point speed is increasing faster than memory access times are decreasing, so floating-point operations are becoming almost free. Algorithms are optimized not by minimizing arithmetic but by minimizing access to slow levels of memory.

Chapter 4

Design of Computer Programs: Writing Your Legacy

In scientific computing, we sometimes begin with a clean slate; we are given a new problem, and we write software to solve it. In this situation, we are inventors. Other times we work on a problem for which considerable software development has been done, often over a period of many years. The existing software may have many authors, some of whom have moved on to other positions. In this situation, our job is more akin to detective work. We study the existing program, run examples to see how it behaves, and come to understand both what it does and how it does it.

In this chapter, we consider the second situation, in which we are asked to work with a **legacy code**, a program that has been in use for a while and now needs maintenance by someone other than its author. We use the MATLAB function `posted.m` from Table 4.1 as an example. Let's consider some principles of documentation and design and see how they apply to `posted`.

4.1 Documentation

Documentation provides you and other potential users of your program an easy source of information about the use and design of the software. Although you completely understand the program you write today, by next year, or even next week, you will be surprised at how difficult it is to reconstruct your ideas if you neglect the documentation.

Ideally, the documentation at the top of the module provides basic information to help a potential user decide whether the software is of interest. It should include the following:

- purpose of the module, since this is certainly the first thing a user wants to know!
- name of author, since this provides someone to whom bugs can be reported and questions asked.
- original date of the module and a list of later modifications, since this gives information such as whether the module is likely to run under the current computer environment and whether it might include the latest advances.

Table 4.1. *A MATLAB function* `posted.m`*, an example of a legacy program.*

```
function [r, q] = posted (C)
[m,n] = size(C);
for k = 1:n
   for j=1:m
       x(j) = C(j,k);
   end
   xn = 0;
   for j=1:m,
       xn = xn + x(j)*x(j);
   end
   r(k,k) = sqrt(xn);
   for j=1:m,
       q(j,k) = C(j,k)/r(k,k);
   end
   for j = k+1:n
      r(k,j) = 0;
      for p=1:m
          r(k,j) = r(k,j) + q(p,k)'*C(p,j);
      end
      for p=1:m
        C(p,j) = C(p,j) - q(p,k)*r(k,j);
      end
   end
end
```

- description of each input parameter, so that a user knows what information needs to be provided and in what format.
- description of each output parameter, so that a user knows what information is produced.
- brief description of the method and references, to help a user decide whether the method fits his/her needs.

In-line documentation identifies the major sections of the module and provides some detail on the method used. It is important in specifying the algorithm, identifying bugs, and providing information to someone who might need to modify the software in order to solve a slightly different problem.

Note that the documentation should be an integral part of the module; in other words, it is not enough to include it in a separate document, because a potential user might not have access to that document. If you find `posted` frustrating, please resolve never to leave your own programs undocumented!

CHALLENGE 4.1.

Guided by the principles listed above, add documentation to `posted`.

4.2 Software Design

Software should be designed according to a principle articulated by Albert Einstein: "Make everything as simple as possible, but not simpler."

- Code should be modular, so that a user can substitute another piece when necessary. For example, a minimization algorithm should call a separate function to get a function value, so that it can be used to minimize any function. Separate modules should also be used when the same computation is repeated in different places, to make the software easier to understand and maintain.

- On the other hand, there is overhead involved in function calls, so ideally each module should involve a substantial computation in order to mask this overhead,

- Input parameters should be tested for validity, and clear error messages should be generated for invalid input, telling a user, for example, if a parameter that must be real and positive has a negative or complex value.

- Data that a function needs should be specified in variables, not constants. For example, a function that solves a linear system should work for any size of the matrix, rather than having a size specified.

- **Spaghetti code** should be avoided. In other words, the sequence of instructions should be top-to-bottom (including loops), without a lot of jumps.

- The names of variables should be chosen to remind the reader of their purpose. For example, `lambda` is better than `l` as the name of a Lagrange multiplier.

- The instructions should be formatted in a way that makes them easy for a reader to understand: the statements within loops or `if` statements should be indented, blank lines should be used to visually partition blocks of code, and lines should be short enough to fit in a window and read at a glance.

- The module should be reasonably efficient. In particular, it should not take an order of magnitude more time or storage than necessary. For example, if a function that solves a linear system with n variables takes $O(n^4)$ operations or $O(n^3)$ storage, then it is not so useful.

- And, of course, the program should be correct.

CHALLENGE 4.2.

Judge `posted` according to each of the first seven design principles listed above. (We'll consider its efficiency and correctness later.)

4.3 Validation and Debugging

It would be comforting to have a proof that each module that we use is correct. Although there has been considerable effort in developing methodologies for proving correctness, there are formidable limitations, both theoretical and practical. If correctness means matching a set of specifications given for the module, how do we know that the specifications are correct? What does correctness mean when we consider the effects of rounding error? And even if each module is correct, can we ensure that the modules interact with each other correctly?

Rather than a proof of correctness, in most situations we settle for a validation of correctness on a limited set of inputs that we believe span the range of possibilities. Proper design of this testing program is just as important as proper design of the modules being tested, and it is a critical part of the debugging process.

- The testing program should be well-documented and easy to read.
- The test should exercise every statement in the target modules, include all typical kinds of correct input, and test all conceivable errors in input.
- The testing program should compare the output against some trusted result and create a log of the test results.
- The testing program should be archived so that it can be used later in case the target modules are modified.

In addition to using a testing program, a powerful way to debug a module is to write it one day and then read it carefully the next, reasoning through each statement to be sure that it performs as intended. Programmers in the 1960s and 1970s had a major incentive to do such **desk checks**: running the program once could mean waiting until the next morning to see the results, so it was important to get it right without too many runs. With today's machines we usually have the luxury of seeing results from our programs much more quickly. It is tempting just to run and modify the program until the answers look good, but this is no substitute for a careful reading.

Let's apply these validation principles to `posted`.

CHALLENGE 4.3.

(a) What does `posted` do?

(b) Develop a testing program for `posted`.

(c) Users complain that `posted` does not seem to behave well when the matrix `C` has more columns than rows. In particular, they expect that `q' * q` is an identity matrix, and this is not true. Investigate this bug complaint and see what can be done.

POINTER 4.1. Further Reading.

Organizations often have their own standards for documentation of programs. One of the most widely admired and underutilized systems is Knuth's Literate Programming [93], used, for example, in his TEXdocument typesetting language. Knuth's programming style is also a model of clarity and good design.

Becoming a good programmer requires practice as well as good models. Bentley's book [14] is an excellent source of deceptively simple problems with beautiful solutions, and Oliveira and Stewart [118] propose problems related to scientific software.

Mastery of the capabilities of the programming language is essential to writing good software, and performance optimization in MATLAB is discussed in [78].

Considerable research was done in the 1970s on proving programs correct, but it is not a mainstream activity. See, for example, [31].

4.4 Efficiency

Finally, we turn our attention to making `posted` more efficient. The main sources of inefficiency in `posted` arise from

- failing to use the vector capabilities of MATLAB. (For example, the programmer used a loop instead of writing `r(k,j) = q(:,k)'*C(:,j)`.)
- failing to use built-in functions like `norm` that make use of BLAS.
- failing to initialize matrices such as `r` to all zeros, and instead forcing MATLAB to allocate new space each time through the loop.

Let's eliminate these inefficiencies.

CHALLENGE 4.4.

Users have run a timing profiler on their software, and they have found that `posted` takes 22% of the total time. Their typical input matrices `C` have about 200 rows and 100 columns. Change `posted` to make it run faster. Test the original and modified versions, graphing the time required for problems with 200 rows and 50, 60, ..., 200 columns.

After redesign, documentation, and validation, you should have a program that runs 100 times faster than `posted` and provides a much more useful legacy.

Unit II

Dense Matrix Computations

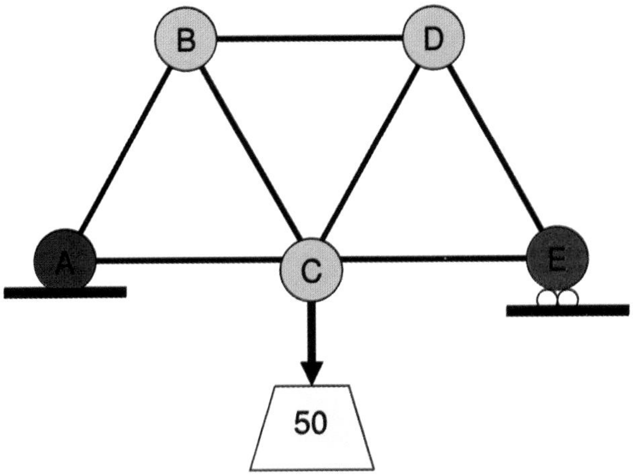

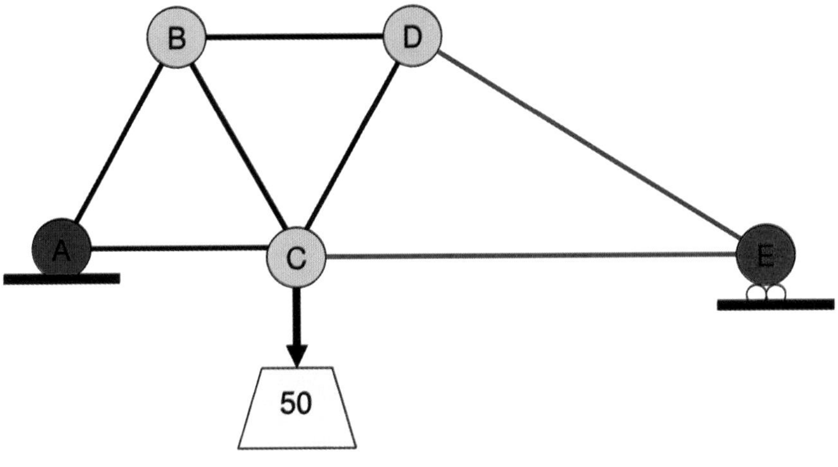

Matrix computations are basic to a wide variety of problems in scientific computing, from function minimization to solving partial differential equations. In this unit, we consider computations involving **dense matrices**, those that don't have a large number of zero elements. Computations involving **sparse matrices**, those that contain a substantial number of zeros, are discussed in Unit VII.

In Chapter 2, we investigated how sensitive matrix problems can be to small changes in the data. In Chapter 3 we learned to efficiently manipulate matrices on computers. Now we focus on being able to choose the right matrix decomposition for a given task.

Our plan is to first, in Chapter 5, discuss some basic tools for matrix manipulation: the BLAS. Then we consider in turn a variety of matrix decompositions and their uses: LU decomposition, QR decomposition, rank-revealing QR decomposition, eigendecomposition, and singular value decomposition (SVD).

One important question is how to adapt a matrix decomposition when the matrix changes slightly. We address this issue in Chapter 7.

We illustrate the use of the SVD in two case studies. We study in Chapter 6 how to deblur images, using the SVD to exploit the Kronecker product structure of a matrix and obtaining an efficient algorithm for quite large problems. In estimating the direction of arrival of signals in Chapter 8, we use the SVD as well as another useful matrix decomposition, the rank-revealing URV.

BASICS: To understand this unit, the following background is helpful:

- **Linear algebra:** matrix multiplication; solution of linear systems; orthogonality; finding a basis for a subspace; definitions of eigenvalues and eigenvectors. See a standard textbook for a first course in linear algebra.

- **Matrix computations:** LU decomposition; Gauss elimination; forward and back substitution for solving $\boldsymbol{Ax} = \boldsymbol{b}$ when $\boldsymbol{A}$ is lower triangular or upper triangular. See, for example, [148, Chapters 5,6,7].

MASTERY: After you have worked through this unit, you should be able to do the following:

- Write a MATLAB algorithm for forward substitution or back substitution (row or column oriented).

- Use BLAS effectively.

- Explain what pivoting is in the LU decomposition and why it is necessary.

- Compute the determinant of a triangular matrix or permutation matrix.

- Recognize when to use the Cholesky decomposition rather than LU.

- Test a matrix for positive definiteness.

- Determine whether a unique solution to a linear system or least squares problem exists and how sensitive it is to perturbations.

- Define orthogonal and unitary matrices and prove simple facts about them (e.g., the product of unitary matrices is unitary; multiplication by them does not change the 2-norm of a vector).
- Compute a Givens transformation to reduce a 2x1 vector to have a zero in its second position.
- Write an algorithm to reduce a matrix to upper-triangular form or upper-Hessenberg form using Givens matrices.
- Write an algorithm to reduce a matrix to upper-triangular form using the Gram–Schmidt algorithm.
- Show that the vectors produced by the QR Gram–Schmidt algorithm are orthogonal and that $\boldsymbol{A} = \boldsymbol{QR}$.
- Understand the difference between QR and RR-QR and state the uses of each.
- Show the relation between the SVD of $\boldsymbol{A}$ and the eigendecomposition of $\boldsymbol{A}^*\boldsymbol{A}$ and $\boldsymbol{A}\boldsymbol{A}^*$ (where $\boldsymbol{A}^*$ is the complex conjugate transpose of $\boldsymbol{A}$).
- Use RR-QR or SVD for tasks such as finding the range and null space of a matrix, solving a least squares problem, and finding a lower-rank matrix near $\boldsymbol{A}$.
- Use MATLAB's `eig` to determine the stability of a matrix.
- Give bounds on sensitivity of eigenvalues to perturbations.
- Choose an appropriate decomposition for a given task.
- Know why matrix inverses are not useful unless the elements themselves are needed.
- Solve linear discrete ill-posed problems.
- Exploit Kronecker product structure in a matrix.
- Update a QR decomposition when a row or a column is added or deleted.
- Know the costs of the matrix decompositions.
- Show the validity of the Sherman–Morrison–Woodbury formula and use it to efficiently solve a modified linear system without using inverses.

Chapter 5

Matrix Factorizations

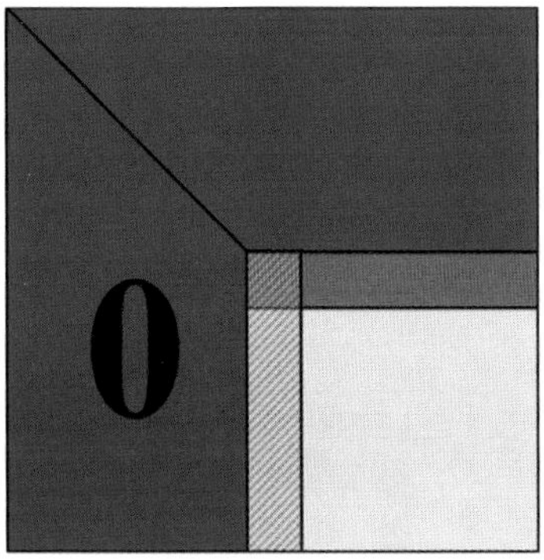

We might have access to many tools for cutting, ranging from a scalpel to a chain saw. In order to perform a given cutting task quickly and safely, we try to choose the appropriate tool. It is good to have access to a variety of tools, since we would not want to have to cut our fingernails with the chain saw, but if we had to choose just one cutting tool, it might be a Swiss Army knife.

Similarly, there are many decompositions for solving matrix problems, and we need to know how to choose the appropriate one for a given problem. In this chapter we develop the basic tools we need to solve matrix problems with speed and stability. The problems include the three most common ones in numerical computation:

Linear system of equations: Given an $n \times n$ matrix $\boldsymbol{A}$ and an $n \times 1$ vector $\boldsymbol{b}$, find $\boldsymbol{x}$ so that

$$\boldsymbol{A}\boldsymbol{x} = \boldsymbol{b}.$$

Least squares problem: Given an $m \times n$ matrix $\boldsymbol{A}$ (usually $m > n$) and an $m \times 1$ vector $\boldsymbol{b}$, solve the problem

$$\min_{\boldsymbol{x}} \|\boldsymbol{b} - \boldsymbol{A}\boldsymbol{x}\|.$$

Eigenvalue problem: Given an $n \times n$ matrix $\boldsymbol{A}$, find a vector $\boldsymbol{x}$ and a scalar λ so that

$$\boldsymbol{A}\boldsymbol{x} = \lambda\boldsymbol{x}.$$

POINTER 5.1. Useful Definitions and Facts from Linear Algebra.

Review the matrix and vector notation defined in Pointer 1.2 and the matrix and vector norms defined in Pointer 2.1. In this chapter we make use of the following facts:

- The **determinant** of a square matrix is the product of its eigenvalues. The usual definition involves an expansion by **minors**, but the determinant of a lower-triangular matrix (one with zeros above the main diagonal) or upper-triangular matrix (one with zeros below the main diagonal) is just the product of the diagonal entries.
- The **rank** of a matrix is the number of linearly independent rows, and it is equal to the number of linearly independent columns.
- An $m \times n$ matrix is **full rank** if its rank is $\min(m,n)$.
- For any nonsingular matrix $\boldsymbol{X}$, the matrix $\boldsymbol{XAX}^{-1}$ is a **similarity transform** of $\boldsymbol{A}$ and the eigenvalues are unchanged.
- The **range** of a matrix $\boldsymbol{A}$ is the set of vectors $\boldsymbol{y}$ that can be expressed as $\boldsymbol{Ax}$ for some vector $\boldsymbol{x}$.
- The **null space** of a matrix $\boldsymbol{A}$ is the set of vectors $\boldsymbol{x}$ for which $\boldsymbol{Ax} = \boldsymbol{0}$.
- A set of vectors forms a **basis** for a linear subspace if every vector in the subspace can be expressed as a linear combination of these vectors and if no set with a smaller number of vectors suffices. The basis is an **orthogonal basis** if the inner product between any two different vectors in the basis is zero. If in addition, the basis vectors have length 1, then they form an **orthonormal basis**. The columns of an orthogonal matrix (see Pointer 5.2) form an orthonormal basis for the range of the matrix.

We discuss in turn the BLAS, the LU decomposition, the QR decomposition, the eigendecomposition, and (the Swiss Army knife) the singular value decomposition. These decompositions, and their uses in computational problems, are summarized in Table 5.1 at the end of the chapter.

5.1 Basic Tools for Matrix Manipulation: The BLAS

There are certain tasks we do all the time. For example, we get dressed, we retrieve our mail, and we travel to work or school or other places. For each of these tasks, we develop shortcuts to minimize the amount of thought and effort that we need to give them.

Similarly, there are certain tasks that are common to many matrix problems, so shortcuts (subroutines) have been developed so that programmers don't need to redo the work each time. These **Basic Linear Algebra Subroutines** or BLAS were mentioned in Chapter 3. Libraries of BLAS are available for all of the standard languages, and using the BLAS has many advantages:

- The functions are reliable, so debugging our own programs is simpler.
- We only need one line to accomplish a complicated task, so our programs are more compact and easier to read and maintain.

POINTER 5.2. Special Kinds of Matrices.

Some matrices have important structure that we can exploit in algorithms. Here are some examples:

- An $n \times n$ matrix is **nonsingular** if it is full rank. A nonsingular matrix has an inverse.
- An $n \times n$ matrix $\boldsymbol{A}$ is **upper-Hessenberg** if it is zero below its first subdiagonal ($a_{ij} = 0$ if $i > j+1$). It is **tridiagonal** if it is zero below its first subdiagonal and above its first superdiagonal ($a_{ij} = 0$ if $i \neq j-1, j, j+1$)
- A real **orthogonal matrix** $\boldsymbol{U}$ satisfies the relation $\boldsymbol{U}^T\boldsymbol{U} = \boldsymbol{I}$, so $\boldsymbol{U}^T = \boldsymbol{U}^{-1}$ if $\boldsymbol{U}$ is square.
- A **permutation matrix** is the identity matrix with the order of its columns scrambled. It is a special case of an orthogonal matrix and it can be "stored" just by recording the column ordering in a vector.
- A **unitary matrix** $\boldsymbol{U}$ satisfies the relation $\boldsymbol{U}^*\boldsymbol{U} = \boldsymbol{I}$, so $\boldsymbol{U}^* = \boldsymbol{U}^{-1}$ if $\boldsymbol{U}$ is square.
- $\boldsymbol{A}$ is **symmetric** if $\boldsymbol{A}^T = \boldsymbol{A}$, and **Hermitian** if $\boldsymbol{A}^* = \boldsymbol{A}$.
- A real symmetric (or complex Hermitian) matrix $\boldsymbol{A}$ is **positive definite** if all of its eigenvalues are positive, or, equivalently, if $\boldsymbol{x}^*\boldsymbol{A}\boldsymbol{x} > 0$ for all $\boldsymbol{x} \neq \boldsymbol{0}$.

- The execution is optimized, so tasks are accomplished faster than we might be able to achieve in a high-level language.

The BLAS are partitioned by level, with a Level k BLAS performing $O(n^k)$ floating-point operations, where n is a dimension of the vector or matrix.

The **Level 1** BLAS perform vector operations. Let $\boldsymbol{x}$ and $\boldsymbol{y}$ be column vectors of length n and let a be a scalar. Examples of Level 1 operations are

- **scal** to compute $a\boldsymbol{x}$,
- **axpy** to compute $a\boldsymbol{x} + \boldsymbol{y}$,
- **dot** to compute $\boldsymbol{x}^*\boldsymbol{y}$ (the **dot-product** or **inner-product** of two vectors).

The **Level 2** BLAS perform operations that cost $O(n^2)$ floating-point computations. These include

- matrix-vector product,
- low-rank updates to a matrix (See the case study of Chapter 7),
- solution of linear systems involving triangular matrices.

The **Level 3** BLAS perform matrix-matrix operations, including

- matrix-matrix product,
- solution of multiple linear systems involving a triangular matrix.

When a BLAS exists for a task you need, it is a good idea to use it in your algorithm! MATLAB automatically uses the BLAS for tasks such as `A * B`. In other languages, you need to call on the BLAS yourself.

CHALLENGE 5.1.

Let A be an $m \times n$ matrix. Write a column-oriented MATLAB algorithm for computing

$$s_i = \sum_{j=1}^{n} |a_{ij}|$$

for $i = 1,2,\ldots,m$. Then use the BLAS, accessed through `norm`, to accomplish the same task, and compare the efficiency of the two algorithms.

5.2 The LU and Cholesky Decompositions

The **LU decomposition** of an $n \times n$ matrix A is defined by $PA = LU$, where P is a **permutation matrix**, L is a **unit lower-triangular** matrix (zero above the main diagonal and ones on the main diagonal) and U is an **upper-triangular matrix** (zero below the main diagonal). These matrices are computed in the process of **Gauss elimination**. We compute the LU decomposition by reducing the matrix A to upper-triangular form. We put zeros below the main diagonal, one column at a time, subtracting a multiple of the current pivot row from all rows below it. The multipliers form the entries of L.

We illustrate using an example with $P = I$. Consider the matrix

$$A = \begin{bmatrix} 4 & 4 & 8 \\ 2 & 8 & 7 \\ 1 & 3 & 6 \end{bmatrix}.$$

In Step 1, we set $U = A$ and then we subtract 1/2 times the first row of U from the second, and 1/4 times the first row from the third. We obtain the matrix

$$U = \begin{bmatrix} 4 & 4 & 8 \\ 0 & 6 & 3 \\ 0 & 2 & 4 \end{bmatrix},$$

In Step 2, we subtract 1/3 times the new second row from the new third row, obtaining

$$U = \begin{bmatrix} 4 & 4 & 8 \\ 0 & 6 & 3 \\ 0 & 0 & 3 \end{bmatrix},$$

POINTER 5.3. Existence, Uniqueness, and Sensitivity of Solutions to Linear Systems of Equations.

- The linear system of equations $\boldsymbol{Ax} = \boldsymbol{b}$, where $\boldsymbol{A}$ is $n \times n$, is guaranteed to have a unique solution if the matrix $\boldsymbol{A}$ is nonsingular.
- If $\boldsymbol{A}$ is singular, then a solution $\boldsymbol{x}_{good}$ exists only if $\boldsymbol{b}$ can be expressed as a linear combination of the columns of $\boldsymbol{A}$, and in that case any vector $\boldsymbol{x} = \boldsymbol{x}_{good} + \boldsymbol{y}$ is also a solution if $\boldsymbol{Ay} = \boldsymbol{0}$.
- Suppose we make a small change in our problem, solving instead a nearby problem

$$(\boldsymbol{A} + \Delta \boldsymbol{A})\boldsymbol{y} = \boldsymbol{b} + \Delta \boldsymbol{b}.$$

We might want to know how close the solution to this problem is to our original problem. To answer this question, define

$$\epsilon_A \equiv \frac{\|\Delta \boldsymbol{A}\|}{\|\boldsymbol{A}\|},$$
$$\epsilon_b \equiv \frac{\|\Delta \boldsymbol{b}\|}{\|\boldsymbol{b}\|},$$
$$\kappa \equiv \|\boldsymbol{A}\| \, \|\boldsymbol{A}^{-1}\|.$$

Then if $\kappa \epsilon_A < 1$ we have the bound

$$\frac{\|\boldsymbol{x} - \boldsymbol{y}\|}{\|\boldsymbol{x}\|} \leq \frac{\kappa}{1 - \kappa \epsilon_A} (\epsilon_A + \epsilon_b).$$

Consult a standard text such as [64, 119, 139] for a proof of this and related results, valid for any of the norms (1, 2, ∞) we have considered. We already used a special case of this inequality, when $\epsilon_b = 0$, in equation (2.1).

---

which is upper triangular. But notice that the multipliers are important, too, since we can express the original matrix as

$$\boldsymbol{A} = \begin{bmatrix} 4 & 4 & 8 \\ 2 & 8 & 7 \\ 1 & 3 & 6 \end{bmatrix} = \begin{bmatrix} 1 & 0 & 0 \\ 1/2 & 1 & 0 \\ 1/4 & 1/3 & 1 \end{bmatrix} \begin{bmatrix} 4 & 4 & 8 \\ 0 & 6 & 3 \\ 0 & 0 & 3 \end{bmatrix} \equiv \boldsymbol{LU}.$$

To obtain a reliable algorithm, it is necessary to **pivot**, or interchange rows in the matrix $\boldsymbol{U}$, as illustrated in Figure 5.1. The most common rule for pivoting is **partial pivoting**. Using this rule, at the kth stage of the decomposition, when we are putting zeros in column k of $\boldsymbol{U}$ below its main diagonal, we first interchange two rows of $\boldsymbol{U}$ (if necessary) so that the maximum magnitude element in column k, among those in rows k through n, appears on the main diagonal. In this way, all of the multipliers have magnitude bounded by 1, and this is sufficient for stability. The record of interchanges defines $\boldsymbol{P}$.

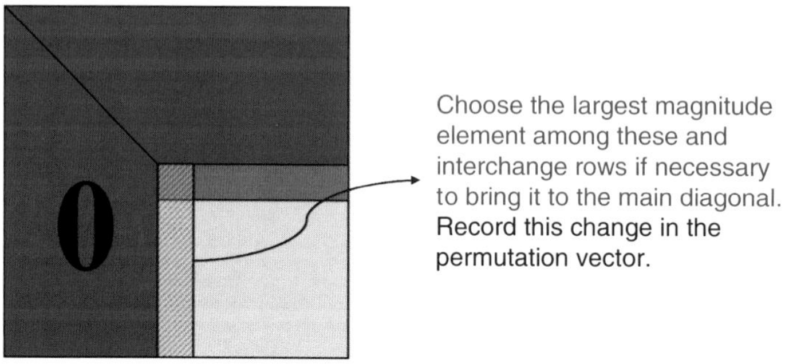

0
Choose the largest magnitude element among these and interchange rows if necessary to bring it to the main diagonal. Record this change in the permutation vector.

Figure 5.1. *Partial pivoting in computing the LU decomposition. We search the cross-hatched part of column k of the* **U** *matrix for the largest magnitude element and interchange rows to move it to the main diagonal position.*

Pivoting is essential for two reasons. First, we need pivoting because the LU decomposition of a nonsingular matrix **A** may not be possible. For example, the matrix

$$\boldsymbol{B} = \begin{bmatrix} 0 & 1 \\ 1 & 1 \end{bmatrix}$$

has no LU decomposition. But for any matrix $\boldsymbol{A}$ we can always find a permutation matrix $\boldsymbol{P}$ so that the matrix $\boldsymbol{PA}$ does have an LU decomposition. For our example, if we premultiply $\boldsymbol{B}$ by a permutation matrix that interchanges the two rows, we obtain

$$\boldsymbol{PB} \equiv \begin{bmatrix} 0 & 1 \\ 1 & 0 \end{bmatrix} \begin{bmatrix} 0 & 1 \\ 1 & 1 \end{bmatrix} = \begin{bmatrix} 1 & 1 \\ 0 & 1 \end{bmatrix},$$

so the LU decomposition of $\boldsymbol{PB}$ is just the identity matrix times $\boldsymbol{PB}$ itself.

Second, as already mentioned, the LU decomposition without pivoting is an unstable algorithm. Let δ be a positive number less than $\frac{1}{2}\epsilon_{mach}$ and consider, for example,

$$\boldsymbol{C} = \begin{bmatrix} \frac{\delta}{4} & 1 \\ 1 & 1 \end{bmatrix} = \begin{bmatrix} 1 & 0 \\ \frac{4}{\delta} & 1 \end{bmatrix} \begin{bmatrix} \frac{\delta}{4} & 1 \\ 0 & -\frac{4}{\delta}+1 \end{bmatrix}.$$

If $\delta/4$ and $4/\delta$ are exactly representable in floating-point arithmetic, then for small positive values of δ, the factors would be approximated in floating-point arithmetic as

$$\begin{bmatrix} 1 & 0 \\ \frac{4}{\delta} & 1 \end{bmatrix} \begin{bmatrix} \frac{\delta}{4} & 1 \\ 0 & -\frac{4}{\delta} \end{bmatrix},$$

and this is the exact decomposition of the matrix

$$\begin{bmatrix} \frac{\delta}{4} & 1 \\ 1 & 0 \end{bmatrix},$$

far from the matrix $\boldsymbol{C}$. Using forward and back substitution to solve the problem $\boldsymbol{Cx} = \boldsymbol{b}$ with these factors can lead to a large residual $\boldsymbol{b} - \boldsymbol{Cx}_c$, where $\boldsymbol{x}_c$ is the computed value.

With pivoting, however, we obtain the decomposition

$$\boldsymbol{PC} = \begin{bmatrix} 1 & 1 \\ \frac{\delta}{4} & 1 \end{bmatrix} = \begin{bmatrix} 1 & 0 \\ \frac{\delta}{4} & 1 \end{bmatrix} \begin{bmatrix} 1 & 1 \\ 0 & 1-\frac{\delta}{4} \end{bmatrix},$$

which would be approximated by

$$\begin{bmatrix} 1 & 0 \\ \frac{\delta}{4} & 1 \end{bmatrix} \begin{bmatrix} 1 & 1 \\ 0 & 1 \end{bmatrix}.$$

This is exact for the matrix

$$\begin{bmatrix} 1 & 1 \\ \frac{\delta}{4} & 1+\frac{\delta}{4} \end{bmatrix},$$

and the residual would be small.

We compute an LU decomposition in MATLAB by typing `[L,U,P] = lu(A)` or `[PtL,U] = lu(A)` to compute $P^T L$ and U. The arithmetic cost is $n^3/3 + O(n^2)$ multiplications.[5] In MATLAB, the **backslash** command `x = A \ b` uses the LU decomposition to solve the problem (or, if `A` is triangular, it uses substitution, discussed below).

The LU decomposition has two main uses: solving linear systems and computing the determinant of a matrix.

- To solve the linear system $\boldsymbol{Ax} = \boldsymbol{b}$, we first solve $\boldsymbol{Ly} = \boldsymbol{Pb}$ by solving equation i for y_i: for $i = 1, \dots, n$,

$$y_i = (\boldsymbol{Pb})_i - \sum_{j=1}^{i-1} \ell_{ij} y_j.$$

This is called **forward substitution**. Then we solve $\boldsymbol{Ux} = \boldsymbol{y}$ by **back substitution**, considering the equations from last to first: for $i = n, n-1, \dots, 1$,

$$x_i = \left(y_i - \sum_{j=i+1}^{n} u_{ij} x_j \right) / u_{ii}.$$

- If $\boldsymbol{PA} = \boldsymbol{LU}$, then

$$\det \boldsymbol{P} \, \det \boldsymbol{A} = \det \boldsymbol{L} \, \det \boldsymbol{U}.$$

The determinant of a triangular matrix is just the product of its main diagonal elements, and $\det(\boldsymbol{P}) = \pm 1$ (positive if the number of interchanges is even and negative otherwise). Therefore, since $\boldsymbol{L}$ has ones on its main diagonal, the determinant of $\boldsymbol{A}$ is just plus or minus the product of the diagonal entries of $\boldsymbol{U}$.

The following challenges give practice with these concepts.

---

[5]For large values of n, only the **high-order term** $n^3/3$ term is significant, so we generally neglect the **low-order terms** n^2, n, and constants.

POINTER 5.4. Testing Positive Definiteness.

By inspection, we can decide whether a matrix is symmetric or Hermitian, but positive definiteness (see Pointer 5.2) is not obvious. Luckily, we often know from the underlying physical problem that a matrix has this property. For example, if $\boldsymbol{A}$ is the matrix of second partial derivatives of a smooth strictly convex function, or if it defines the energy function for a variational problem, then it is positive definite. Another easy case is if a nonsingular $\boldsymbol{A}$ is formed as $\boldsymbol{B}^*\boldsymbol{B}$. Then $\boldsymbol{x}^*\boldsymbol{A}\boldsymbol{x} = \boldsymbol{x}^*\boldsymbol{B}^*\boldsymbol{B}\boldsymbol{x} = \|\boldsymbol{B}\boldsymbol{x}\|_2^2$ and this quantity is positive for nonzero $\boldsymbol{x}$ as long as $\boldsymbol{B}$ has linearly independent columns.

A test based on the **Gerschgorin circle theorem** [64] is sometimes useful. For any matrix $\boldsymbol{A}$, all of the eigenvalues of $\boldsymbol{A}$ lie in the union of the n circles in the complex plane defined by

$$\{z : |a_{ii} - z| \le \sum_{\substack{j=1 \\ j \ne i}}^{n} |a_{ij}|\}$$

for $i = 1, \ldots, n$. Therefore, if our matrix is real symmetric or complex Hermitian, and none of these circles touches the negative real axis, then the matrix is positive definite.

CHALLENGE 5.2.

Use back substitution to solve the linear system

$$\begin{bmatrix} 2 & 5 \\ 0 & 3 \end{bmatrix} \begin{bmatrix} x_1 \\ x_2 \end{bmatrix} = \begin{bmatrix} 8 \\ 6 \end{bmatrix}.$$

What is the determinant of the matrix?

CHALLENGE 5.3.

Write a column oriented algorithm to compute the determinant of $\boldsymbol{A}$ and to solve the linear system $\boldsymbol{A}\boldsymbol{x} = \boldsymbol{b}$ by forward substitution when $\boldsymbol{A}$ is an $n \times n$ nonsingular lower-triangular matrix.

One special case of triangular decomposition is important. If the matrix $\boldsymbol{A}$ is real symmetric and positive definite then we can use the **Cholesky decomposition** in place of LU. This gives a decomposition as

- $\boldsymbol{L}\boldsymbol{L}^T$, where $\boldsymbol{L}$ is lower triangular (see MATLAB's `chol`)
- or $\boldsymbol{L}\boldsymbol{D}\boldsymbol{L}^T$, where $\boldsymbol{L}$ is unit lower triangular and $\boldsymbol{D}$ is diagonal

at half the arithmetic cost of LU. The restriction to positive definite matrices assures that the decomposition is stable without pivoting. A similar decomposition exists for a Hermitian positive definite matrix.

CHALLENGE 5.4.

Suppose $\boldsymbol{A}$ is real $n \times n$ symmetric positive definite, and let $\boldsymbol{A} = \boldsymbol{L}\boldsymbol{L}^T$ be its Cholesky decomposition.

(a) Let $n = 3$ and write formulas for a_{ij}, $i, j = 1, 2, 3$, in terms of the entries of $\boldsymbol{L}$.

(b) Notice that if we consider the formulas in column order, we can compute the elements $\ell_{11}, \ell_{21}, \ell_{31}, \ell_{22}, \ell_{32}$, and ℓ_{33}. Use this insight to write a MATLAB function to compute the Cholesky decomposition for a general value of n. Check the correctness of your function by comparing with `chol`.

5.3 The QR Decomposition

The **QR decomposition** of a full-rank $m \times n$ matrix $\boldsymbol{A}$ ($m \geq n$) is defined by

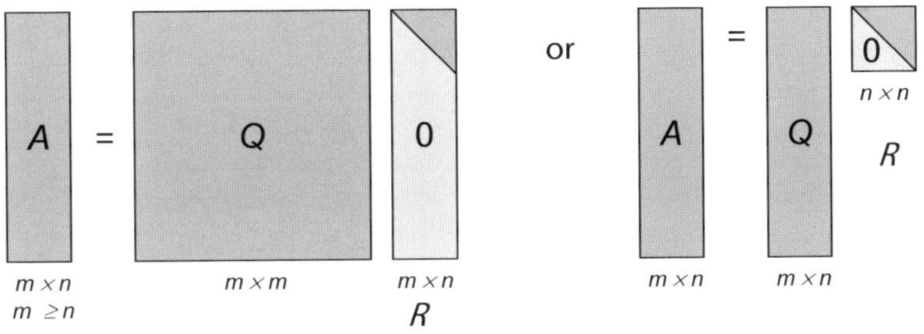

where

- $\boldsymbol{Q}$ is an $m \times m$ unitary matrix (orthogonal, if $\boldsymbol{A}$ is real) and $\boldsymbol{R}$ is an $m \times n$ matrix with zeros below the main diagonal (upper-trapezoidal) or
- $\boldsymbol{Q}$ is an $m \times n$ unitary matrix (orthogonal, if $\boldsymbol{A}$ is real) and $\boldsymbol{R}$ is an $n \times n$ upper-triangular matrix.

The compact $m \times n$ decomposition arises because part of the $m \times m$ matrix $\boldsymbol{Q}$ is not needed in the decomposition:

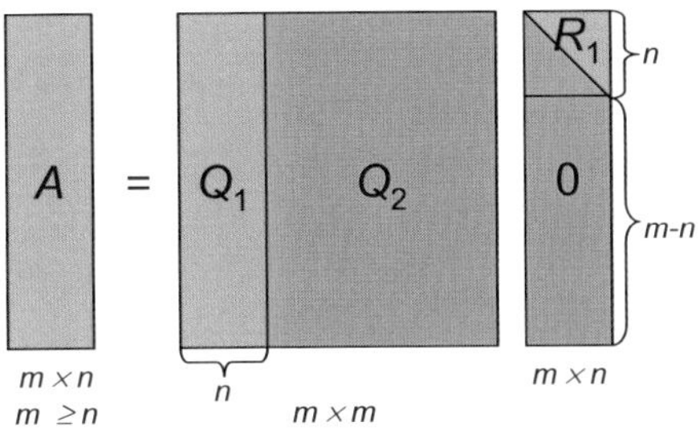

$$= Q_1 R_1 + Q_2 0 = Q_1 R_1$$

Although Q_2 is not needed to reconstruct A, it contains useful information, an orthogonal basis for the null space of A^*, as we will see in Section 5.3.3.

We'll discuss two algorithms for computing the QR decomposition:

- Givens rotations (good for Q $m \times m$),
- Gram-Schmidt orthogonalization (good for Q $m \times n$).

A third algorithm using **Householder transformations**, also useful, is presented as Challenge 5.10.

5.3.1 QR Decomposition by Givens Rotations

A simple orthogonal matrix, a **rotation**, can be used to introduce one zero at a time into a real matrix A. (We'll discuss the extension of the algorithm to complex matrices in Challenge 5.7.)

We'll write the **Givens rotation matrix** as

$$G = \begin{bmatrix} c & s \\ -s & c \end{bmatrix},$$

where $c^2 + s^2 = 1$. Thus, c and s have the geometric interpretation of the cosine and sine of an angle, and the product Gx rotates the vector x through that angle.

We use Givens rotations to rotate a given vector to point in the direction of the positive x-axis, thus putting a zero in its second entry. Given a vector $z \neq 0$ of dimension 2×1, we want to find G so that $Gz = xe_1$ where $x = \|z\|$ and e_1 is the first column of the identity matrix.

To solve this, note that

$$Gz = \begin{bmatrix} cz_1 + sz_2 \\ -sz_1 + cz_2 \end{bmatrix} = xe_1.$$

Multiplying the first equation by c and the second by s and then subtracting yields

$$(c^2 + s^2)z_1 = cx\,,$$

so

$$c = z_1/x\,.$$

Similarly, we can determine that $s = z_2/x$. Since $c^2 + s^2 = 1$, we conclude that

$$z_1^2 + z_2^2 = x^2,$$

so

$$c = \frac{z_1}{\sqrt{z_1^2 + z_2^2}},$$

$$s = \frac{z_2}{\sqrt{z_1^2 + z_2^2}}.$$

The next challenge gives practice in computing a real-valued Givens rotation.

CHALLENGE 5.5.

Find an orthogonal matrix $\boldsymbol{G}$ and a number w so that

$$\boldsymbol{G}\begin{bmatrix} 3 \\ 4 \end{bmatrix} = \begin{bmatrix} w \\ 0 \end{bmatrix}.$$

We can use Givens matrices to zero out single components of a matrix. We use the notation $\boldsymbol{G}_{ij}$ to denote an $n \times n$ identity matrix with its ith and jth rows modified to include the Givens rotation. For example, if $n = 6$, then

$$\boldsymbol{G}_{25} = \begin{bmatrix} 1 & 0 & 0 & 0 & 0 & 0 \\ 0 & c & 0 & 0 & s & 0 \\ 0 & 0 & 1 & 0 & 0 & 0 \\ 0 & 0 & 0 & 1 & 0 & 0 \\ 0 & -s & 0 & 0 & c & 0 \\ 0 & 0 & 0 & 0 & 0 & 1 \end{bmatrix}.$$

Multiplication of a vector by this matrix leaves all but rows 2 and 5 of the vector unchanged. If we choose the angle appropriately, multiplying a matrix by $\boldsymbol{G}_{25}$ can put a zero in some entry in row 5.

We can reduce a matrix $\boldsymbol{A}$ to upper-trapezoidal form by the sequence of rotations indicated in Algorithm 5.1. Since the product of orthogonal matrices is an orthogonal matrix, the product of the Givens rotation matrices is an $m \times m$ orthogonal matrix that we call $\boldsymbol{Q}^T$. We use a variant of this algorithm in the next challenge.

Algorithm 5.1 QR via Givens Rotations

Initialize $\boldsymbol{Q}$ to be the $m \times m$ identity matrix.
Initialize $\boldsymbol{R}$ to be the $m \times n$ matrix $\boldsymbol{A}$.
for $i = 1, \ldots, n$,
  for $j = i+1, \ldots, m$,
    Choose the matrix $\boldsymbol{G}_{ij}$ to put a zero in position (j,i) of the matrix $\boldsymbol{R}$, using the current value in position (i,i), and set $\boldsymbol{R} = \boldsymbol{G}_{ij}\boldsymbol{R}$.
    $\boldsymbol{Q} = \boldsymbol{Q}\boldsymbol{G}_{ij}^*$.
  end
end

CHALLENGE 5.6.

MATLAB's function `W = planerot(y)` takes a 2×1 vector $\boldsymbol{y}$ as input and returns a Givens matrix $\boldsymbol{W}$ so that $\boldsymbol{W}\boldsymbol{y}$ has a zero in its 2nd position.

Write a MATLAB program that uses `planerot` to reduce a matrix of the form

$$\boldsymbol{A} = \begin{bmatrix} \times & \times & \times & \times \\ \times & \times & \times & \times \\ 0 & \times & \times & \times \\ 0 & 0 & \times & \times \end{bmatrix}$$

to upper-triangular form, where $\times$ indicates a nonzero element. (In other words, do a QR decomposition of this matrix, but don't worry about saving $\boldsymbol{Q}$.)

A derivation similar to the one above determines (complex) Givens matrices to reduce a complex matrix $\boldsymbol{A}$ to triangular form.

CHALLENGE 5.7.

For a given 2×1 vector z with complex entries, let $c = |z_1|/\|z\|$. Determine s in the matrix

$$\boldsymbol{G} = \begin{bmatrix} c & s \\ -\bar{s} & c \end{bmatrix}$$

so that $\boldsymbol{G}$ is unitary and $\boldsymbol{G}z = \|z\|_2 \boldsymbol{e}_1$. (Note that $\bar{s}$ denotes the complex conjugate of the number s.) This is the complex extension of the idea behind the Givens matrix and it can be used in applying Algorithm 5.1 to complex matrices.

5.3.2 QR by Gram–Schmidt Orthogonalization

An alternative to the Givens QR uses **Gram–Schmidt orthogonalization** to compute the decomposition. From the columns $[\boldsymbol{a}_1, \ldots, \boldsymbol{a}_n]$ of the matrix $\boldsymbol{A}$, we create an orthonormal basis $\{\boldsymbol{q}_1, \ldots, \boldsymbol{q}_n\}$ (the columns of $\boldsymbol{Q}$) and save the coefficients that compute this basis in an upper-triangular matrix $\boldsymbol{R}$. We summarize this in Algorithm 5.2.

Algorithm 5.2 Gram–Schmidt Orthogonalization

$r_{11} = \|\boldsymbol{a}_1\|$
$\boldsymbol{q}_1 = \boldsymbol{a}_1 / r_{11}$
(Comment: Compute the component of $\boldsymbol{q}_{k+1}$ orthogonal to all previous vectors $\boldsymbol{q}_1, \ldots, \boldsymbol{q}_k$.)
for $k = 1, \ldots, n-1$,
  $\boldsymbol{q}_{k+1} = \boldsymbol{a}_{k+1}$
  for $i = 1, \ldots, k$,
    $r_{i,k+1} = \boldsymbol{q}_i^* \boldsymbol{q}_{k+1}$
    $\boldsymbol{q}_{k+1} = \boldsymbol{q}_{k+1} - r_{i,k+1} \boldsymbol{q}_i$
  end
  (Comment: Normalize $\boldsymbol{q}_{k+1}$ to length 1.)
  $r_{k+1,k+1} = \|\boldsymbol{q}_{k+1}\|$
  $\boldsymbol{q}_{k+1} = \boldsymbol{q}_{k+1} / r_{k+1,k+1}$
end

Convince yourself that Algorithm 5.2 computes a matrix decomposition $\boldsymbol{A} = \boldsymbol{QR}$, where $\boldsymbol{Q}$ is $m \times n$. The geometry of the transformation is illustrated in Figure 5.2, and you can try the algebra in the next challenge.

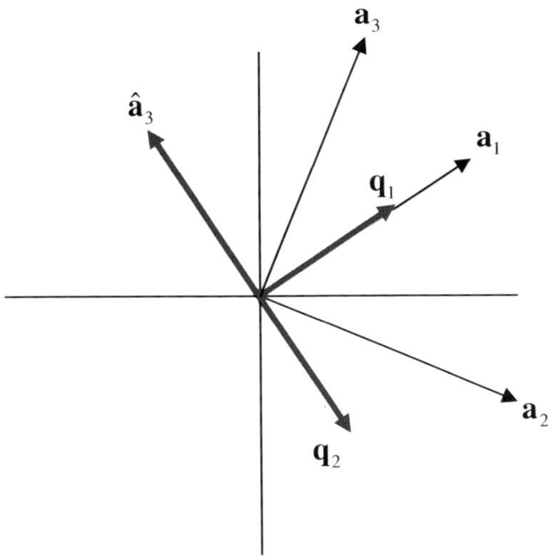

Figure 5.2. *The geometry of the QR decomposition. The vector* $\mathbf{a}_1$ *is normalized to length 1 to obtain the first basis vector* $\mathbf{q}_1$*. We find the component of* $\mathbf{a}_2$ *orthogonal to this vector and normalize to obtain* $\mathbf{q}_2$*. If we have a third vector* $\mathbf{a}_3$*, we find that its component orthogonal to* $\mathbf{q}_1$ *is* $\widehat{\mathbf{a}}_3$*, which has no component orthogonal to* $\mathbf{q}_2$*, since in* $\mathcal{R}^2$*, we can have at most two linearly independent basis vectors.*

CHALLENGE 5.8.
Compute a QR decomposition of the matrix

$$\begin{bmatrix} 3 & 3 \\ 3 & 1 \end{bmatrix}$$

using

(a) a Givens rotation,

(b) Gram–Schmidt orthogonalization.

Let's make sure we believe that the Gram–Schmidt algorithm produces orthogonal vectors.

CHALLENGE 5.9.
For the Gram–Schmidt algorithm, show (by finite induction) that $q_i^* q_k = 0$ for $i < k$.

Finally, in the next challenge we consider a third QR algorithm, based on **Householder transformations**.

CHALLENGE 5.10. (Challenging)

(a) Suppose we are given an $n \times 1$ vector z. Represent its first coordinate in polar notation as $z_1 = e^{i\theta}\zeta$, where ζ is real. Define $v = z - \alpha e_1$, where $\alpha = -e^{i\theta}\|z\|$. Let $u = v/\|v\|$, and define

$$Q = I - 2uu^*.$$

Verify that Q is a unitary matrix and that $Qz = \alpha e_1$. The matrix Q is one version of a **Householder transformation**.

(b) Given an $m \times n$ matrix A, we can determine a Householder transformation Q_1 so that $A_1 = Q_1 A$ has zeros in its first column below the main diagonal. Show how to determine Q_2 in the form

$$Q_2 = \begin{bmatrix} 1 & \mathbf{0} \\ \mathbf{0} & I - 2u_2u_2^* \end{bmatrix}$$

so that $A_2 = Q_2 A_1$ has zeros in its first and second column below the main diagonal.

(c) Continuing this process, write an algorithm to reduce a matrix A to upper-triangular form by multiplying by a series of Householder transformations.

(d) How many floating-point multiplications does your algorithm take? (Hint: When $m = n$, your answer should reduce to $2n^3/3 + O(n^2)$ multiplications to form R.)

5.3.3 Computing and Using the QR Decomposition

In MATLAB, we compute the QR decomposition of the $m \times n$ matrix A ($m \geq n$) by typing `[Q,R] = qr(A)`. Alternatively, `qr(A,0)` returns the compact matrix Q of dimension $m \times n$ (although first the full $m \times m$ matrix is computed). Householder transformations are used. The function `orth` can be used if only the compact Q is needed and not R.

The arithmetic cost of the QR decomposition is approximately

- $2mn^2 - 2/3n^3$ multiplications, using Givens rotations (without explicit formation of Q),
- $mn^2 - 1/3n^3$ multiplications, using Householder reflections (without explicit formation of Q),
- mn^2 multiplications, using Gram–Schmidt orthogonalization.

A nice feature of the QR decomposition is that in general, we don't need to pivot to preserve numerical stability. This makes QR a useful alternative to LU for solving linear systems. Although the operations count is twice as big, the data handling is simpler. To solve $Ax = b$, we use back substitution to solve $Rx = Q^*b$, taking advantage of the fact that $Q^{-1} = Q^*$.

If the columns of A are linearly dependent, or are close to being linearly dependent, then the QR decomposition does not behave well, and the rank-revealing QR of Section 5.4 or the SVD of Section 5.6 should be used instead.

An important use of the QR decomposition is to obtain **orthogonal bases** for the range of a full-rank matrix $\boldsymbol{A}$ and the null space of $\boldsymbol{A}^*$, i.e., a set of orthonormal vectors spanning these spaces. Suppose $\boldsymbol{A} = \boldsymbol{QR}$. Then

$$\boldsymbol{A} = \boldsymbol{QR} = \begin{bmatrix} \boldsymbol{Q}_1 & \boldsymbol{Q}_2 \end{bmatrix} \begin{bmatrix} \boldsymbol{R}_1 \\ \boldsymbol{0} \end{bmatrix} = \begin{bmatrix} \boldsymbol{q}_1 \dots \boldsymbol{q}_n & \boldsymbol{q}_{n+1} \dots \boldsymbol{q}_m \end{bmatrix} \begin{bmatrix} \boldsymbol{R}_1 \\ \boldsymbol{0} \end{bmatrix}.$$

Since $\boldsymbol{Q}$ has linearly independent columns, we can express any m-vector z as

$$z = \sum_{i=1}^{m} c_i \boldsymbol{q}_i$$

for some coefficients c_i. Note that

$$\boldsymbol{Q}^* z = \begin{bmatrix} \boldsymbol{q}_1^* z \\ \vdots \\ \boldsymbol{q}_m^* z \end{bmatrix} = \begin{bmatrix} c_1 \\ \vdots \\ c_m \end{bmatrix}$$

and

$$\boldsymbol{A}^* z = \boldsymbol{R}^* \boldsymbol{Q}^* z = \begin{bmatrix} \boldsymbol{R}_1^* & \boldsymbol{0} \end{bmatrix} \begin{bmatrix} c_1 \\ \vdots \\ c_m \end{bmatrix} = \boldsymbol{R}_1^* \begin{bmatrix} c_1 \\ \vdots \\ c_n \end{bmatrix}.$$

Therefore, if $\boldsymbol{R}_1$ is full rank, then $\boldsymbol{A}^* z = \boldsymbol{0}$ if and only if $c_1 = \ldots = c_n = 0$, which happens if and only if z can be expressed as a linear combination of the columns of $\boldsymbol{Q}_2$, so these columns are an orthogonal basis for the null space of $\boldsymbol{A}^*$.

Similarly, since

$$\boldsymbol{Ax} = \boldsymbol{Q}_1 \boldsymbol{R}_1 \boldsymbol{x},$$

every vector in the range of $\boldsymbol{A}$ can be expressed as a linear combination of columns of $\boldsymbol{Q}_1$. Thus the first n columns of $\boldsymbol{Q}$ form an orthogonal basis for the range of $\boldsymbol{A}$ and the last $m - n$ columns of $\boldsymbol{Q}$ form an orthogonal basis for the null space of $\boldsymbol{A}^*$.

By far the most common use of the QR decomposition is to solve linear least squares problems

$$\min_{\boldsymbol{x}} \|\boldsymbol{b} - \boldsymbol{Ax}\|$$

when $\boldsymbol{A}$ has more rows than columns. The MATLAB backslash command, `x = A \ b`, which solves a linear system when $\boldsymbol{A}$ is square and nonsingular, uses the QR decomposition to solve the problem in the least squares sense when $m > n$. The **residual** is defined as

$$\boldsymbol{r} = \boldsymbol{b} - \boldsymbol{Ax}. \tag{5.1}$$

The solution process is based on three fundamental facts, whose proof we leave to the next challenge.

CHALLENGE 5.11.
Prove the following facts:

- Minimizing the norm of $\boldsymbol{b} - \boldsymbol{Ax}$ gives the same solution as minimizing the square of the norm.
- The norm of a vector is invariant under multiplication by $\boldsymbol{Q}^*$, so $\|\boldsymbol{y}\| = \|\boldsymbol{Q}^*\boldsymbol{y}\|$ for any vector $\boldsymbol{y}$.
- Suppose we partition the vector $\boldsymbol{y}$ into two pieces:

$$\boldsymbol{y} = \begin{bmatrix} \boldsymbol{y}_1 \\ \boldsymbol{y}_2 \end{bmatrix}.$$

Then $\|\boldsymbol{y}\|^2 = \|\boldsymbol{y}_1\|^2 + \|\boldsymbol{y}_2\|^2$.

Now suppose we have factored the full-rank matrix $\boldsymbol{A} = \boldsymbol{QR}$, and define

$$\boldsymbol{c} = \boldsymbol{Q}^*\boldsymbol{b} = \begin{bmatrix} \boldsymbol{c}_1 \\ \boldsymbol{c}_2 \end{bmatrix},$$

where $\boldsymbol{c}_1$ is $n \times 1$ and $\boldsymbol{c}_2$ is $(m-n) \times 1$. Then

$$\begin{aligned} \|\boldsymbol{b} - \boldsymbol{Ax}\|^2 &= \|\boldsymbol{Q}^*(\boldsymbol{b} - \boldsymbol{Ax})\|^2 \\ &= \|\boldsymbol{c} - \boldsymbol{Rx}\|^2 \\ &= \|\boldsymbol{c}_1 - \boldsymbol{R}_1\boldsymbol{x}\|^2 + \|\boldsymbol{c}_2 - \boldsymbol{0x}\|^2 \\ &= \|\boldsymbol{c}_1 - \boldsymbol{R}_1\boldsymbol{x}\|^2 + \|\boldsymbol{c}_2\|^2. \end{aligned}$$

To minimize this quantity, we make the first term zero by taking $\boldsymbol{x}$ to be the solution to the $n \times n$ linear system $\boldsymbol{R}_1\boldsymbol{x} = \boldsymbol{c}_1$. This leads to Algorithm 5.3. The MATLAB command `x = A \ b` performs the first three steps.

Algorithm 5.3 Linear Least Squares via QR for a Full-Rank Matrix *A*

Factor $\boldsymbol{A} = \boldsymbol{QR}$.
Partition $\boldsymbol{Q} = [\boldsymbol{Q}_1, \boldsymbol{Q}_2]$, where $\boldsymbol{Q}_1$ contains the first n columns.
Compute $\boldsymbol{c}_1 = \boldsymbol{Q}_1^*\boldsymbol{b}$.
Solve $\boldsymbol{R}_1\boldsymbol{x} = \boldsymbol{c}_1$ by back substitution.
The norm of the residual can be computed as $\|\boldsymbol{b} - \boldsymbol{Ax}\|$ or as $\|\boldsymbol{c}_2\| = \|\boldsymbol{Q}_2^*\boldsymbol{b}\|$.

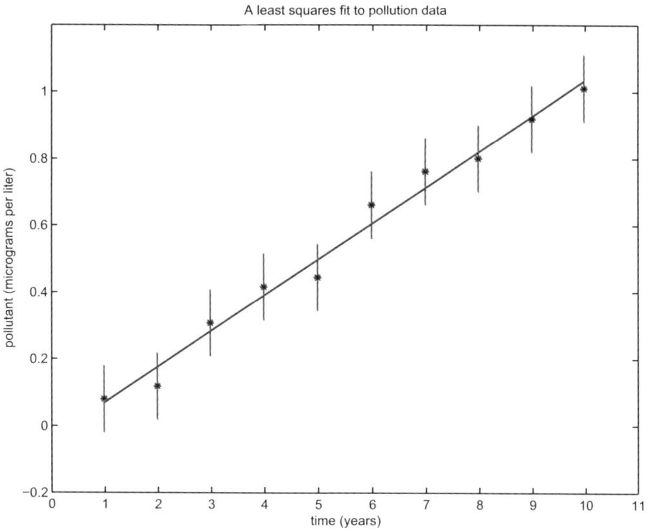

Figure 5.3. *Is a straight line a good model to this data? The measured values of the pollutant are indicated by stars, and the vertical lines indicate the error bars, an uncertainty of two standard deviations plus or minus.*

CHALLENGE 5.12.

Try justifying the QR algorithm for least squares without referring to the previous derivation. Suppose we have factored the full-rank $m \times n$ matrix $\boldsymbol{A} = \boldsymbol{QR}$ ($m \geq n$), and let $\widehat{\boldsymbol{x}}$ be the solution to the least squares problem

$$\min_{\boldsymbol{x}} \|\boldsymbol{b} - \boldsymbol{A}\boldsymbol{x}\|.$$

Show that $\|\boldsymbol{b} - \boldsymbol{A}\widehat{\boldsymbol{x}}\|^2 = \|\boldsymbol{c}_2\|^2$, where $\boldsymbol{c}_2$ is the vector consisting of the last $m-n$ components of $\boldsymbol{Q}^*\boldsymbol{b}$.

5.3.4 Mini Case Study: Least Squares Data Fitting

We consider the problem of fitting a model to data in order to reduce the effects of noise in the measurements.

As an example, we consider the data (t_i, f_i), $i = 1, \ldots, 10$, in Figure 5.3, representing the amount (μg per liter) of a pollutant in a river, measured once a year. We want to know whether a straight line is a good fit to this data. We suppose that the errors in the measurements f_i are independent, with mean 0 and standard deviation equal to $\sigma = 0.05$. We express the line as $x_1 + tx_2$ and determine the parameters x_1 and x_2 by solving

$$\min_{\boldsymbol{x}} \|\boldsymbol{b} - \boldsymbol{A}\boldsymbol{x}\|,$$

POINTER 5.5. Existence, Uniqueness, and Sensitivity of Solutions to Linear Least Squares Problems.

- A solution to the problem

$$\min_{\boldsymbol{x}} \|\boldsymbol{b} - \boldsymbol{A}\boldsymbol{x}\|$$

always exists, and it is unique if the $m \times n$ matrix $\boldsymbol{A}$ has rank n. If the rank of $\boldsymbol{A}$ is less than n, and if $\boldsymbol{x}_{good}$ solves the problem, then any vector $\boldsymbol{x} = \boldsymbol{x}_{good} + \boldsymbol{y}$ is also a solution if $\boldsymbol{A}\boldsymbol{y} = \boldsymbol{0}$.

- Suppose we make a small change in our problem, solving instead a nearby problem

$$\min_{\boldsymbol{y}} \|(\boldsymbol{A} + \Delta\boldsymbol{A})\boldsymbol{y} - (\boldsymbol{b} + \Delta\boldsymbol{b})\|.$$

We might want to know how close the solution to this problem is to our original problem. To answer this question, define

$$\epsilon_A \equiv \frac{\|\Delta\boldsymbol{A}\|}{\|\boldsymbol{A}\|},$$
$$\epsilon_b \equiv \frac{\|\Delta\boldsymbol{b}\|}{\|\boldsymbol{b}\|}.$$

Then if the rank of $\boldsymbol{A}$ is equal to the rank of $\boldsymbol{A} + \Delta\boldsymbol{A}$, and if $\kappa\epsilon_A < 1$ (where κ is the square root of the ratio of the largest to the smallest eigenvalue of $\boldsymbol{A}^*\boldsymbol{A}$), the solution to the problem can change by at most

$$\|\boldsymbol{x} - \boldsymbol{y}\| \le \frac{\kappa}{1 - \kappa\epsilon_A}\left(\epsilon_A\|\boldsymbol{x}\| + \epsilon_b\frac{\|\boldsymbol{b}\|}{\|\boldsymbol{A}\|} + \epsilon_A\kappa\frac{\|\boldsymbol{r}_x\|}{\|\boldsymbol{A}\|}\right) + \epsilon_A\kappa\|\boldsymbol{x}\|.$$

Similarly, the residuals $\boldsymbol{r}_x = \boldsymbol{b} - \boldsymbol{A}\boldsymbol{x}$ and $\boldsymbol{r}_y = (\boldsymbol{b} + \Delta\boldsymbol{b}) - (\boldsymbol{A} + \Delta\boldsymbol{A})\boldsymbol{y}$ are related by

$$\|\boldsymbol{r}_y - \boldsymbol{r}_x\| \le \epsilon_A\|\boldsymbol{x}\|\|\boldsymbol{A}\| + \epsilon_b\|\boldsymbol{b}\| + \epsilon_A\kappa\|\boldsymbol{r}_x\|.$$

See the excellent book by Björck [16, Thm. 1.4.6] for the proof of these results.

---

where

$$\boldsymbol{A} = \begin{bmatrix} 1 & t_1 \\ \cdot & \cdot \\ \cdot & \cdot \\ \cdot & \cdot \\ 1 & t_{10} \end{bmatrix}, \quad \boldsymbol{b} = \begin{bmatrix} f_1 \\ \cdot \\ \cdot \\ \cdot \\ f_{10} \end{bmatrix}.$$

We compute and plot the straight-line fit using the following MATLAB program:

```
sigma=.05
t = [1:10];
b = ...    % the values of f
plot(t,b,'g*')
hold on
for i=1:10,
    plot([t(i),t(i)],[b(i)+2*sigma,b(i)-2*sigma])
end
axis([0 11 -.2 1.2])
A = [ones(10,1),t'];
x = A \ b';
plot(t,A*x,'m')
xlabel('time (years)')
ylabel('pollutant (micrograms per liter)')
title('A least squares fit to pollution data')
```

Figure 5.3 shows the data (with error bars covering the range $\pm 2\sigma$) and the straight line produced by least squares. The fit is quite good, with the straight line passing through all of the error bars.

5.4 The Rank-Revealing QR Decomposition (RR-QR)

The **RR-QR decomposition** of an $m \times n$ matrix $\boldsymbol{A}$ is defined by $\boldsymbol{AP} = \boldsymbol{QR}$, where $\boldsymbol{P}$ is a permutation matrix chosen so that the leading principal submatrix of $\boldsymbol{R}$ of dimension $p \times p$ is well conditioned and the other diagonal block (of dimension $(n-p) \times (n-p)$) has entries of small magnitude. This means that

$$\boldsymbol{R} = \begin{bmatrix} \boldsymbol{R}_1 & \boldsymbol{F} \\ \boldsymbol{0} & \boldsymbol{R}_2 \\ \boldsymbol{0} & \boldsymbol{0} \end{bmatrix},$$

where $\boldsymbol{R}_1$ is $p \times p$ and upper triangular, $\boldsymbol{R}_2$ is $(n-p) \times (n-p)$ and upper triangular, and $\|\boldsymbol{R}_2\|$ is small relative to the main diagonal elements of $\boldsymbol{R}_1$. The **numerical rank** of $\boldsymbol{A}$ is p, indicating that $\boldsymbol{A}$ is quite close to the matrix $\boldsymbol{Q}\widehat{\boldsymbol{R}}$ of rank p, where $\widehat{\boldsymbol{R}}$ is obtained by replacing $\boldsymbol{R}_2$ by $\boldsymbol{0}$.

We can compute this decomposition by a pivoted version of our Givens QR decomposition, Algorithm 5.4. A similar modification can be made to the Gram–Schmidt and Householder versions. The number of multiplications in computing the RR-QR decomposition is the same as for the original decomposition, whichever version is used, but the pivoting introduces $O(mn)$ extra overhead.

In MATLAB, `[Q,R,P] = qr(A)` produces a rank-revealing QR decomposition.

The RR-QR decomposition has two main uses in addition to those of the QR decomposition. We can use the RR-QR decomposition to

- Determine whether $\boldsymbol{A}$ is rank-deficient. As discussed above, we can even use the RR-QR to construct a matrix of lower rank that is close to $\boldsymbol{A}$.

Algorithm 5.4 Givens QR with Pivoting

for $i = 1 : n$,
  Among columns $i : n$ of rows $i : m$ of the current $\boldsymbol{A}$ matrix, choose the column with largest norm.
  Permute so that this column becomes the ith column.
  Perform Givens rotations to put zeros below row i in column i of $\boldsymbol{A}$.
end

- Compute a "poor man's" principal component analysis (PCA). If we are modeling $\boldsymbol{b}$ as $\boldsymbol{b} \approx \boldsymbol{A}\boldsymbol{x}$, then the RR-QR decomposition tells us that we can produce almost as good a fit using p parameters instead of n, and the first p elements of the permutation tell us which parameters to use.

The RR-QR decomposition can be fooled into overestimating the numerical rank of the matrix (but it never underestimates it). If reliability is essential, use the somewhat more expensive SVD of Section 5.6.

5.5 Eigendecomposition

The **eigendecomposition** of a matrix $\boldsymbol{A}$ of dimension $n \times n$ is

$$\boldsymbol{A} = \boldsymbol{U}\boldsymbol{\Sigma}\boldsymbol{U}^{-1}, \qquad (5.2)$$

where $\boldsymbol{\Sigma}$ is a diagonal matrix with the **eigenvalues** λ_i as its entries. The columns of $\boldsymbol{U}$ are the **right eigenvectors**,

$$\boldsymbol{A}\boldsymbol{u}_i = \lambda_i \boldsymbol{u}_i\,,$$

and the rows of $\boldsymbol{U}^{-1}$ are the **left eigenvectors** z_i^*, satisfying

$$z_i^* \boldsymbol{A} = \lambda_i z_i^*\,.$$

The decomposition is guaranteed to exist if

- $\boldsymbol{A}$ is real symmetric or complex Hermitian, or
- the eigenvalues of $\boldsymbol{A}$ are distinct.

Otherwise, the decomposition may fail to exist, although it always exists for a nearby matrix.

5.5.1 Computing the Eigendecomposition

The basic algorithm to compute the eigendecomposition is simple, but the refinements that make it work really well are not. We'll just focus on the basics. When you do an eigendecomposition, make sure you choose high-quality software to compute it. The algorithm is a two-step process.

POINTER 5.6. Existence, Uniqueness, and Sensitivity of Solutions to the Eigenproblem.

Here are some basic results, found, for example, in [64, 139]. Other properties of the eigenproblem are presented in Chapter 31.

- Every $n \times n$ matrix has n eigenvalues (counting multiplicities), the roots of the equation $\det(\boldsymbol{A} - \lambda \boldsymbol{I}) = 0$, but computing the roots using this polynomial equation is unstable.
- Almost every square matrix has a full set of linearly independent eigenvectors; the only exceptions are matrices which have a Jordan block of size greater than one; see Section 5.7. If the matrix is real symmetric or complex Hermitian, there is a full set of orthogonal eigenvectors.
- If the multiplicity of the eigenvalue λ is one, then the corresponding eigenvector is unique, up to multiplication by a nonzero constant.
- If the multiplicity of an eigenvalue λ is greater than 1, then although the eigenvectors corresponding to this eigenvalue are not unique, the subspace that they span, known as an **invariant subspace** of the matrix, is unique.
- For general matrices $\boldsymbol{A}$ and $\boldsymbol{E}$, suppose $\boldsymbol{A}$ has a full set of n linearly independent eigenvectors, the columns of the matrix $\boldsymbol{U}$, and let $\kappa(\boldsymbol{U}) = \|\boldsymbol{U}\| \|\boldsymbol{U}^{-1}\|$ using either the $1, 2$, or ∞ norm.
  - (Bauer–Fike theorem) The matrix $\boldsymbol{A}+\boldsymbol{E}$ has an eigenvalue within $\kappa(\boldsymbol{U})\|\boldsymbol{E}\|$ of each eigenvalue of $\boldsymbol{A}$.
  - If we have a number $\tilde{\lambda}$ and a vector $\boldsymbol{z}$ of norm 1 satisfying $\|\boldsymbol{A}\boldsymbol{z} - \tilde{\lambda}\boldsymbol{z}\| = \epsilon$, then $\tilde{\lambda}$ is within $\epsilon\,\kappa(\boldsymbol{U})$ of an eigenvalue of $\boldsymbol{A}$.
- Suppose that $\boldsymbol{A}$ and $\boldsymbol{E}$ are real symmetric or Hermitian matrices. Denote the kth largest eigenvalue of a matrix $\boldsymbol{B}$ by $\lambda_k(\boldsymbol{B})$, the largest by $\lambda_{\max}$, and the smallest by $\lambda_{\min}$. Then
  - for $k = 1, \ldots, n$,

$$\lambda_k(\boldsymbol{A}) + \lambda_{\min}(\boldsymbol{E}) \leq \lambda_k(\boldsymbol{A}+\boldsymbol{E}) \leq \lambda_k(\boldsymbol{A}) + \lambda_{\max}(\boldsymbol{E}).$$

  - (Weilandt–Hoffman theorem)

$$\sum_{k=1}^{n} (\lambda_k(\boldsymbol{A}+\boldsymbol{E}) - \lambda_k(\boldsymbol{A}))^2 \leq \|\boldsymbol{E}\|_F^2.$$

Results on the sensitivity of eigenvectors can be found in the references mentioned above.

- Step 1: Reduce the matrix $\boldsymbol{A}$ to compact form, so that it is easy to manipulate. To do this, find a unitary matrix $\boldsymbol{V}$ so that

$$\boldsymbol{V}^*\boldsymbol{A}\boldsymbol{V} = \boldsymbol{H},$$

  where $\boldsymbol{H}$ is

  - **tridiagonal** if $\boldsymbol{A}$ is Hermitian (or real symmetric),
  - **upper-Hessenberg** otherwise.

  This can be done in $O(n^3)$ operations using Givens rotations. For example, if $\boldsymbol{A}$ is 4×4, then we choose $\boldsymbol{G}_{24}$ to reduce the $(4,1)$ element of $\boldsymbol{A}$ to zero, $\boldsymbol{G}_{23}$ to reduce the $(3,1)$ element of $\boldsymbol{G}_{24}\boldsymbol{A}\boldsymbol{G}_{24}^*$ to zero, and $\boldsymbol{G}_{34}$ to reduce the $(4,2)$ element of $\boldsymbol{G}_{23}\boldsymbol{G}_{24}\boldsymbol{A}\boldsymbol{G}_{24}^*\boldsymbol{G}_{23}^*$ to zero, obtaining $\boldsymbol{V} = \boldsymbol{G}_{24}^*\boldsymbol{G}_{23}^*\boldsymbol{G}_{34}^*$ and $\boldsymbol{H} = \boldsymbol{G}_{34}\boldsymbol{G}_{23}\boldsymbol{G}_{24}\boldsymbol{A}\boldsymbol{G}_{24}^*\boldsymbol{G}_{23}^*\boldsymbol{G}_{34}^* = \boldsymbol{V}^*\boldsymbol{A}\boldsymbol{V}$. Note that we have applied a **similarity transform**, so if we find an eigendecomposition of $\boldsymbol{H}$ as

$$\boldsymbol{H} = \boldsymbol{U}\boldsymbol{\Sigma}\boldsymbol{U}^{-1}$$

  then the eigendecomposition of $\boldsymbol{A}$ is

$$\boldsymbol{A} = (\boldsymbol{V}\boldsymbol{U})\boldsymbol{\Sigma}(\boldsymbol{V}\boldsymbol{U})^{-1}.$$

  The arithmetic cost of this step is $O(n^3)$.

- Step 2: Find the eigendecomposition of $\boldsymbol{H}$ by **QR iteration**:

  - Form $\boldsymbol{H} = \boldsymbol{Q}\boldsymbol{R}$.
  - Replace $\boldsymbol{H}$ by $\boldsymbol{R}\boldsymbol{Q}$.

  Note that since $\boldsymbol{Q}^*\boldsymbol{Q} = \boldsymbol{I}$ and $\boldsymbol{H} = \boldsymbol{Q}\boldsymbol{R}$,

$$\boldsymbol{R}\boldsymbol{Q} = (\boldsymbol{Q}^*\boldsymbol{Q})\boldsymbol{R}\boldsymbol{Q} = \boldsymbol{Q}^*\boldsymbol{H}\boldsymbol{Q}.$$

  Therefore we have performed a similarity transform, so the new $\boldsymbol{H}$ has the same eigenvalues as the old one, and if we find an eigendecomposition of $\boldsymbol{R}\boldsymbol{Q}$, then we have an eigendecomposition of $\boldsymbol{H}$.

  The arithmetic cost of this step is $O(n^2)$.

We repeat Step 2 $O(n)$ times. Although it is not obvious, some subdiagonal elements of $\boldsymbol{H}$ often converge to zero. Once that happens, we can read some eigenvalues off the diagonal.

In order to ensure convergence of all subdiagonal elements and make it faster, there are many refinements to the algorithm, mostly involving shifting the matrix (by subtracting a multiple γ of the identity matrix, factoring $\boldsymbol{H} - \gamma\boldsymbol{I} = \boldsymbol{Q}\boldsymbol{R}$ and forming the new $\boldsymbol{H}$ by $\boldsymbol{R}\boldsymbol{Q} + \gamma\boldsymbol{I}$) in order to emphasize one eigenvalue for the iteration. The resulting algorithm typically requires about $2n$ iterations of Step 2 [37, p. 173]), so the total arithmetic cost is $O(n^3)$.

In MATLAB, we compute the eigendecomposition as `[U,Lambda] = eig(A)`.

In general, the eigenvalues with large magnitude are computed more accurately than those of small magnitude. This happens because, due to rounding error, a stable algorithm computes the exact eigendecomposition for some matrix close to the given matrix $\boldsymbol{A}$. The eigenvalues cannot be changed by more than the norm of the perturbation, so the large eigenvalues are computed quite accurately. But if $\boldsymbol{A}$ has an eigenvalue close to zero, the relative error in its computed value may be large.

Applications of the eigendecomposition include determining modes of resonance of structures (see Chapter 31), determining the long-term behavior of a Markov chain (used implicitly in Chapters 17 and 19), determining the solution (or stability) of a system of linear differential equations (Section 20.1.2), and analyzing the stability of control systems, as in the next section.

5.5.2 Mini Case Study: Stability Analysis of a Linear Control System

Any automatic system must accomplish its goal without going out of control. For example, the antilock braking system on an automobile must be able to bring the car to a stop as quickly as possible without causing it to skid. **Control theory** is a complicated area, but the first thing to know about it is the importance of **stability** of the system.[6] Without stability, small changes in the controls create arbitrarily large changes in the response of the system, and this can be dangerous.

Suppose we have a model of our system:

$$\begin{aligned} \boldsymbol{x}_{k+1} &= \boldsymbol{A}\boldsymbol{x}_k + \boldsymbol{B}\boldsymbol{u}_k, \\ \boldsymbol{v}_k &= \boldsymbol{C}\boldsymbol{x}_k + \boldsymbol{D}\boldsymbol{u}_k, \end{aligned}$$

where

- $\boldsymbol{A}$ $(n \times n)$, $\boldsymbol{B}$ $(n \times p)$, $\boldsymbol{C}$ $(q \times n)$, and $\boldsymbol{D}$ $(q \times p)$ are design matrices,
- $\boldsymbol{x}_k$ is the **state vector** of the system at time k,
- $\boldsymbol{u}_k$ defines the **controls** used on the system at time k,
- $\boldsymbol{v}_k$ is some observation of the system at time k.

For stability, i.e., for small changes in $\boldsymbol{u}$ to make small changes in $\boldsymbol{x}$, it is sufficient that $\|\boldsymbol{A}\|_2 < 1$.

For example, we might want to design a system with

$$\boldsymbol{A} = \begin{bmatrix} .5 & .4 & a \\ a & .3 & .4 \\ .3 & .3 & .3 \end{bmatrix}, \tag{5.3}$$

where a is a parameter to be determined. We might wish to determine the range of values of a for which $\|\boldsymbol{A}\|_2 < 1$. This is one type of stability problem; other matrix stability problems involve determining whether all of the eigenvalues of a matrix lie in the left half-plane.

CHALLENGE 5.13.

Determine the range of values of a in (5.3) for which $\|\boldsymbol{A}\|_2 < 1$.

---

[6]Using the terminology in Chapter 1, we would call the system well-conditioned or ill-conditioned, but the jargon of the field is stable or unstable.

5.5.3 Other Uses for Eigendecompositions

The eigendecomposition of a matrix determines its invariant subspaces and is also used to study convergence of iterations, as demonstrated in the next two challenges.

CHALLENGE 5.14. (Challenging: requires considerable comfort with linear algebra.)

Let A be an $n \times n$ matrix with linearly independent eigenvectors $\boldsymbol{u}_1, \ldots, \boldsymbol{u}_n$. Let S be a subspace of $\mathcal{R}^n$ such that for any $\boldsymbol{x} \in S$, the vector $\boldsymbol{Ax}$ is also in S. We call S an **invariant subspace** for $\boldsymbol{A}$.

Show that S is an invariant subspace if a subset of the eigenvectors of A contains a basis for it. (This is actually an *if and only if statement*, but proving the other direction is more challenging.)

CHALLENGE 5.15.

Let $\boldsymbol{A}$ be an $n \times n$ matrix with a full set of linearly independent eigenvectors. Given a vector $\boldsymbol{x}^{(0)}$, consider the iteration

$$\boldsymbol{x}^{(k+1)} = \boldsymbol{A}\boldsymbol{x}^{(k)} + \boldsymbol{b},$$

for $k = 0, 1, \ldots$. (This is called a **stationary iterative method** or **SIM**.)

(a) Suppose we have a vector $\boldsymbol{x}_{true}$ satisfying

$$\boldsymbol{x}_{true} = \boldsymbol{A}\boldsymbol{x}_{true} + \boldsymbol{b}.$$

Such a vector is called a **fixed point** of the iteration. Show that if $\boldsymbol{e}^{(k)} = \boldsymbol{x}^{(k)} - \boldsymbol{x}_{true}$, then

$$\boldsymbol{e}^{(k+1)} = \boldsymbol{A}\boldsymbol{e}^{(k)}.$$

(b) Show by induction that

$$\boldsymbol{e}^{(k)} = \boldsymbol{A}^k \boldsymbol{e}^{(0)}.$$

(c) Show that $\boldsymbol{e}^{(k)} \to \boldsymbol{0}$ as $k \to \infty$ for any initial vector $\boldsymbol{x}^{(0)}$ if and only if all eigenvalues of $\boldsymbol{A}$ lie within the unit circle; i.e., if and only if $|\lambda_j| < 1$ for $j = 1, \ldots, n$. Hint: To do this, consider expressing $\boldsymbol{e}^{(0)}$ as

$$\boldsymbol{e}^{(0)} = \sum_{j=1}^{n} \alpha_j \boldsymbol{u}_j,$$

where $\boldsymbol{u}_j$ are the eigenvectors of $\boldsymbol{A}$ and the values α_j are appropriate coefficients. Now compute $\boldsymbol{A}^k \boldsymbol{e}^{(0)}$ and study its convergence.

POINTER 5.7. The Relation between the 2-Norm and the SVD.

The largest singular value of a matrix is its 2-norm. The 2-norm condition number $\kappa(\boldsymbol{A})$ is the ratio between the largest singular value and the smallest. Equivalently, it is the square root of the ratio between the largest and smallest eigenvalues of $\boldsymbol{A}^*\boldsymbol{A}$.

5.6 The Singular Value Decomposition (SVD)

Every matrix $\boldsymbol{A}$ of dimension $m \times n$ $(m \geq n)$ can be decomposed as

$$\boldsymbol{A} = \boldsymbol{U}\boldsymbol{\Sigma}\boldsymbol{V}^*,$$

where

- $\boldsymbol{U}$ has dimension $m \times m$ and $\boldsymbol{U}^*\boldsymbol{U} = \boldsymbol{I}$,
- $\boldsymbol{\Sigma}$ has dimension $m \times n$, the only nonzeros are on the main diagonal, and they are nonnegative real numbers $\sigma_1 \geq \sigma_2 \geq \cdots \geq \sigma_n \geq 0$,
- $\boldsymbol{V}$ has dimension $n \times n$ and $\boldsymbol{V}^*\boldsymbol{V} = \boldsymbol{I}$.

This is the SVD. Observe that

$$\begin{aligned} \boldsymbol{A}^*\boldsymbol{A} &= (\boldsymbol{U}\boldsymbol{\Sigma}\boldsymbol{V}^*)^*\boldsymbol{U}\boldsymbol{\Sigma}\boldsymbol{V}^* \\ &= \boldsymbol{V}\boldsymbol{\Sigma}^*\boldsymbol{U}^*\boldsymbol{U}\boldsymbol{\Sigma}\boldsymbol{V}^* = \boldsymbol{V}\boldsymbol{\Sigma}^*\boldsymbol{\Sigma}\boldsymbol{V}^*. \end{aligned}$$

So we have a decomposition of the real symmetric matrix $\boldsymbol{A}^*\boldsymbol{A}$ as a unitary matrix times a diagonal matrix times the conjugate transpose of the unitary, and the diagonal matrix has entries σ_i^2. Therefore, by the eigendecomposition (5.2), we have the following facts:

- The **singular values** σ_i of $\boldsymbol{A}$ are the square roots of the eigenvalues of $\boldsymbol{A}^*\boldsymbol{A}$.
- The columns of $\boldsymbol{V}$ are the **right singular vectors** of $\boldsymbol{A}$ and the eigenvectors of $\boldsymbol{A}^*\boldsymbol{A}$.
- By forming $\boldsymbol{A}\boldsymbol{A}^*$, we would see that the columns of $\boldsymbol{U}$, which are the **left singular vectors** of $\boldsymbol{A}$, are the eigenvectors of $\boldsymbol{A}\boldsymbol{A}^*$.

5.6.1 Computing and Using the SVD

Because of the SVD-eigendecomposition relation, algorithms to compute the SVD are variants on algorithms for computing eigendecompositions. We won't give details, but note that it is rather complicated, so you would want to use a high-quality existing function rather than writing your own. Also because of this relationship, the singular values are quite stable with respect to perturbations in the matrix $\boldsymbol{A}$.

In MATLAB we compute `[U,S,V] = svd(A)`. The arithmetic cost of computing the SVD is $O(mn^2)$ when $m \geq n$. The constant is usually of order 10.

Uses of the SVD include solving ill-conditioned least squares problems and solving discretized ill-posed problems (See the case study in Chapter 6.).

The SVD is expensive but the information it provides is quite reliable. Its main virtue is that it can be used to solve almost any of the problems we have discussed: solving linear systems of equations, determining the range and null space of a matrix, and solving least squares problems and other problems, as we see in the following challenges.

CHALLENGE 5.16.

Suppose we have an SVD of a nonsingular matrix $\boldsymbol{A}$ of dimension $n \times n$: $\boldsymbol{A} = \boldsymbol{U}\boldsymbol{\Sigma}\boldsymbol{V}^*$. Given the SVD, how many multiplications would it take to solve the linear system $\boldsymbol{A}\boldsymbol{x} = \boldsymbol{b}$? (In your answer, make clear how you would use the factors to solve the problem.)

CHALLENGE 5.17.

If $\boldsymbol{A} = \boldsymbol{U}\boldsymbol{\Sigma}\boldsymbol{V}^*$, find an orthogonal basis for

(a) the range of $\boldsymbol{A}$,

(b) the null space of $\boldsymbol{A}^*$.

CHALLENGE 5.18.

(a) Suppose the matrix $\boldsymbol{A}$ is $m \times n$ and suppose that $\boldsymbol{b}$ is in the range of $\boldsymbol{A}$. Give a formula for all solutions to the equation $\boldsymbol{A}\boldsymbol{x} = \boldsymbol{b}$ in terms of the SVD of $\boldsymbol{A}$, when $m \geq n$ and $\boldsymbol{A}$ has rank $p < n$.

(b) Suppose the matrix $\boldsymbol{A}$ is $m \times n$ with $\text{rank}(A) = n < m$. Give a formula for all solutions to the equation $\boldsymbol{A}^*\boldsymbol{x} = \boldsymbol{b}$ in terms of the SVD of $\boldsymbol{A}$.

So if you are stranded on a desert island and allowed only one piece of linear algebra software, which should you choose?

5.6.2 Mini Case Study: Solving Ill-Conditioned and Rank-Deficient Least Squares Problems

Suppose we want to solve the least squares problem

$$\min_{\boldsymbol{x}} \|\boldsymbol{b} - \boldsymbol{A}\boldsymbol{x}\|_2 .$$

As in (5.1), we define the residual to be $\boldsymbol{r} = \boldsymbol{b} - \boldsymbol{A}\boldsymbol{x}$. Since $\|\boldsymbol{r}\|^2 = \boldsymbol{r}^*\boldsymbol{r} = (\boldsymbol{U}^*\boldsymbol{r})^*(\boldsymbol{U}^*\boldsymbol{r}) = \|\boldsymbol{U}^*\boldsymbol{r}\|^2$, we can minimize the norm of $\boldsymbol{r}$ by minimizing the norm of $\boldsymbol{U}^*\boldsymbol{r} = \boldsymbol{U}^*\boldsymbol{b} - \boldsymbol{U}^*\boldsymbol{A}\boldsymbol{x} = \boldsymbol{c} - \boldsymbol{\Sigma}\boldsymbol{V}^*\boldsymbol{x}$, where

$$c_i = \boldsymbol{u}_i^*\boldsymbol{b}, \qquad i = 1, \ldots, m,$$

and $\boldsymbol{u}_i$ is the ith column of $\boldsymbol{U}$. If we change the coordinate system (as in Section 5.3.3) by letting $\boldsymbol{w} = \boldsymbol{V}^*\boldsymbol{x}$, then our problem is to minimize

$$\|\boldsymbol{U}^*\boldsymbol{r}\|^2 = |c_1 - \sigma_1 w_1|^2 + \cdots + |c_n - \sigma_n w_n|^2 + |c_{n+1}|^2 + \cdots + |c_m|^2 .$$

POINTER 5.8. Matrix Approximation.

Two other facts about the SVD are useful:

- The SVD expresses the matrix $\boldsymbol{A}$ as a sum of rank-1 matrices

$$\boldsymbol{A} = \boldsymbol{U}\boldsymbol{\Sigma}\boldsymbol{V}^* = \sum_{j=1}^{n} \sigma_j \boldsymbol{u}_j \boldsymbol{v}_j^*,$$

and it is worthwhile to take the time to verify this.

- The best rank p approximation to $\boldsymbol{A}$, in the sense of minimizing both $\|\boldsymbol{A} - \tilde{\boldsymbol{A}}\|_2$ and $\|\boldsymbol{A} - \tilde{\boldsymbol{A}}\|_F$, is

$$\tilde{\boldsymbol{A}} = \sum_{j=1}^{p} \sigma_j \boldsymbol{u}_j \boldsymbol{v}_j^*.$$

Thus the SVD can be useful in matrix approximation problems.

In the next two challenges we see that the SVD gives us not only an algorithm for solving the linear least squares problem, but also a measure of the sensitivity of the solution $\boldsymbol{x}$ to small changes in the data $\boldsymbol{b}$.

CHALLENGE 5.19.

Our data vector $\boldsymbol{b}$ has been measured with some error. Let $\boldsymbol{b}_{true}$ be the true but unknown data, and let $\boldsymbol{A}\boldsymbol{x}_{true} = \boldsymbol{b}_{true}$.

(a) The columns of the matrix $\boldsymbol{V} = [\boldsymbol{v}_1, \ldots, \boldsymbol{v}_n]$ form an orthonormal basis for n-dimensional space. Let's express the solution $\boldsymbol{x}_{true}$ to the least squares problem as

$$\boldsymbol{x}_{true} = w_1 \boldsymbol{v}_1 + \cdots + w_n \boldsymbol{v}_n .$$

Determine a formula for w_i $(i = 1, \ldots, n)$ in terms of $\boldsymbol{U}$, $\boldsymbol{b}_{true}$, and the singular values of $\boldsymbol{A}$.

(b) Justify the reasoning behind these two statements.

$$\boldsymbol{A}(\boldsymbol{x} - \boldsymbol{x}_{true}) = \boldsymbol{b} - \boldsymbol{b}_{true} - \boldsymbol{r} \text{ means } \|\boldsymbol{x} - \boldsymbol{x}_{true}\| \le \frac{1}{\sigma_n}(\|\boldsymbol{b} - \boldsymbol{b}_{true} - \boldsymbol{r}\|),$$

$$\boldsymbol{b}_{true} = \boldsymbol{A}\boldsymbol{x}_{true} \text{ means } \|\boldsymbol{b}_{true}\| = \|\boldsymbol{A}\boldsymbol{x}_{true}\| \le \|\boldsymbol{A}\|\,\|\boldsymbol{x}_{true}\|.$$

(c) Use the two statements above and the fact that $\|\boldsymbol{A}\| = \sigma_1$ to derive an upper bound on $\|\boldsymbol{x} - \boldsymbol{x}_{true}\|/\|\boldsymbol{x}_{true}\|$ in terms of the condition number $\kappa(\boldsymbol{A}) \equiv \sigma_1/\sigma_n$ and $\|\boldsymbol{b} - \boldsymbol{b}_{true} - \boldsymbol{r}\|/\|\boldsymbol{b}_{true}\|$.

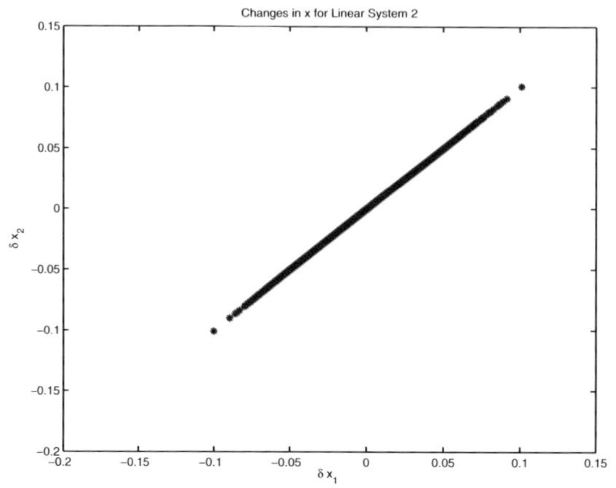

Figure 5.4. *Perturbations in the two coordinates of the solution to an ill-conditioned linear system.*

CHALLENGE 5.20.

Denote the SVD of the 2×2 matrix $\boldsymbol{A}$ by $\boldsymbol{U\Sigma V}^*$.

(a) Express the solution to the linear system $\boldsymbol{Ax} = \boldsymbol{b}$ as $\boldsymbol{x} = \alpha_1 \boldsymbol{v}_1 + \alpha_2 \boldsymbol{v}_2$, where $\boldsymbol{V} = [\boldsymbol{v}_1, \boldsymbol{v}_2]$.

(b) Consider the linear system $\boldsymbol{Ax} = \boldsymbol{b}$ with

$$\boldsymbol{A} = \begin{bmatrix} 1+\delta & \delta-1 \\ \delta-1 & 1+\delta \end{bmatrix}, \qquad \boldsymbol{b} = \begin{bmatrix} 2 \\ -2 \end{bmatrix},$$

and $\delta = 0.002$, and suppose we compute the solution to the nearby systems

$$(\boldsymbol{A} + \boldsymbol{E}^{(i)})\boldsymbol{x}^{(i)} = \boldsymbol{b}$$

for $i = 1, \ldots, 1000$, where the elements of $\boldsymbol{E}^{(i)}$ are independent and normally distributed with mean 0 and standard deviation $\tau = .0001$. Note that this system is very ill-conditioned; the graphs of the two equations lie almost on top of each other. Using part (a), explain why the resulting solutions all fall near a straight line, as shown in Figure 5.4.

Sometimes we want to solve least squares problems in which the rank of the matrix $\boldsymbol{A}$ is $p < n$. Convince yourself that the problem has many solutions, but a solution $\boldsymbol{x}$ of minimal norm is computed by Algorithm 5.5.

5.7 Some Matrix Tasks to Avoid

There are some matrix computations that should be avoided, either because there are better alternatives or they are too expensive. Here are two examples:

Algorithm 5.5 Minimum-Norm Solution to a Least Squares Problem

Compute $c = U^* b$, where $A = U\Sigma V^*$.
Let p be the number of nonzero singular values of A.
for $j = 1, \ldots, p$,
  Set $w_j = c_j/\sigma_j$.
end
The minimum norm solution is $x = V(:, 1:p)w$.
The norm of the residual is $(|c_{p+1}|^2 + \cdots + |c_m|^2)^{1/2}$.

- Computing a matrix inverse. We can solve $Ax = b$ by multiplying both sides of the equation by A^{-1}:

$$A^{-1}Ax = x = A^{-1}b.$$

  Therefore, we can solve linear systems by multiplying the right-hand side b by A^{-1}.

  This is generally a BAD idea. It is more expensive than the LU decomposition and it generally computes an answer that has larger error. Using the LU decomposition with pivoting in floating-point arithmetic is guaranteed to produce a vector x that solves a nearby problem $Ax = b - r$, where r is small. There is no such guarantee when you solve using A^{-1}.

  So, whenever you see a matrix inverse in a formula, think "LU decomposition."

  Because of this, never compute a matrix inverse unless you really want to look at the entries in it. Otherwise, find a decomposition that accomplishes your task. Your answer is then generally more accurate and less expensive than if you used the inverse matrix. If you choose to use A^{-1}, at least be aware of the trade-offs.

- Computing the Jordan canonical form of a matrix. As we mentioned above, some matrices do not have an eigendecomposition. For example,

$$A = \begin{bmatrix} 1 & 1 & 0 & 0 & 0 \\ 0 & 1 & 1 & 0 & 0 \\ 0 & 0 & 1 & 1 & 0 \\ 0 & 0 & 0 & 1 & 1 \\ 0 & 0 & 0 & 0 & 1 \end{bmatrix}$$

  has an eigenvalue 1 of multiplicity 5, but its only right-eigenvector is e_1, the first column of the identity matrix. Therefore, the full-rank matrix of eigenvectors does not exist. (The matrix is called **defective** since it fails to have a full set of eigenvectors.) This is an example of a **Jordan block**, with a multiple eigenvalue and only one linearly independent eigenvector.

  Every matrix can be decomposed into **Jordan canonical form** as

$$A = WJW^{-1},$$

  where W is nonsingular and J is almost diagonal, with the eigenvalues on the main diagonal and ones in positions on the superdiagonal defining the Jordan blocks. If the eigendecomposition exists, then this decomposition is just the eigendecomposition

POINTER 5.9. Software for Matrix Decompositions.

For computing matrix decompositions and solving matrix problems in FORTRAN or C, look for LAPACK software [2] (more than 20 million downloads!). For Java, see `http://math.nist.gov/javanumerics/`. These systems provide

- numerically stable algorithms,
- a uniform interface, making them easy to use,
- row or column oriented implementation, appropriate for the matrix storage scheme used by the language,
- efficient software built using BLAS.

(More precisely, they are efficient when n is large (100 or more). The overhead for small n is rather big.)

($\boldsymbol{J}$ is diagonal) and the computation can be done as in Section 5.5. If not, rounding error introduced in computing the eigendecomposition generally changes the matrix enough to make it nondefective so that an eigendecomposition exists.

Note that the $\boldsymbol{A}$ in our example is arbitrarily close to a matrix that has a full set of eigenvectors:

$$\begin{bmatrix} 1 & 1 & 0 & 0 & 0 \\ 0 & 1+\epsilon & 1 & 0 & 0 \\ 0 & 0 & 1-\epsilon & 1 & 0 \\ 0 & 0 & 0 & 1+2\epsilon & 1 \\ 0 & 0 & 0 & 0 & 1-2\epsilon \end{bmatrix},$$

for example, where $\epsilon > 0$ is arbitrarily small. This is true for any defective matrix, and for this reason it is difficult to determine whether a given matrix is defective.

There are alternatives to Jordan canonical form that are generally just as useful computationally; in particular, the **Schur decomposition**, which factors $\boldsymbol{A}$ as $\boldsymbol{URU}^*$ with $\boldsymbol{U}$ unitary and $\boldsymbol{R}$ upper triangular, also displays the eigenvalues, and its computation is stable [139]. See Chapter 30 for a use of this decomposition.

5.8 Summary

We summarize in Table 5.1 our matrix decompositions and their uses. We review this material in the following two challenges.

CHALLENGE 5.21.

Suppose we have a matrix $\boldsymbol{A}$ of dimension $n \times n$ of rank $n-1$. Explain two numerically stable ways to find a nonzero vector $\boldsymbol{z}$ so that $\boldsymbol{Az} = \boldsymbol{0}$.

POINTER 5.10. Further Reading.

More information on the LAPACK software can be found in the users' guide [2]. Matrix decompositions are covered in more detail in [64] and [139].

Table 5.1. *A summary of our matrix decompositions.*

Decomposition	Multiplications	Examples of uses
LU	$n^3/3$	Solving linear systems. Computing determinants.
QR	mn^2 (Gram–Schmidt)	Solving well-conditioned linear least squares problems. Representing the range or null space of a matrix.
rank-revealing QR	mn^2 (Gram–Schmidt)	Determining whether a matrix is rank-deficient. Representing the range or null space of a matrix. Solving rank-deficient least squares problems. Fitting a model with a reduced number of parameters.
eigen-decomposition	$O(n^3)$	Determining eigenvalues or eigenvectors of a matrix. Determining invariant subspaces. Determining stability of a control system. Determining convergence of $\mathbf{A}^p$ as $p \to \infty$.
SVD	$O(mn^2)$	Solving ill-conditioned linear least squares problems. Solving discretizations of ill-posed problems. Fitting a model with a reduced number of parameters. Representing the range or null space of a matrix. Computing an approximation to a matrix.

CHALLENGE 5.22.

Choose a matrix decomposition that can be used to efficiently solve each of the following problems in a stable manner, and review how the solution can be computed given the decomposition.

(a) Find the null space of a matrix.

(b) Solve a least squares problem when the matrix is well conditioned.

(c) Determine the rank of a matrix.

(d) Find the determinant of a matrix.

(e) Determine whether a symmetric matrix is positive definite.

Chapter 6 / Case Study

Image Deblurring: I Can See Clearly Now

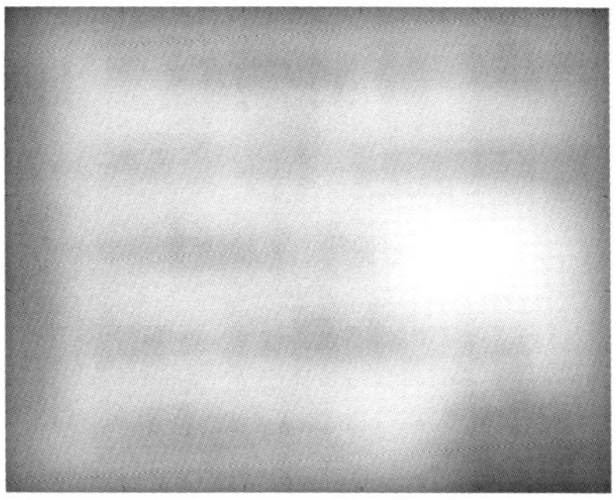

(coauthored by James G. Nagy)

Inverse problems are among the most challenging computations in science and engineering. They involve determining the parameters of a system that is only observed indirectly. For example, we may have a spectrum and want to determine the species that produced it, as well as their relative proportions. Or we may have taken sonar measurements of a containment tank and want to decide whether it has a hidden crack.

In this case study we consider such an inverse problem: given a blurred image and a linear model for the blurring, reconstruct the original image. This is a linear inverse problem (a **discrete ill-posed problem** that comes from an **integral equation of the first kind** (See Pointer 6.1). It illustrates the impact of **ill-conditioning** on the choice of algorithms. Although deblurring your vacation pictures might be important to you, the techniques we study are even more important for applications such as interpreting CAT scans or astronomical images.

Consider a linear system of equations

$$\boldsymbol{K}\boldsymbol{f} = \boldsymbol{g},$$

where $\boldsymbol{K}$ is a real $n \times n$ matrix, and $\boldsymbol{f}$ and $\boldsymbol{g}$ are vectors. Such a system turns out to be a good model for the process of blurring an image $\boldsymbol{f}$ by a blurring matrix $\boldsymbol{K}$ to obtain an observed image $\boldsymbol{g}$; see Pointers 6.1 and 6.2 for the connection. Now suppose that $\boldsymbol{K}$ is scaled so that its largest singular value is $\sigma_1 = 1$. If the smallest singular value is $\sigma_n \approx 0$, then $\boldsymbol{K}$ is ill-conditioned. We distinguish two types of ill-conditioning:

- The matrix $\boldsymbol{K}$ is considered **numerically rank deficient** if there is a j such that

  $$\sigma_j \gg \sigma_{j+1} \approx \cdots \approx \sigma_n \approx 0.$$

  That is, there is an obvious gap between large and small singular values.

- If the singular values decay to zero with no particular gap in the spectrum, then we say the linear system $\boldsymbol{K}\boldsymbol{f} = \boldsymbol{g}$ is a **discrete ill-posed problem**.

Figure 6.1. *Can you deblur this image?*

POINTER 6.1. Where Does the Image Data Come From?

In our model $\boldsymbol{Kf} + \boldsymbol{\eta} = \boldsymbol{g}$, the vector $\boldsymbol{g}$ contains the recorded image, and the noise $\boldsymbol{\eta}$ arises from imperfections in recording the data.

Let f be the true image. Then f is actually a function over some two-dimensional domain that we call Ω, pehaps $[0,1] \times [0,1]$. The function values are the intensities of the image at each coordinate (s_1, s_2) in the domain.

Let g be the recorded image. Again, g is actually a function over the two-dimensional domain, but we have only a few samples of this function, perhaps an $n_r \times n_c$ array of pixel values which we may assume are measured at points $\boldsymbol{s}_{jk} = (j/n_r, k/n_c)$ for $j = 1, \ldots, n_r$, $k = 1, \ldots, n_c$. To form the vector we call $\boldsymbol{g}$, we stack the columns of this two-dimensional array of values to form a single column with $n_r n_c$ entries. We use these recorded values to estimate a vector $\boldsymbol{f}$, which might correspond to samples at the same points $\boldsymbol{s}_{jk}$ as those that determine $\boldsymbol{g}$.

It is very difficult to compute accurate approximate solutions of discrete ill-posed problems, especially because in most real applications, the right-hand-side vector $\boldsymbol{g}$ is not known exactly. Rather, it is more typical that the collected data have the form

$$\boldsymbol{g} = \boldsymbol{Kf} + \boldsymbol{\eta},$$

where $\boldsymbol{\eta}$ is a vector representing (unknown) noise or measurement errors. The goal, then, is: Given an ill-conditioned matrix $\boldsymbol{K}$ and a vector $\boldsymbol{g}$, compute an approximation of the unknown vector $\boldsymbol{f}$. Naïvely solving $\boldsymbol{Kf} = \boldsymbol{g}$ usually does not work, since the matrix $\boldsymbol{K}$ is so ill-conditioned. Instead, one usually uses something called **regularization** to make the problem less sensitive to the noise.

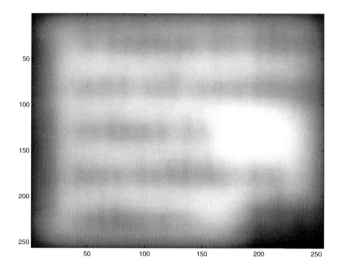

POINTER 6.2. Where Does the Matrix Problem Come From?

For simplicity, we consider the noise-free case with $\boldsymbol{\eta} = \mathbf{0}$. Then the recorded image g is the result of the **convolution** of the true image f with a recording device specified by a **kernel function** K so that

$$g(\boldsymbol{s}) = \int_{\Omega} K(\boldsymbol{s},\boldsymbol{t}) f(\boldsymbol{t}) d\boldsymbol{t}.$$

If $K(\boldsymbol{s},\boldsymbol{t}) = \delta(\|\boldsymbol{s}-\boldsymbol{t}\|)$, where δ is the Dirac δ function, then $g(\boldsymbol{s}) = f(\boldsymbol{s})$; this is the ideal case, and K is zero almost everywhere.

In practical situations, K is not this nice, although it often has **small support**, so that $K(\boldsymbol{s},\boldsymbol{t})$ is zero when $\boldsymbol{t}$ and $\boldsymbol{s}$ are not close to each other. In this case, the value of the integral is a weighted average of values of f in a neighborhood of $\boldsymbol{s}$.

We obtain the matrix equation $\boldsymbol{g} = \boldsymbol{K}\boldsymbol{f}$ by discretizing the integral. The row of this equation corresponding to s_{jk} approximates the relation

$$g(\boldsymbol{s}_{jk}) = \int_{\Omega} K(\boldsymbol{s}_{jk},\boldsymbol{t}) f(\boldsymbol{t}) d\boldsymbol{t} \approx \sum_{\ell=1}^{n_r} \sum_{p=1}^{n_c} w_{\ell p} K(\boldsymbol{s}_{jk},\boldsymbol{t}_{\ell p}) f(\boldsymbol{t}_{\ell p}),$$

where the values $w_{\ell p}$ are chosen to make the approximation as accurate as desired. For example, choosing $w_{\ell p} = 1/(n_r n_c)$ for all values of ℓ and p gives a **rectangle rule** for integration. If we use our sample values of $\boldsymbol{s}$ as sample values for $\boldsymbol{t}$, then the entry in the row of $\boldsymbol{K}$ corresponding to $\boldsymbol{s}_{jk}$ and the column corresponding to $\boldsymbol{s}_{\ell p}$ is $w_{\ell p} K(\boldsymbol{s}_{jk},\boldsymbol{s}_{\ell p})$, and this defines our matrix problem.

If the kernel function has the property that $K(\boldsymbol{s},\boldsymbol{t})$ depends only on the difference $\boldsymbol{s}-\boldsymbol{t}$, then ordering the pixels row-by-row (or column-by-column) gives a matrix with Kronecker product structure.

One additional issue needs to be addressed: how do we determine K? Usually it is either modeled by some mathematical function or measured by aiming the camera at a point source: a picture that is black except for a single white pixel. By moving that white pixel and repeating the measurement—or by assuming that the image is unchanged except for translation as we move the white pixel—we can approximately determine all of the values $K(\boldsymbol{s}_{jk},\boldsymbol{s}_{\ell p})$.

Method 1: Tikhonov Regularization

The best-known regularization procedure, called **Tikhonov regularization**, computes a solution of the **damped least squares** problem

$$\min_{\boldsymbol{f}} \{\|\boldsymbol{g} - \boldsymbol{K}\boldsymbol{f}\|_2^2 + \alpha^2 \|\boldsymbol{f}\|_2^2\} \tag{6.1}$$

The extra term $\alpha^2\|\boldsymbol{f}\|_2^2$ imposes a penalty for making the norm of the solution too big, and this means that the effects of small singular values are reduced. This regularized problem is also a least squares problem.

CHALLENGE 6.1.

Show that (6.1) is equivalent to the linear least squares problem

$$\min_{f} \left\| \begin{bmatrix} g \\ 0 \end{bmatrix} - \begin{bmatrix} K \\ \alpha I \end{bmatrix} f \right\|_2^2. \tag{6.2}$$

The scalar α (called a **regularization parameter**) controls the degree of smoothness of the solution. Note that $\alpha = 0$ implies no regularization, and for a discrete ill-posed problem, the computed solution of (6.2) with $\alpha = 0$ is likely to be horribly corrupted with noise. On the other hand, if α is large, then the computed solution cannot be a good approximation to the exact f. It is not a trivial matter to choose an appropriate value for α. Various algorithms are discussed in the literature [69, 131], but we use a manual approach here.

We turn to the problem of solving the least squares problem (6.2).

CHALLENGE 6.2.

Show that if K has a singular value decomposition $K = U\Sigma V^T$, then (6.2) can be transformed into the equivalent least squares problem

$$\min_{\widehat{f}} \left\| \begin{bmatrix} \widehat{g} \\ 0 \end{bmatrix} - \begin{bmatrix} \Sigma \\ \alpha I \end{bmatrix} \widehat{f} \right\|_2^2, \tag{6.3}$$

where $\widehat{f} = V^T f$ and $\widehat{g} = U^T g$.

CHALLENGE 6.3.

Derive a linear system of equations whose solution is the solution to (6.3). Hint: Set the derivative of the minimization function to zero and solve for $\widehat{f}$.

This gives us an algorithm to determine the Tikhonov solution to a discrete ill-posed problem. Next we consider a second method for regularization.

Method 2: Truncated SVD

Another way to regularize the problem is to truncate the singular value decomposition. The next challenge demonstrates how the solution to the least squares problem can be expressed in terms of the SVD.

CHALLENGE 6.4.

Show that the solution to the problem

$$\min_{f} ||g - Kf||_2^2$$

can be written as

$$\boldsymbol{f}_{\ell s} = \boldsymbol{V}\boldsymbol{\Sigma}^{\dagger}\boldsymbol{U}^T\boldsymbol{g} \equiv \sum_{i=1}^{n} \frac{\boldsymbol{u}_i^T\boldsymbol{g}}{\sigma_i}\boldsymbol{v}_i\,,$$

where $\boldsymbol{u}_i$ is the ith column of $\boldsymbol{U}$ and $\boldsymbol{v}_i$ is the ith column of $\boldsymbol{V}$.

We see that trouble occurs in $\boldsymbol{f}_{\ell s}$ if we have a small value of σ_i dividing a term $\boldsymbol{u}_i^T\boldsymbol{g}$ that is dominated by error. In that case, $\boldsymbol{f}_{\ell s}$ is dominated by error, too.

To overcome this, Golub and Kahan [62] suggested truncating the expansion above:

$$\boldsymbol{f}_t = \sum_{i=1}^{p} \frac{\boldsymbol{u}_i^T\boldsymbol{g}}{\sigma_i}\boldsymbol{v}_i$$

for some value of $p < n$.

Using Regularization for Image Deblurring

Now we have all the tools in place to solve a problem in image processing: **image deblurring**. Suppose we have a blurred, noisy image $\boldsymbol{G}$, as in Figure 6.1, and some knowledge of the blurring operator, and we want to reconstruct the true original image $\boldsymbol{F}$. This is an example of a discrete ill-posed problem, where the vectors in the linear system $\boldsymbol{g} = \boldsymbol{K}\boldsymbol{f} + \boldsymbol{\eta}$ represent the image arrays stacked by columns to form vectors. In MATLAB notation,

- $\boldsymbol{f} = \texttt{reshape}(F, n, 1)$,
- $\boldsymbol{g} = \texttt{reshape}(G, n, 1)$.

The goal in this problem is, given $\boldsymbol{K}$ and $\boldsymbol{G}$, reconstruct an approximation of the unknown image $\boldsymbol{F}$.

If we assume $\boldsymbol{F}$ and $\boldsymbol{G}$ contain $\sqrt{n} \times \sqrt{n}$ pixels, then $\boldsymbol{f}$ and $\boldsymbol{g}$ are vectors of length n, and $\boldsymbol{K}$ is an $n \times n$ matrix representing the blurring operation. In general this matrix is too large to use the SVD. However, in some cases $\boldsymbol{K}$ can be written as a **Kronecker product**, $\boldsymbol{K} = \boldsymbol{A} \otimes \boldsymbol{B}$, and the SVD can be used.

A Few Facts on Kronecker Products

The **Kronecker product** $\boldsymbol{A} \otimes \boldsymbol{B}$, where $\boldsymbol{A}$ is an $m \times m$ matrix, is defined to be

$$\boldsymbol{A} \otimes \boldsymbol{B} = \begin{bmatrix} a_{11}\boldsymbol{B} & a_{12}\boldsymbol{B} & \dots & a_{1m}\boldsymbol{B} \\ a_{21}\boldsymbol{B} & a_{22}\boldsymbol{B} & \dots & a_{2m}\boldsymbol{B} \\ \vdots & \vdots & \ddots & \vdots \\ a_{m1}\boldsymbol{B} & a_{m2}\boldsymbol{B} & \dots & a_{mm}\boldsymbol{B} \end{bmatrix}.$$

Kronecker products have a very convenient property: If $\boldsymbol{A} = \boldsymbol{U}_A\boldsymbol{\Sigma}_A\boldsymbol{V}_A^T$, $\boldsymbol{B} = \boldsymbol{U}_B\boldsymbol{\Sigma}_B\boldsymbol{V}_B^T$, then

$$\boldsymbol{K} = \boldsymbol{A} \otimes \boldsymbol{B}\boldsymbol{K} = \boldsymbol{U}\boldsymbol{\Sigma}\boldsymbol{V}^T,$$

where $\boldsymbol{U} = \boldsymbol{U}_A \otimes \boldsymbol{U}_B$, $\boldsymbol{\Sigma} = \boldsymbol{\Sigma}_A \otimes \boldsymbol{\Sigma}_B$, and $\boldsymbol{V} = \boldsymbol{V}_A \otimes \boldsymbol{V}_B$.

POINTER 6.3. Further Reading.

Regularization methods are considered in detail by Hansen [68]. Kronecker products are discussed by Horn and Johnson [81]. Hansen, Nagy, and O'Leary [69] give more detail on image deblurring using matrix techniques.

Therefore, it is possible to compute the SVD of a rather large matrix if it is the Kronecker product of two smaller ones. The website for this book contains a sample MATLAB program, `projdemo.m`, illustrating this property.

In order to solve our image deblurring problem, we need to operate rather carefully with the small matrices; otherwise, storage quickly becomes an issue. Again, see the sample program for guidance. With the Kronecker product as a tool, we are ready to compute.

CHALLENGE 6.5.

Write a program that takes matrices $\boldsymbol{A}$ and $\boldsymbol{B}$ and an image $\boldsymbol{G}$ and computes an image $\boldsymbol{F}$ using

- Tikhonov regularization,
- truncated SVD.

For each of these two algorithms, experiment to find the value of the regularization parameter (α for Tikhonov or p for truncated SVD) that gives the clearest image.

Sample data (i.e., a blurred image $\boldsymbol{G}$, and the matrices $\boldsymbol{A}$ and $\boldsymbol{B}$) are given on the book's website. Your job is to restore the image well enough that you can read the text in it.

Report how you decided what parameters to try. Comment on the usefulness of the two methods for this particular example.

Chapter 7 / Case Study

Updating and Downdating Matrix Factorizations: A Change in Plans

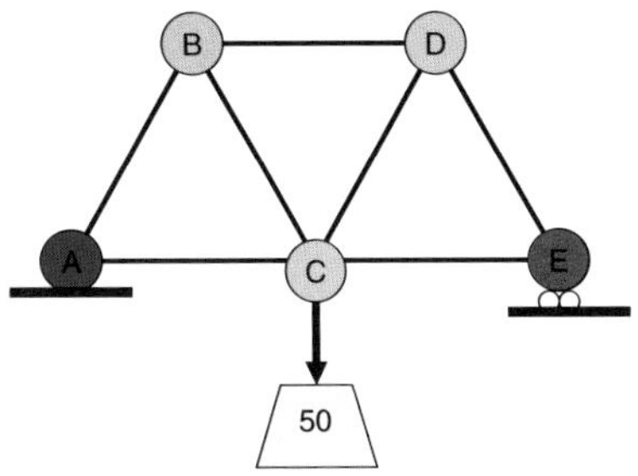

We seldom get it right the first time. Whether we are composing an important e-mail, seasoning a stew, painting a picture, or planning an experiment, we almost always make improvements on our original thought. The same is true of engineering design; we draft a plan, but changes are almost always made. Perhaps the customer changes the performance specifications, or perhaps a substitution of building materials leads to a redesign. In this case study, we consider numerical methods that make it easier to reanalyze the design after small changes are made. For definiteness, we focus on truss design, but the same principles apply to any linear model.

Consider the truss at the top of Figure 7.1. Beginning engineering students learn to compute the force acting on each truss member (beam) by considering equations for each node in the truss, ensuring that the sum of the horizontal forces is zero, the sum of the vertical forces is zero, and the moment is zero. They can "march" through the equations, solving for the horizontal and vertical forces by a clever ordering. If the design is changed—by moving a node, for example, as in the second truss in the figure—then the resulting forces are just as easy to determine.

For more complex models, for example, a finite element model of the forces on the bridge of Figure 1.1, "marching" no longer works, and the system of equations must be solved using a method such as Gauss elimination. We would like to have an algorithm that would enable us to easily recompute the forces if the design of the bridge is changed slightly.

To introduce methods for solving modified models, let's return to the Figure 7.1 example, forget the marching trick, and write a system of equations $\boldsymbol{Af} = \boldsymbol{\ell}$ for the unknown force on each member. The matrix of the system has one column for each unknown force and two rows per node (for horizontal and vertical forces). The load on node C is put in the right-hand side. For the very simple top truss of Figure 7.1, for example, there are $n = 10$

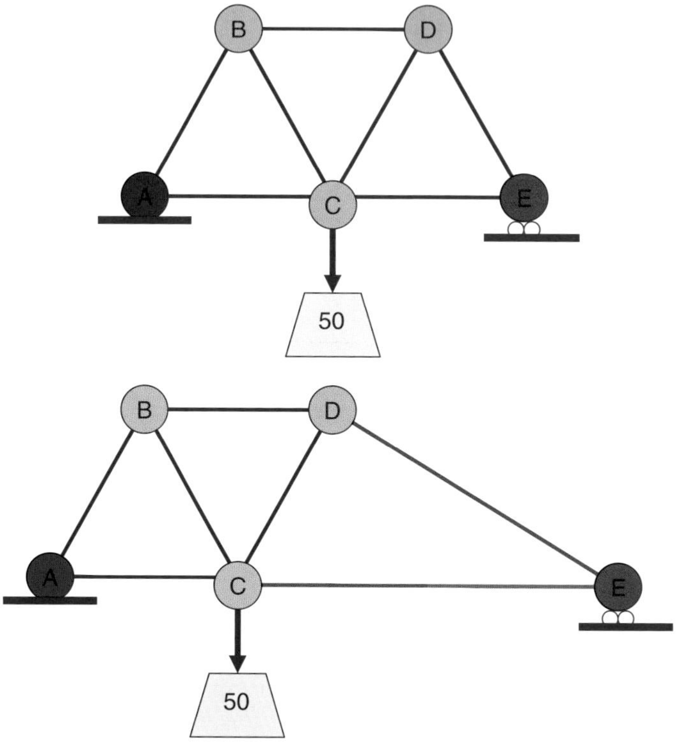

Figure 7.1. *(Top) A truss with* 5 *nodes and* 7 *equal-length members, loaded with a force of* 50 *Newtons. Node A is fixed (supported horizontally and vertically) and node E is rolling (supported only vertically). (Bottom) A change in the truss. Member CE is now two times its original length.*

forces to be computed, the solution to the linear system

$$\begin{bmatrix} -1 & 0 & c & 0 & 0 & 0 & 0 & 1 & 0 & 0 \\ 0 & 1 & s & 0 & 0 & 0 & 0 & 0 & 0 & 0 \\ 0 & 0 & -c & c & 0 & 0 & 1 & 0 & 0 & 0 \\ 0 & 0 & s & s & 0 & 0 & 0 & 0 & 0 & 0 \\ 0 & 0 & 0 & -c & c & 0 & 0 & -1 & 1 & 0 \\ 0 & 0 & 0 & s & s & 0 & 0 & 0 & 0 & 0 \\ 0 & 0 & 0 & 0 & -c & c & -1 & 0 & 0 & 0 \\ 0 & 0 & 0 & 0 & s & s & 0 & 0 & 0 & 0 \\ 0 & 0 & 0 & 0 & 0 & -c & 0 & 0 & -1 & 0 \\ 0 & 0 & 0 & 0 & 0 & s & 0 & 0 & 0 & -1 \end{bmatrix} \begin{bmatrix} f_{Ah} \\ f_{Av} \\ f_{AB} \\ f_{BC} \\ f_{CD} \\ f_{DE} \\ f_{BD} \\ f_{AC} \\ f_{CE} \\ f_{Ev} \end{bmatrix} = \begin{bmatrix} \ell_{Ah} \\ \ell_{Av} \\ \ell_{Bh} \\ \ell_{Bv} \\ \ell_{Ch} \\ \ell_{Cv} \\ \ell_{Dh} \\ \ell_{Dv} \\ \ell_{Eh} \\ \ell_{Ev} \end{bmatrix}. \quad (7.1)$$

The subscript Ah, for example, denotes a horizontal force at node A, and the subscript BC refers to the member connecting nodes B and C. The right-hand-side vector is zero except for the entry ℓ_{Cv}, which is -50, corresponding to the vertical load on node C. The truss members are all equal length, so $c = \cos(\pi/3)$ and $s = \sin(\pi/3)$.

There are several simple kinds of changes to the truss design that produce "small" changes in the system of equations. For example:

- If we change the loading on the truss, then we keep the same matrix but change the right-hand-side $\boldsymbol{\ell}$.
- If we add a new node along with two new truss members, then our new matrix has two additional rows and columns and contains the old matrix as a submatrix.
- If we remove a set of nodes and their truss members, then we delete the columns of the matrix corresponding to the forces exerted by the members and we delete the pair of rows corresponding to each removed node.
- If we move a node, then we change the two rows corresponding to the horizontal and vertical forces on that node, and we change the columns corresponding to its members.

For a problem with $n = 10$ unknowns, we could easily recompute the answer after any of these changes, but if $n = 1,000,000$, then we might want to take advantage of our solution to the original problem to more quickly obtain a solution to the modified problem. In this case study, we develop the tools to do this.

Changes to the Right-Hand Side

If we need to analyze the truss for several different loadings, then it is a good idea to compute a decomposition of the matrix $\boldsymbol{A}$ once and save it for multiple uses. For example, suppose that we compute the LU decomposition with partial pivoting

$$\boldsymbol{PA} = \boldsymbol{LU}, \tag{7.2}$$

where $\boldsymbol{P}$ is a permutation matrix that interchanges rows of $\boldsymbol{A}$, $\boldsymbol{L}$ is a lower-triangular matrix, and $\boldsymbol{U}$ is an upper-triangular matrix. Then each of the loads can be processed by solving

$$\boldsymbol{Ly} = \boldsymbol{P}^T \boldsymbol{\ell}$$

by forward substitution and then computing

$$\boldsymbol{Uf} = \boldsymbol{y}$$

by back substitution. If $\boldsymbol{A}$ is a dense matrix, with very few nonzeros, then the initial LU decomposition costs $O(n^3)$ operations, while forward and back substitution costs only $O(n^2)$. Taking advantage of the sparsity of $\boldsymbol{A}$ can reduce the cost of both the LU decomposition and the substitution (see Chapter 27), but substitution is still significantly less costly than factorization, especially when n is large.

Changes to the Matrix

The Sherman–Morrison–Woodbury Formula

Sometimes our matrix changes in a rather simple way, and we want to reconsider our problem, making use of the original decomposition without explicitly forming the update.

POINTER 7.1. Other Uses for Updated Models.

The problem of efficiently handling small changes in the model matrix arises in many situations other than engineering design.

- Suppose we are solving a system of linear inequalities

  $$\boldsymbol{Ax} \geq \boldsymbol{b}$$

  with $\boldsymbol{A}$ of dimension $m \times n$ ($m \geq n$) and we think that the first n of them should be **active**:

  $$\boldsymbol{a}_i^T \boldsymbol{x} = b_i, \qquad i = 1, \ldots, n,$$

  where $\boldsymbol{a}_i^T$ is the ith row of $\boldsymbol{A}$. Suppose then that the solution to these equations violates the kth inequality ($k > n$), and we want to add this inequality to our current system of equations and delete the jth equation. Can we solve our new linear system easily?

  This problem routinely arises in minimization problems when we have linear inequality constraints (see Chapter 10), and is the basic computation in the simplex algorithm for solving linear programming problems.

- Suppose that we are solving a linear least squares problem

  $$\boldsymbol{Ax} \approx \boldsymbol{b}$$

  or

  $$\min_{\boldsymbol{x}} \|\boldsymbol{b} - \boldsymbol{Ax}\|,$$

  and we get some new measurements. This adds rows to $\boldsymbol{A}$ and $\boldsymbol{b}$. Can we solve our new least squares problem easily?

- Suppose we have computed the eigenvalues and eigenvectors of $\boldsymbol{A}$, and then $\boldsymbol{A}$ is changed by addition of a **rank-1 matrix**

  $$\widehat{\boldsymbol{A}} = \boldsymbol{A} + \boldsymbol{c}\boldsymbol{r}^T .$$

  What are the eigenvalues of $\widehat{\boldsymbol{A}}$?

Each of these problems (and similar ones) can be solved cheaply using the techniques discussed in this case study.

As an example, suppose we have the decomposition from equation (7.2) and now we want to solve the linear system

$$(\boldsymbol{A} - \boldsymbol{z}\boldsymbol{v}^T)\boldsymbol{f} = \boldsymbol{\ell},$$

where $\boldsymbol{z}$ and $\boldsymbol{v}$ are column vectors of length n, so that $\boldsymbol{z}\boldsymbol{v}^T$ is an $n \times n$ matrix. For example, if we decide to move node E as in the truss of Figure 7.1, then we need to change column

6 in our matrix. To do this, we set $\boldsymbol{v}$ to be the sixth column of the identity matrix and set $\boldsymbol{z}$ to be the difference between the old column and the new one.

In the next challenge, we see how to apply this principle to more than one set of changes in the matrix.

CHALLENGE 7.1.

(a) Suppose we want to change columns 6 and 7 in our matrix $\boldsymbol{A}$. Express the new matrix as $\boldsymbol{A} - \boldsymbol{Z}\boldsymbol{V}^T$, where $\boldsymbol{Z}$ and $\boldsymbol{V}$ have dimension $n \times 2$.

(b) Suppose we want to change both column 6 and row 4 of $\boldsymbol{A}$. Find $\boldsymbol{Z}$ and $\boldsymbol{V}$ so that our new matrix is $\boldsymbol{A} - \boldsymbol{Z}\boldsymbol{V}^T$.

In the next challenge, we see how this formulation of our new matrix as a small-rank change in our old matrix leads to an efficient computational algorithm.

CHALLENGE 7.2.

(a) Assume that $\boldsymbol{A}$ and $\boldsymbol{A} - \boldsymbol{Z}\boldsymbol{V}^T$ are both nonsingular. Show that

$$(\boldsymbol{A} - \boldsymbol{Z}\boldsymbol{V}^T)^{-1} = \boldsymbol{A}^{-1} + \boldsymbol{A}^{-1}\boldsymbol{Z}(\boldsymbol{I} - \boldsymbol{V}^T\boldsymbol{A}^{-1}\boldsymbol{Z})^{-1}\boldsymbol{V}^T\boldsymbol{A}^{-1}$$

by verifying that the product of this matrix with $\boldsymbol{A} - \boldsymbol{Z}\boldsymbol{V}^T$ is the identity matrix $\boldsymbol{I}$. This is called the **Sherman–Morrison–Woodbury formula**.

(b) Suppose we have an LU decomposition of $\boldsymbol{A}$ as in (7.2). Assume that $\boldsymbol{Z}$ and $\boldsymbol{V}$ are $n \times k$ and $k << n$. Show that we can use this decomposition and the Sherman–Morrison–Woodbury formula to solve the linear system $(\boldsymbol{A} - \boldsymbol{Z}\boldsymbol{V}^T)\boldsymbol{f} = \boldsymbol{\ell}$ without forming any matrix inverses. (If $\boldsymbol{A}$ is dense, then we perform $O(kn^2)$ operations using Sherman–Morrison–Woodbury, rather than the $O(n^3)$ operations needed to solve the linear system from scratch.) Hint: Remember that $\boldsymbol{A}^{-1}\boldsymbol{y}$ can be computed by solving a linear system using forward and back substitution with the factors L and U.

In the following two challenges, we experiment with the Sherman–Morrison–Woodbury algorithm to see when it can be useful.

CHALLENGE 7.3.

Implement the Sherman–Morrison–Woodbury algorithm from Challenge 7.2(b). Debug it by factoring the matrix in equation (7.1) modeling the top truss in Figure 7.1 and then changing the model to the second truss.

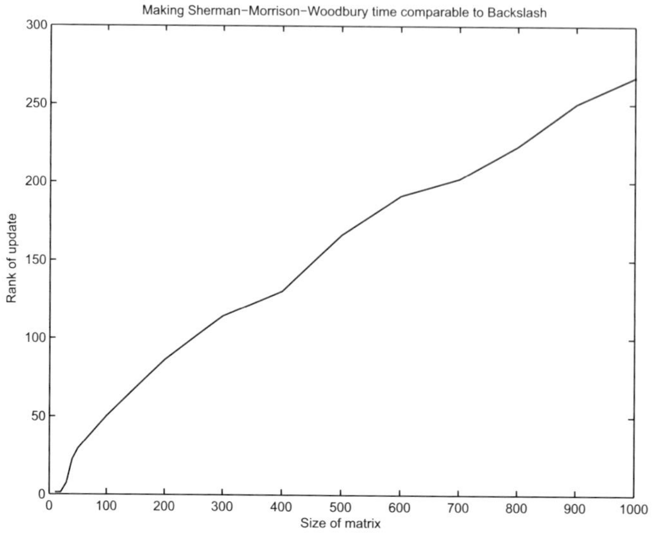

Figure 7.2. *Results of Challenge* 7.4. *The plot shows the rank* k_0 *of the update for which the time for using the Sherman–Morrison–Woodbury formula was approximately the same as the time for solving using factorization. Sherman–Morrison–Woodbury was faster for* $n \geq 40$ *when the rank of the update was less than* $0.25n$.

CHALLENGE 7.4.

For n taken to be various numbers between 10 and 1000, generate a random $n \times n$ matrix $\boldsymbol{A}$. Find the number of updates k_0 that makes the time for solving a linear system using the Sherman–Morrison–Woodbury method comparable to the time for computing $\boldsymbol{A} \setminus \boldsymbol{\ell}$. Plot k_0 as a function of n. Results for one machine are shown in Figure 7.2. Are yours similar?

Updating a Matrix Decomposition

The Sherman–Morrison–Woodbury formula enables us to solve modified linear systems without explicitly modifying our matrix decomposition. It is very efficient when only a few changes are to be made.

In some problems, though, we need to do a long series of updates to the matrix, and it is better to explicitly update the decomposition. We could consider updating an LU decomposition, but because pivoting is necessary to preserve stability, this can be complicated. Instead let's use a decomposition that is stable without pivoting. This also enables us to consider matrices that have more rows than columns, such as those that arise in least squares problems.

Suppose we have factored

$$\boldsymbol{A} = \boldsymbol{Q}\boldsymbol{R},$$

where $\boldsymbol{A}$ is $m \times n$ with $m \geq n$, $\boldsymbol{Q}$ is $m \times m$ and orthogonal ($\boldsymbol{Q}^T\boldsymbol{Q} = \boldsymbol{I}$), and $\boldsymbol{R}$ is $m \times n$ and has zeros below its main diagonal. For definiteness, we'll let $\boldsymbol{A}$ have dimensions 5×3.

As examples, we'll consider two kinds of common changes:

- Adding a row. In least squares, this happens when new data comes in; in our truss problem it means that we have a new node.
- Deleting a column. In least squares, this happens when we decide to reduce the number of parameters in the model; in our truss problem, it could result from removing a member.

Adding a row. Denote the new matrix as $\widehat{\boldsymbol{A}}$. Our decomposition can be written as

$$\widehat{\boldsymbol{A}} = \begin{bmatrix} \boldsymbol{Q} & \boldsymbol{0} \\ \boldsymbol{0} & 1 \end{bmatrix} \begin{bmatrix} r_{11} & r_{12} & r_{13} \\ 0 & r_{22} & r_{23} \\ 0 & 0 & r_{33} \\ 0 & 0 & 0 \\ 0 & 0 & 0 \\ a_{61} & a_{62} & a_{63} \end{bmatrix}. \tag{7.3}$$

In order to complete the decomposition, we need to reduce the as to zeros. We can do this using n Givens rotations (see Section 5.3.1), which is much faster than recomputing the entire decomposition.

Let's see how we can use Givens rotation matrices.

CHALLENGE 7.5.

(a) Given a vector $z \neq \boldsymbol{0}$ of dimension 2×1, find a Givens matrix $\boldsymbol{G}$ so that $\boldsymbol{G}z = x\boldsymbol{e}_1$, where $x = \|z\|$ and $\boldsymbol{e}_1$ is the vector with a 1 in the first position and zero in the second.

(b) We use the notation $\boldsymbol{G}_{ij}$ to denote an $m \times m$ identity matrix with its ith and jth rows modified to include the Givens rotation: for example, if $m = 6$, then

$$\boldsymbol{G}_{25} = \begin{bmatrix} 1 & 0 & 0 & 0 & 0 & 0 \\ 0 & c & 0 & 0 & s & 0 \\ 0 & 0 & 1 & 0 & 0 & 0 \\ 0 & 0 & 0 & 1 & 0 & 0 \\ 0 & -s & 0 & 0 & c & 0 \\ 0 & 0 & 0 & 0 & 0 & 1 \end{bmatrix},$$

where $c^2 + s^2 = 1$. Multiplication of a vector by this matrix leaves all but rows 2 and 5 of the vector unchanged. Show that we can finish our QR decomposition in equation (7.3) by (left) multiplying the $\boldsymbol{R}$ matrix first by $\boldsymbol{G}_{16}$, then by $\boldsymbol{G}_{26}$, and finally by $\boldsymbol{G}_{36}$, where the angle defining each of these matrices is suitably chosen. In order to preserve the equality, we multiply the $\boldsymbol{Q}$-matrix by $\boldsymbol{G}_{16}^T\boldsymbol{G}_{26}^T\boldsymbol{G}_{36}^T$ on the right, and we have the updated decomposition.

Deleting a column. As an example, if we delete column 1 from $\boldsymbol{A}$, we can write the decomposition as

$$\widehat{\boldsymbol{A}} = \boldsymbol{Q} \begin{bmatrix} r_{12} & r_{13} \\ r_{22} & r_{23} \\ 0 & r_{33} \\ 0 & 0 \\ 0 & 0 \end{bmatrix}.$$

The resulting $\boldsymbol{R}$ is almost upper triangular; we just need rotations to reduce the elements labeled r_{22} and r_{33} to zero.

In general, we need $n-k$ rotations when column k is deleted.

Algorithms for deleting a row and adding a column are similar to those that we just discussed. In the next challenge, we construct algorithms for changing a column.

CHALLENGE 7.6.

Write a MATLAB function that updates a QR decomposition of a matrix $\boldsymbol{A}$ when the entries in a single column are changed. Apply it to the truss examples in Challenge 7.3.

The Point of Updating

It may seem silly to worry so much about whether to update or recompute; computers are fast, and if we make one change to the matrix, it really doesn't matter which we do. But when we need to do the task over and over again, perhaps in a loop that solves a more complicated problem, it is essential to use appropriate updating techniques to reduce the cost.

As a final example, we consider one more use of matrix updating, adding both a row and a column.

CHALLENGE 7.7.

Suppose we have factored the $n \times n$ matrix $\boldsymbol{A}$ as $\boldsymbol{PA} = \boldsymbol{LU}$ and now we want to solve the linear system formed by adding one row and one column to $\boldsymbol{A}$ to make a matrix

$$\boldsymbol{A}_{new} = \begin{bmatrix} a_{1,1} & \dots & a_{1,n} & a_{1,n+1} \\ a_{2,1} & \dots & a_{2,n} & a_{2,n+1} \\ \vdots & \vdots & \vdots & \vdots \\ a_{n,1} & \dots & a_{n,n} & a_{n,n+1} \\ a_{n+1,1} & \dots & a_{n+1,n} & a_{n+1,n+1} \end{bmatrix}.$$

Express $\boldsymbol{A}_{new}$ as

$$\boldsymbol{A}_{new} = \begin{bmatrix} \boldsymbol{A} & \boldsymbol{0} \\ \boldsymbol{0} & 1 \end{bmatrix} - \boldsymbol{Z}\boldsymbol{V}^T$$

(where $\boldsymbol{Z}$ and $\boldsymbol{V}$ are rank-2 matrices) so that the Sherman–Morrison–Woodbury formula can be applied.

POINTER 7.2. Further Reading.

The QR decomposition and the use of Givens rotations is discussed in Section 5.3. Other stable methods for modifying matrix decompositions are considered by Gill, Golub, Murray, and Saunders [58], and Golub [60] and Golub and Van Loan [64] discuss the solution of eigenvalue problems when the matrix is modified.

Note that there are many unstable updating algorithms in the literature, so it is important to understand the importance of **stability** and use only trusted algorithms such as the ones given here and in the references in Pointer 7.2.

Chapter 8 / Case Study

The Direction-of-Arrival Problem: Coming at You

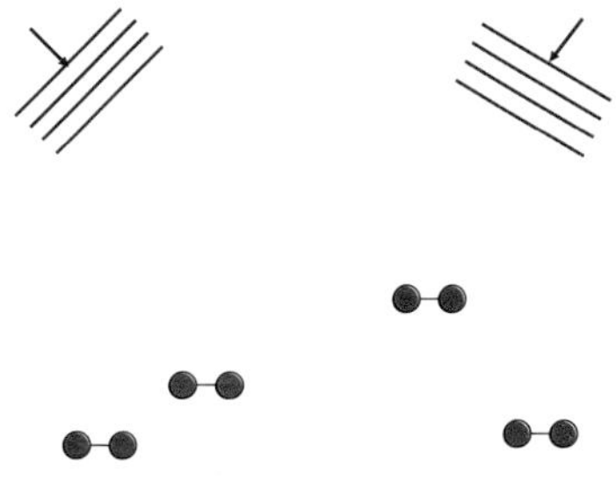

If you break your leg on a mountain but have a cell phone or other transmitter with you, you will hope that a rescuer will be able to determine the direction in which to travel in order to reach you. Similarly, when a navy detects a transmission from a submarine, they want to determine the signal's **direction of arrival** (DOA) in order to locate the sub. The problem is complicated if more than one signal needs to be processed, especially if the number of signals is unknown, and even more complicated if the submarines are moving.

Surprisingly, we will see that your rescuer can solve an eigenvalue problem—involving the product of two unknown matrices—and use that information to find you! This DOA-finding algorithm is known as **ESPRIT**. To understand the process, we use several matrix decompositions and illustrate the necessity of using **update techniques**, as presented in Chapter 7, for real-time computations.

The DOA Problem Definition

Suppose we have d signal sources, each a long distance from the sensors. Suppose that the signals are **narrowband**, so that we can approximate each by a plane wave of fixed frequency ω.

Let's take m sensor pairs whose locations are arbitrary except that the spacing δ and the orientation of the two sensors in each pair is constant. We measure the signal reaching each sensor as a function of time. Figure 8.1 illustrates a sample configuration with $m = 4$ sensor pairs and $d = 2$ signal sources. Let $s_k(t)$ be the signal emitted from the kth source at time t, $k = 1, \dots, d$, and let $\boldsymbol{s}(t)$ be the vector made up of these components. The vector function $\boldsymbol{x}_1(t)$ denotes the m signal measurements $x_{1j}(t)$, $j = 1, \dots, m$, received by the first sensor in each pair at time t, and $\boldsymbol{x}_2(t)$ denotes the corresponding measurements for the second sensor in each pair.

After we take measurements for some time, we want to determine the DOAs: the angles between each plane wave and a line parallel to the lines connecting each sensor pair. We call these d angles θ_k, $k = 1, \dots, d$. In Figure 8.1, the angles are 45° and −30°.

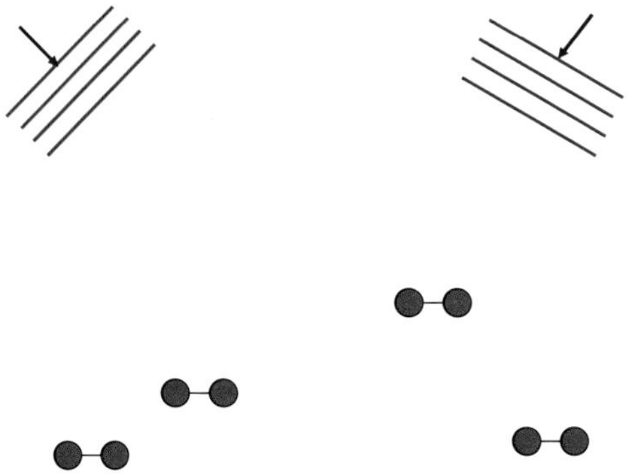

Figure 8.1. *Two signals (plane waves) received by four sensor pairs.*

We model the sensor measurements as a function of signal as

$$\boldsymbol{x}_1(t) = \boldsymbol{A}\boldsymbol{s}(t) + \boldsymbol{\epsilon}_1(t),$$
$$\boldsymbol{x}_2(t) = \boldsymbol{A}\boldsymbol{\Phi}\boldsymbol{s}(t) + \boldsymbol{\epsilon}_2(t).$$

The matrix $\boldsymbol{A}$ of size $m \times d$ is an unknown matrix of **array steering vectors**, and the matrix $\boldsymbol{\Phi}$ is a diagonal matrix that accounts for the phase delays between the sensors in each pair. The kth diagonal entry is

$$\phi_k = e^{i\omega\delta \sin\theta_k / c}, \qquad k = 1, \dots, d,$$

where $i = \sqrt{-1}$ and c is the speed of sound. Our problem, then, is to determine $\boldsymbol{\Phi}$, given $\boldsymbol{x}$, δ, and ω, without knowing $\boldsymbol{A}$, $\boldsymbol{s}(t)$, the measurement noise $\boldsymbol{\epsilon}_1(t)$ and $\boldsymbol{\epsilon}_2(t)$, or even d. As an added complication, if the sources are moving, then $\boldsymbol{\Phi}$ is also a function of t.

We build algorithms on a very clever observation for extracting $\boldsymbol{\Phi}$. Let's temporarily assume that we know the number of signals $d < m$. We observe the system for n timesteps, obtaining $\boldsymbol{X}_1$ with m rows and n columns containing the data values $\boldsymbol{x}_1(t)$. Similarly, we construct $\boldsymbol{X}_2$ from the data $\boldsymbol{x}_2(t)$. This way of collecting the data is called **rectangular windowing** because we just look at the data within a time window of size n. If we neglect the errors $\boldsymbol{\epsilon}(t)$, then our system becomes

$$\boldsymbol{X} = \begin{bmatrix} \boldsymbol{X}_1 \\ \boldsymbol{X}_2 \end{bmatrix} = \begin{bmatrix} \boldsymbol{A} \\ \boldsymbol{A}\boldsymbol{\Phi} \end{bmatrix} \boldsymbol{S},$$

where the columns of $\boldsymbol{S}$ are $\boldsymbol{s}(t)$. If $\boldsymbol{A}$ is full rank, then its rank is d, and this is also the maximal rank of $\boldsymbol{X}$. Consider the following recipe:

- Find a matrix $\boldsymbol{B}$ of size $d \times m$ so that $\boldsymbol{BA}$ is $d \times d$ and full rank.
- Find a matrix $\boldsymbol{C}$ of size $n \times d$ so that $\boldsymbol{SC}$ is $d \times d$ and full rank.
- Find d vectors z_k and d values λ_k so that $\boldsymbol{BA\Phi SC}z_k = \lambda_k \boldsymbol{BASC}z_k$.

Note that λ_k is an **eigenvalue** of the **generalized eigenproblem**

$$\widehat{M}z = \lambda \widehat{A}z$$

involving the matrices $\widehat{M} \equiv BA\Phi SC$ and $\widehat{A} \equiv BASC$, and z_k is the corresponding **eigenvector**.

CHALLENGE 8.1.

Let $w_k = SCz_k$ and show that $\Phi w_k = \lambda_k w_k$. Conclude that the eigenvalues λ_k are equal to the diagonal entries of Φ.

We've accomplished something rather surprising: without knowing A or Φ, we can choose matrices B and C of the proper dimensions and construct the matrices for the eigenproblem just by knowing X, since $BASC = BX_1C$ and $BA\Phi SC = BX_2C$. But we need to make sure that BA and SC have full rank.

SVD-ESPRIT and Rectangular Windowing

In order to ensure the full-rank conditions, we use the **singular value decomposition** (SVD) of a matrix: any matrix F of dimension $p \times q$ can be factored as

$$F = U\Sigma W^*,$$

where U is $p \times p$, W is $q \times q$, $U^*U = I$, and $W^*W = I$. The real diagonal matrix Σ of dimension $p \times q$ has nonzeros $\sigma_1 \geq \sigma_2 \geq \cdots \geq \sigma_{\hat{q}}$, where $\hat{q} = \min(p,q)$. Because U and W are unitary matrices, their columns (and rows) are well-conditioned bases for the subspaces they span, and using unitary matrices leads to numerically stable choices for B and C, which we define in the next challenge.

CHALLENGE 8.2.

Suppose that the SVD of X is $U\Sigma W^*$, where $\sigma_j = 0$, $j > d$. Let Σ_1 be the square diagonal matrix with entries $\sigma_1, \ldots, \sigma_d$, and partition U into

$$U = \begin{bmatrix} U_1 & U_3 \\ U_2 & U_4 \end{bmatrix},$$

where U_1 and U_2 have m rows and d columns, so that

$$X_1 = AS = U_1[\Sigma_1, \mathbf{0}_{d\times(n-d)}]W^*,$$
$$X_2 = A\Phi S = U_2[\Sigma_1, \mathbf{0}_{d\times(n-d)}]W^*,$$

where $\mathbf{0}_{d\times(n-d)}$ is the zero matrix of size $d \times (n-d)$. Let $\widehat{U} = [U_1, U_2]$ have SVD $T\Delta V^*$, and denote the leading $d \times d$ submatrix of Δ by Δ_1. Partition

$$V = \begin{bmatrix} V_1 & V_3 \\ V_2 & V_4 \end{bmatrix}$$

so that V_1 and V_2 have dimension $d \times d$. Let

$$B = [\Delta_1{}^{-1}, \mathbf{0}_{d\times(m-d)}]T^*$$

and

$$C = W \begin{bmatrix} \Sigma_1{}^{-1} \\ \mathbf{0}_{(n-d)\times d} \end{bmatrix}.$$

Show that the eigenvalues λ_k that satisfy the equation $V_2^* z_k = \lambda_k V_1^* z_k$ are ϕ_k.

Thus we have our first algorithm for solving the DOA problem:

- Compute the SVD of $X = U\Sigma W^*$.
- Compute the SVD of $\widehat{U} = [U_1, U_2] = T\Delta V^*$.
- Solve the generalized eigenvalue problem $V_2^* z = \lambda V_1^* z$ for the values $\lambda_k = \phi_k$, $k = 1, \ldots, d$.

In the next challenge, we investigate how this algorithm performs.

CHALLENGE 8.3.

Program the SVD algorithm and experiment with rectangularly windowed data and a window size of $n = 10$. Note that we need to compute U and V, but we do not need B or C. You can find sample data for X and Φ on the website. Plot the true and computed DOAs as a function of time and compute the average absolute error in your DOA estimates (absolute value of true value minus computed value) and the average relative error (absolute error divided by absolute value of true value). Your results should resemble those in Figure 8.2.

Eigen-ESPRIT and Exponential Windowing

Experimenting further with the data of Challenge 8.3, we would discover that rectangular windowing has a drawback: if the window size n is too small, then the DOAs are very sensitive to errors in the measurements and our estimates can change abruptly. But if the window size is too large, then very old data contribute to our measurements, and our estimates will be bad if the sources are moving too fast. The cure for this is to use old data but give more weight to newer data. We do this in **exponential windowing** by multiplying all of our old data by a **forgetting factor** f (with $0 < f \le 1$) every time we add new data. Thus, after n observations, column ℓ of X_1 contains data from observation ℓ multiplied by $f^{n-\ell}$, and similarly for X_2.

Using exponential windowing, the number of columns in the matrix X can grow very large, so the SVD is too expensive. We could avoid an SVD (for either exponential windowing or rectangular windowing) but still use an orthogonal basis by noting that $XX^* = U\Sigma\Sigma^T U^*$, so U can be computed from the eigendecomposition of the $2m \times 2m$ matrix XX^*. The next challenge shows how XX^* can be formed quickly.

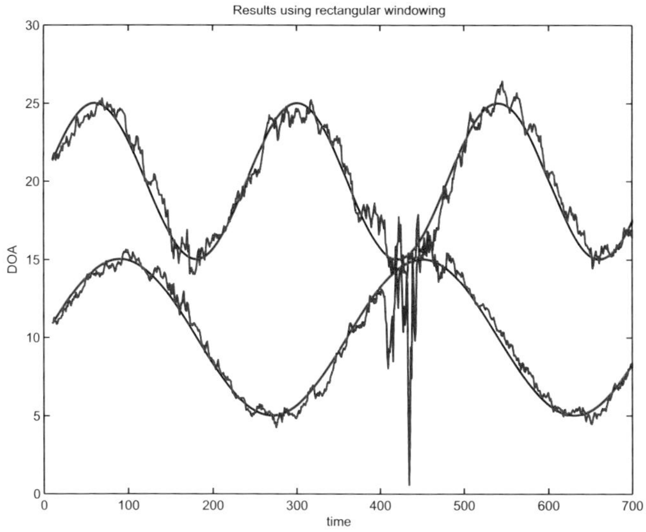

Figure 8.2. *Results of Challenge* 8.3*: the true DOA (blue) and the DOA estimated by rectangular windowing (red) as a function of time.*

CHALLENGE 8.4.

Suppose that the matrix $\boldsymbol{X}$ contains the exponential windowing data and that a new data vector $\boldsymbol{x}$ arrives. Give a formula for the new exponential windowing data matrix $\boldsymbol{X}\boldsymbol{X}^*$ and show that the cost of computing it from $\boldsymbol{X}$ and $\boldsymbol{x}$ is $O(m^2)$ multiplications.

Now we try this eigenvalue variant of ESPRIT, computing an eigendecomposition of $\boldsymbol{X}\boldsymbol{X}^*$ in place of an SVD of $\boldsymbol{X}$.

CHALLENGE 8.5.

Program the Eigen-ESPRIT algorithm and experiment with exponential windowing for the data of Challenge 8.3. Use the forgetting factor $f = 0.9$, and compare the results with those of Challenge 8.3.

Determining the Number of Sources

Now suppose we do not know how many source signals we are receiving. Recall that if $m \geq d$ and $\boldsymbol{A}$ is full rank, then its rank is d, and this is also the maximal rank of $\boldsymbol{X}$. Thus we can estimate d experimentally by taking it to be the rank of the matrix $\boldsymbol{X}$. This works fine in the absence of error if the signals are all stationary, but if they are moving, or if there is error in our measurements, then the matrix $\boldsymbol{X}$ has some small nonzero singular values in addition to d nonzeros. We need to be able to distinguish between real signals and noise. If you know some statistics, you can solve the next challenge to predict how large the noise will be.

CHALLENGE 8.6.

Suppose that we have a matrix X of size $m \times n$, $m \leq n$, and each element of X is normally distributed with mean 0 and standard deviation ψ.

(a) Show that the random variable equal to the sum of the squares of the entries of X is equal to the sum of the squares of the singular values of X.

(b) Show, therefore, that for rectangular windowing of this data, the expected value of $\sigma_1^2 + \cdots + \sigma_m^2$ is $\psi^2 mn$, where σ_j is a singular value of X.

(c) Using a similar argument, show that for exponential windowing, the expected value of $\sigma_1^2 + \cdots + \sigma_m^2$ is approximately $mf^2\psi^2/(1-f^2)$, where σ_i is a singular value of FX. Here, F is a diagonal matrix, with jth entry equal to f^j.

Solving this challenge relies on a formula for $a + a^2 + \cdots + a^k$ when $0 < a < 1$, and information in any basic statistics textbook that discusses the normal distribution.

This gives us a way to experimentally determine d: choose it to be the smallest value for which the remaining singular values satisfy

$$\sigma_{d+1}^2 + \cdots + \sigma_m^2 < n(m-d)\kappa\psi^2$$

for rectangular windowing and

$$\sigma_{d+1}^2 + \cdots + \sigma_m^2 < (m-d)\kappa f^2\psi^2/(1-f^2)$$

for exponential windowing, where $\kappa > 1$ is a user-chosen tolerance.

Now we experiment with this algorithm for determining the number of signals and their directions of arrival.

CHALLENGE 8.7.

Modify the programs so that they also determine d, and explore the methods' sensitivity to the choice of n, f, and κ.

Using URV for Efficiency

Computing SVDs and eigendecompositions from scratch can be too computationally intensive to keep up with the incoming data; the operations counts are proportional to m^2n for the SVD and m^3 for the eigendecomposition. In order to keep up with incoming data, we need to find ways to update our DOA estimates at lower cost. Unfortunately, there is no easy way to update SVDs or eigendecompositions, but there is a closely related decomposition, the **rank-revealing URV decomposition**, that can be updated. If we substitute this

POINTER 8.1. Further Reading.

In this case study, we use three matrix decompositions: the SVD and the eigendecomposition from Chapter 5, and the rank-revealing URV decomposition with updating [141].

The **generalized eigenvalue problem** used in Challenge 8.1 is discussed in linear algebra textbooks and numerical linear algebra textbooks such as [64].

The ESPRIT algorithm was proposed in [127]. The URV variant was proposed in [104], which is also the source of the data we use in Challenge 8.3.

For Challenge 8.8, you can find a detailed discussion of the URV-ESPRIT algorithm in [104].

A more common algorithm for determining DOAs is the **Music algorithm**. This algorithm can also be formulated in terms of matrix decompositions. Standard textbooks such as that of Haykin [70, Chap. 12] provide further information.

---

for the SVD of $\boldsymbol{X}$ or the eigendecomposition of $\boldsymbol{X}\boldsymbol{X}^*$, then our algorithm has a cost proportional to $d^3 + m^2 + n^2$ and is suitable for real-time applications as long as the number of incoming signals is not too great.

The rank-revealing URV decomposition of $\boldsymbol{X}^*$ is

$$\boldsymbol{X}^* = \boldsymbol{U}\boldsymbol{R}\boldsymbol{V}^* = \boldsymbol{U}\begin{bmatrix} \bar{\boldsymbol{R}} & \boldsymbol{F} \\ \boldsymbol{0} & \boldsymbol{G} \end{bmatrix}\boldsymbol{V}^*,$$

where $\boldsymbol{U}$ and $\boldsymbol{V}$ are square unitary matrices ($\boldsymbol{U}^*\boldsymbol{U} = \boldsymbol{I}$, $\boldsymbol{V}^*\boldsymbol{V} = \boldsymbol{I}$), $\bar{\boldsymbol{R}}$ is an upper-triangular matrix of size $d \times d$, and $\boldsymbol{G}$ is an upper-triangular matrix of size $(n-d) \times (2m-d)$. In addition, the norms of the matrices $\boldsymbol{F}$ and $\boldsymbol{G}$ should be small. Therefore, $\boldsymbol{X}$ is within $\sqrt{||\boldsymbol{F}||^2 + ||\boldsymbol{G}||^2}$ of the matrix of rank d obtained by setting these two blocks to zero. The SVD is a special case of this decomposition in which $\boldsymbol{F}$ is zero and $\bar{\boldsymbol{R}}$ and $\boldsymbol{G}$ are diagonal, but by allowing the more general case, we gain the ability to update the decomposition inexpensively as new data arrives.

The rank-revealing QR decomposition of Section 5.4 is related to the rank-revealing URV decomposition. In the QR case, we set $\boldsymbol{V} = \boldsymbol{I}$ and ask that $\boldsymbol{G}$ be small, but we do not restrict $\boldsymbol{F}$.

CHALLENGE 8.8. (Extra)

Implement the URV updating algorithm, or use available software, and use it on the matrix $\boldsymbol{X}$ to solve the DOA problem for rectangular windowing and exponential windowing.

Unit III

Optimization and Data Fitting

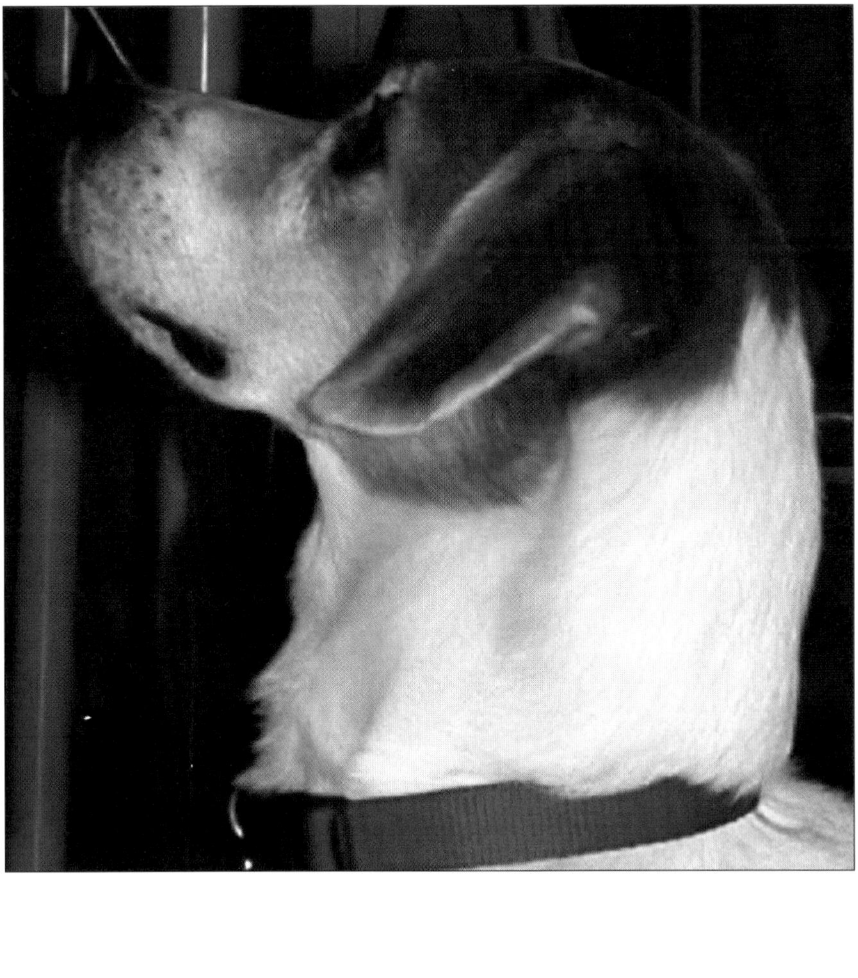

For some problems we are interested in finding any solution, but in other cases we want the best solution. "Best" can have many meanings, but we usually achieve it by solving an optimization problem.

In this unit we focus on two types of optimization problems: unconstrained problems, in which any point in $\mathcal{R}^n$ can be considered as a candidate (Chapter 9), and constrained problems, in which candidate points must satisfy extra conditions (Chapter 10). We discuss algorithms as well as the criteria that characterize a solution.

Chapter 11 considers a difficult class of optimization problems, dividing a set of data points into appropriate clusters. In Chapter 12, we solve perhaps the most common optimization problem, the problem of fitting a model to data using nonlinear least squares. We investigate this further in Chapter 13, in which we introduce more efficient algorithms for the special case of solving separable least squares problems. In Chapters 14 and 15, we fit a model to spectroscopy data, assuming that all of the data in the model has some uncertainty, first using (total) least squares and then using minimization in other norms. Constrained optimization is also used in Chapter 29.

BASICS: To understand this unit, the following background is helpful:

- Taylor series and the use of Lagrange multipliers for equality constraints in optimization problems. This can be found in a standard calculus textbook.
- Methods for minimizing a function of a single variable, including Newton's method and some alternatives to it. This can be found in standard scientific computing textbooks such as [148].
- Methods for fitting a function, such as a polynomial or a spline, to a set of data points. Basic textbooks such as [32, 148] are good references.

MASTERY: After you have worked through this unit, you should be able to do the following:

- Recognize a solution to an optimization problem (necessary and sufficient conditions).
- Compute a descent direction to reduce a function $f(\boldsymbol{x})$ starting from a given point.
- Compute the Newton direction and test whether it is downhill.
- Define linear and superlinear rates of convergence. Define quadratic rate of convergence and recognize that Newton's method usually exhibits this.
- Explain why the Newton direction might need to be modified, and be able to define and use the Levenberg–Marquardt method and explain the reasoning behind a modified Cholesky strategy.
- Explain what a backtracking linesearch is, contrast it with an exact linesearch, and explain why we use backtracking.
- Write a program that uses a trust region method (using the Euclidean norm) to solve an optimization problem. Explain the relation between trust region methods and Levenberg–Marquardt.

- State the secant condition and where it comes from. Verify that a given quasi-Newton matrix update formula satisfies the secant condition, and test whether it satisfies the no-change conditions.
- Give advantages and disadvantages of storing a quasi-Newton approximation to the Hessian matrix $\boldsymbol{H}$, to $\boldsymbol{H}^{-1}$, or to factors of $\boldsymbol{H}$.
- Give the convergence rates, storage requirements, and number of f, $\boldsymbol{g}$, and $\boldsymbol{H}$ evaluations per iteration for the optimization methods studied.
- Write a program that uses a conjugate gradient method or a quasi-Newton method to solve an optimization problem, using a given linesearch algorithm.
- Write a program for the truncated Newton method, given a program for the linear conjugate gradient method.
- Explain what automatic differentiation is used for and what information you need to provide to it.
- Explain the simplex-based algorithms for unconstrained optimization.
- Define a positive spanning set. Given a positive spanning set and a linesearch routine, write a program that does pattern search.
- Give advantages and disadvantages of our methods for minimizing without derivatives: automatic differentiation with our gradient-based methods, finite difference methods, simplex algorithm, pattern search, and Monte Carlo minimization (Chapter 17).
- Use Lagrange multipliers to determine the sensitivity of solutions to constrained optimization problems.
- Use a feasible direction method to solve an optimization problem with linear equality constraints.
- Use QR to get a basis for the feasible directions.
- Program and use a barrier method or a penalty method to solve a constrained optimization problem.
- Determine the dual problem and the central path for a linear programming, second-order cone programming, or semidefinite programming problem.
- Use the k-means algorithm for clustering data.
- Minimize the root mean squared distance between points in two objects.
- Solve nonlinear least squares problems, and recognize separable ones.
- Solve total least squares and total least norm problems.

Chapter 9

Numerical Methods for Unconstrained Optimization

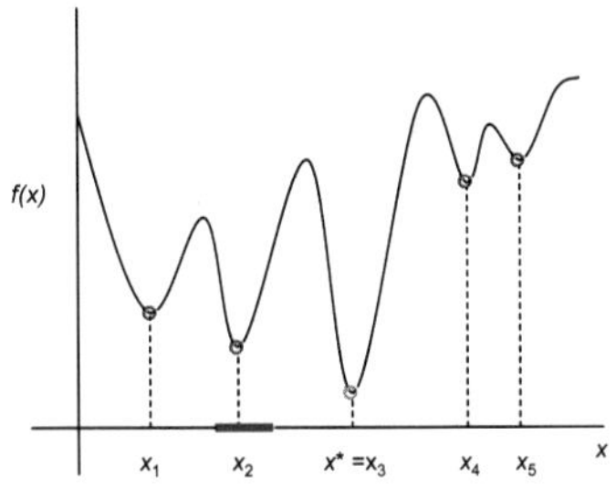

Our goal in this chapter is to develop efficient algorithms to solve the following problem:

Unconstrained Optimization Problem: Given a function $f : S \to \mathcal{R}$ with $S \subseteq \mathcal{R}^n$, find $\boldsymbol{x}_{opt}$ so that

$$f(\boldsymbol{x}_{opt}) = \min_{\boldsymbol{x} \in S} f(\boldsymbol{x}).$$

The point $\boldsymbol{x}_{opt}$ is called the **minimizer**, and the value $f(\boldsymbol{x}_{opt})$ is the **minimum**.

For **unconstrained optimization**, the set S is usually taken to be $\mathcal{R}^n$, but sometimes we make use of **lower and upper bounds** on the variables, restricting our search to a box

$$S = \{\boldsymbol{x} : \boldsymbol{\ell} \leq \boldsymbol{x} \leq \boldsymbol{u}\}$$

for some given vectors $\boldsymbol{\ell}, \boldsymbol{u} \in \mathcal{R}^n$.

We begin in Section 9.1 by discussing the basics of how to recognize a solution algebraically and geometrically. Then we present a template algorithm and consider in Section 9.2 how to use Newton's method in this framework. We discuss how close to Newton's method we must be to have rapid convergence. We discuss making methods safe through the use of descent directions and linesearches or the use of trust regions. Finally, in Section 9.5 we discuss the families of alternatives to Newton's method.

9.1 Fundamentals for Unconstrained Optimization

In this section we address some basic issues in minimizing a function. First, we determine how we can recognize a solution and the geometry that determines a solution.

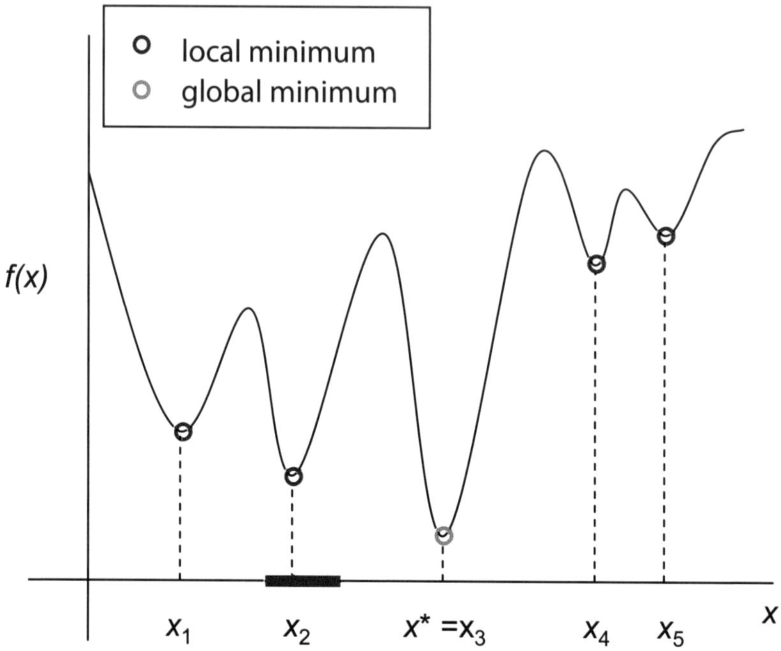

Figure 9.1. *Local and global minimizers. The function $f(x)$ has five local minimizers $x_1,\ldots,x_5$. The point x_2, for example, is a local minimizer because $f(x_2)$ is less than or equal to the function values for all of the points in a neighborhood that includes the blue line segment. The global minimizer is $x_{opt} = x_3$, since its function value is the lowest.*

9.1.1 How Do We Recognize a Solution?

The point $\boldsymbol{x}_{opt}$ is a **local minimizer** for our unconstrained optimization problem if there is a $\delta > 0$ so that if $\boldsymbol{x} \in S$ and $\|\boldsymbol{x} - \boldsymbol{x}_{opt}\| < \delta$, then $f(\boldsymbol{x}_{opt}) \le f(\boldsymbol{x})$. In other words, $\boldsymbol{x}_{opt}$ is at least as good as any point in its neighborhood. The point $\boldsymbol{x}_{opt}$ is a **global minimizer** if $f(\boldsymbol{x}_{opt}) \le f(\boldsymbol{x})$ for any $\boldsymbol{x} \in S$. Figure 9.1 illustrates these different types of solutions.

It would be nice if every local solution was guaranteed to be global. This is true when f is **convex**, which means that for all points $\boldsymbol{x}$ and $\boldsymbol{y}$ in its domain and for all values α between 0 and 1,

$$f(\alpha\boldsymbol{x} + (1-\alpha)\boldsymbol{y}) \le \alpha f(\boldsymbol{x}) + (1-\alpha) f(\boldsymbol{y}).$$

(Geometrically, this means that the surface $f(\boldsymbol{x})$ lies on or below any line connecting two points on the surface.) The function in Figure 9.1 is convex on the interval marked with the blue line, for example.

We assume throughout this chapter that f is smooth enough that it has as many continuous derivatives as we need. For this section, that means two continuous derivatives plus one more derivative, possibly discontinuous.

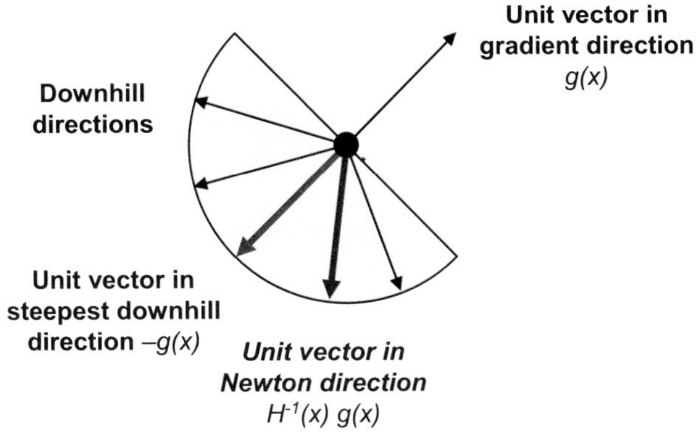

Figure 9.2. *Given the gradient vector* $\mathbf{g}(\mathbf{x})$ *at some point* $\mathbf{x}$*, the downhill directions are those in the shaded region. The steepest downhill direction (red) is in the direction of* $-\mathbf{g}(\mathbf{x})$*. The Newton direction is downhill if* $\mathbf{H}(\mathbf{x})$ *is positive definite.*

POINTER 9.1. Some Hessian Matrices Are Easy.

Although the Hessian matrix has n^2 elements, it is sometimes quite inexpensive to compute. For example,

- If each component of the gradient depends on only a few of the variables, then each row of the Hessian has only a few nonzero components. In this case we would exploit the sparsity of the Hessian matrix by using special techniques to solve linear systems of equations involving the matrix; see Chapters 27 and 28.
- Terms in f involving sines, cosines, or exponentials (as well as some other functions) have second derivatives that reuse these values.

The **gradient** and **Hessian** of f help us to recognize local solutions. The **gradient** of f at $\boldsymbol{x}$ is defined to be the vector

$$\boldsymbol{g}(\boldsymbol{x}) = \nabla f(\boldsymbol{x}) = \begin{bmatrix} \partial f(\boldsymbol{x})/\partial x_1 \\ \vdots \\ \partial f(\boldsymbol{x})/\partial x_n \end{bmatrix}.$$

The **Hessian** of f at $\boldsymbol{x}$ is the derivative of the gradient:

$$\boldsymbol{H}(\boldsymbol{x}) = \nabla^2 f(\boldsymbol{x}), \text{ with } h_{ij} = \frac{\partial^2 f(\boldsymbol{x})}{\partial x_i \, \partial x_j}, \quad i, j = 1, \ldots, n.$$

Under our smoothness assumption, the Hessian is symmetric.

First-Order Necessary Condition for Optimality

Taylor series expansions, explained in calculus courses, give us our principal tool for recognizing a solution. Suppose we have a point $\boldsymbol{x}_{opt}$, a small positive scalar h, and a vector $\boldsymbol{p} \in \mathcal{R}^n$ with $\|\boldsymbol{p}\| = 1$. Then Taylor series tell us that

$$f(\boldsymbol{x}_{opt} + h\boldsymbol{p}) = f(\boldsymbol{x}_{opt}) + h\boldsymbol{p}^T \boldsymbol{g}(\boldsymbol{x}_{opt}) + \frac{1}{2}h^2 \boldsymbol{p}^T \boldsymbol{H}(\boldsymbol{x}_{opt})\boldsymbol{p} + O(h^3).$$

Now suppose that $\boldsymbol{g}(\boldsymbol{x}_{opt})$ is nonzero. Then referring to Figure 9.2, we can always find a **descent** or **downhill direction** $\boldsymbol{p}$ so that

$$\boldsymbol{p}^T \boldsymbol{g}(\boldsymbol{x}_{opt}) < 0.$$

Take, for example, $\boldsymbol{p} = -\boldsymbol{g}(\boldsymbol{x}_{opt})/\|\boldsymbol{g}(\boldsymbol{x}_{opt})\|$. Using this direction with a small enough stepsize h, we can make $\frac{1}{2}h^2 \boldsymbol{p}^T \boldsymbol{H}(\boldsymbol{x}_{opt})\boldsymbol{p}$ small enough that

$$f(\boldsymbol{x}_{opt} + h\boldsymbol{p}) < f(\boldsymbol{x}_{opt}).$$

Therefore, a necessary condition for $\boldsymbol{x}_{opt}$ to be a minimizer is that $\boldsymbol{g}(\boldsymbol{x}_{opt}) = \boldsymbol{0}$. This is called the **first-order optimality condition**, since it depends on first derivatives.

Second-Order Necessary Condition for Optimality

We now know that if $\boldsymbol{x}_{opt}$ is a minimizer, then $\boldsymbol{g}(\boldsymbol{x}_{opt}) = \boldsymbol{0}$, so

$$f(\boldsymbol{x}_{opt} + h\boldsymbol{p}) = f(\boldsymbol{x}_{opt}) + \frac{1}{2}h^2 \boldsymbol{p}^T \boldsymbol{H}(\boldsymbol{x}_{opt})\boldsymbol{p} + O(h^3).$$

Now suppose that we have a direction $\boldsymbol{p}$ so that $\boldsymbol{p}^T \boldsymbol{H}(\boldsymbol{x}_{opt})\boldsymbol{p} < 0$. (We call this a **direction of negative curvature**.) Then again, for small enough h, we could make $f(\boldsymbol{x}_{opt} + h\boldsymbol{p}) < f(\boldsymbol{x}_{opt})$.

This leads us to the **second-order condition for optimality**: a necessary condition for $\boldsymbol{x}_{opt}$ to be a minimizer is that there be no direction of negative curvature. This is equivalent to saying that the matrix $\boldsymbol{H}(\boldsymbol{x}_{opt})$ must be **positive semidefinite**. In other words, all of the eigenvalues of $\boldsymbol{H}(\boldsymbol{x}_{opt})$ must be nonnegative.

Are these conditions sufficient? Not quite. For example, consider

$$f(x) = x^3, \quad \text{where } x \in \mathcal{R}^1.$$

Then $f'(x) = 3x^2$ and $f''(x) = 6x$, so $f'(0) = 0$ and $f''(0) = 0$, so $x = 0$ satisfies the first- and second-order necessary conditions for optimality, but it is not a minimizer of f.

We are very close to sufficiency, though: Recall that a symmetric matrix is **positive definite** if all of its eigenvalues are positive. If $\boldsymbol{g}(\boldsymbol{x}) = \boldsymbol{0}$ and $\boldsymbol{H}(\boldsymbol{x})$ is positive definite, then $\boldsymbol{x}$ is a local minimizer.

9.1.2 Geometric Conditions for Optimality

Imagine we are at point $\boldsymbol{x}$ on a mountain whose surface is described by the graph of the function $f(\boldsymbol{x})$. (So $\boldsymbol{x} \in \mathcal{R}^2$.) Now imagine that it is foggy, so that we cannot see very far in front of us.

The direction $\boldsymbol{g}(\boldsymbol{x})$ is the **direction of steepest ascent**. So if we want to climb the mountain, it is the best direction to walk. Similarly, the direction $-\boldsymbol{g}(\boldsymbol{x})$ is the **direction of steepest descent**, the fastest way down.

Further, any direction $\boldsymbol{p}$ that makes a positive inner product with the gradient is an **uphill direction**, and any direction that makes a negative inner product is **downhill**.

If we are standing at a point where the gradient is zero, then there is no ascent direction and no descent direction, but a **direction of positive curvature** leads us to a point where we can go uphill, and a **direction of negative curvature** leads us to a point where we can descend.

If there is no descent direction and no direction of negative curvature, then we are at the bottom of a valley, a point that (locally) has minimum altitude.

9.1.3 The Basic Minimization Algorithm

Our basic strategy for finding a local minimizer of a function is inspired by the foggy mountain and is given in Algorithm 9.1. Initially, we study algorithms for which the stepsize $\alpha_k = 1$. We sometimes omit the indices k on $\boldsymbol{x}, \boldsymbol{p}$, and α, and the argument $\boldsymbol{x}$ for $\boldsymbol{H}$ and $\boldsymbol{g}$, for brevity.

Algorithm 9.1 Basic Minimization Algorithm

Take an initial guess $\boldsymbol{x}^{(0)}$ for the solution.
Set $k = 0$.
while $\boldsymbol{x}^{(k)}$ is not a good enough solution,
  Find a search direction $\boldsymbol{p}^{(k)}$.
  Set $\boldsymbol{x}^{(k+1)} = \boldsymbol{x}^{(k)} + \alpha_k \boldsymbol{p}^{(k)}$, where α_k is a scalar chosen to guarantee that progress is made.
  Set $k = k+1$.
end

There are several "details" that we need to resolve to make our algorithm practical: testing for convergence, finding a search direction, and computing the stepsize α_k.

The test for determining that $\boldsymbol{x}^{(k)}$ is good enough depends on precisely what we are looking for. It is really a matter of forward and backward error analysis, discussed in Section 1.6, and in the following discussion we assume that all arithmetic is exact. If we are most concerned about finding a point for which f is close to its (local) minimum, then, using the Taylor series expansion

$$f(\boldsymbol{x}_{opt}) \approx f(\boldsymbol{x}^{(k)}) + (\boldsymbol{x}_{opt} - \boldsymbol{x}^{(k)})^T \boldsymbol{g}(\boldsymbol{x}^{(k)}),$$

we stop when $\|\boldsymbol{g}(\boldsymbol{x}^{(k)})\|$ is small enough, since that forces $f(\boldsymbol{x}^{(k)}) \approx f(\boldsymbol{x}_{opt})$ unless $\|\boldsymbol{x}_{opt} - \boldsymbol{x}^{(k)}\|$ is very large. This is a backward error bound approach. Estimating the distance from $\boldsymbol{x}^{(k)}$ to a local minimizer $\boldsymbol{x}_{opt}$ gives the forward error bound, and this is more difficult. Some programs stop when $\alpha_k \boldsymbol{p}^{(k)}$ is small, but really this just means that our algorithm is not making much progress. Instead we might try to estimate the distance to the true solution. Again, by Taylor series,

$$\boldsymbol{0} = \boldsymbol{g}(\boldsymbol{x}_{opt}) \approx \boldsymbol{g}(\boldsymbol{x}^{(k)}) + \boldsymbol{H}(\boldsymbol{x}^{(k)})(\boldsymbol{x}_{opt} - \boldsymbol{x}^{(k)}),$$

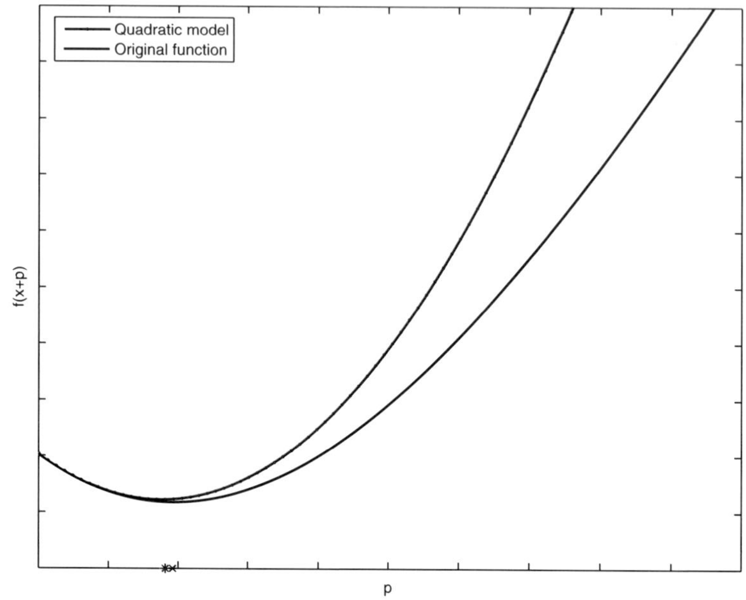

Figure 9.3. *Newton's method fits a quadratic model $q(p)$ (blue curve) to the function $f(x+p)$ (black curve) so that $q(p)$ matches F in function value and first two derivatives at $p=0$. The minimizer of the quadratic model (star) is close to the minimizer of f but not identical.*

so

$$\|\boldsymbol{H}(\boldsymbol{x}^{(k)})^{-1}\boldsymbol{g}(\boldsymbol{x}^{(k)})\| \approx \|\boldsymbol{x}_{opt} - \boldsymbol{x}^{(k)}\|, \tag{9.1}$$

and therefore we can expect that $\|\boldsymbol{x}^{(k)} - \boldsymbol{x}_{opt}\|$ is less than $\|\boldsymbol{g}(\boldsymbol{x}^{(k)})\|/\sigma_{min}$, where σ_{min} denotes the smallest singular value of $\boldsymbol{H}(\boldsymbol{x}^{(k)})$. We will see in the next section that the vector $\boldsymbol{H}(\boldsymbol{x}^{(k)})^{-1}\boldsymbol{g}(\boldsymbol{x}^{(k)})$ is computed in Newton's method, so in that case we can evaluate the left-hand side of (9.1) directly.

We consider methods for finding a search direction and determining the stepsize in the next section.

9.2 The Model Method: Newton

Newton's method is one way to determine the search direction $\boldsymbol{p}^{(k)}$. It is inspired by our Taylor series expansion

$$f(\boldsymbol{x}+\boldsymbol{p}) \approx f(x) + \boldsymbol{p}^T \boldsymbol{g}(x) + \frac{1}{2}\boldsymbol{p}^T \boldsymbol{H}(x)\boldsymbol{p} \equiv q(\boldsymbol{p}). \tag{9.2}$$

Suppose we replace $f(\boldsymbol{x}+\boldsymbol{p})$ by the **quadratic model** $q(\boldsymbol{p})$ and minimize that, as illustrated in Figure 9.3. In general, the model won't fit f well at all—except in a neighborhood of the point $\boldsymbol{x}$ for which it is built. But if our step $\boldsymbol{p}$ is not too big, that is all right! So let's try to minimize q with respect to $\boldsymbol{p}$. If we set the derivative of q equal to zero we obtain the equation

$$g(x) + H(x)p = 0.$$

We see that we need the vector p defined by

$$H(x)p = -g(x). \tag{9.3}$$

This vector is called the **Newton direction**, and it is obtained by solving the linear system involving the Hessian matrix and the negative gradient.

The next three challenges give us some practice with Newton's method.

CHALLENGE 9.1.

Apply one step of Newton's method (with stepsize $\alpha_k = 1$) to the problem

$$\min_{x} x_1^4 + x_2(x_2 - 1),$$

starting at the point $x_1 = 2$, $x_2 = -1$.

CHALLENGE 9.2.

Let $f(x) = e^{x_1+x_2}x_1 + x_2^2$ and consider the point $x_1 = 1$, $x_2 = 0.3863$. Compute the Newton direction and determine whether it is downhill.

Helpful fact: $e^{1.3863} = 4.0000$.

CHALLENGE 9.3.

Write a MATLAB program to apply 5 iterations of Newton's method to the problem

$$\min_{x} (x_1 - 5)^4 + (x_2 + 1)^4 - x_1 x_2$$

with a stepsize of $\alpha_k = 1$ and with an initial starting guess of $x = [1, 2]^T$.

Note that if the Hessian $H(x)$ is positive definite, then the linear system defining p is guaranteed to have a unique solution (since $H(x)$ is therefore nonsingular) and, in addition,

$$0 < p^T H(x)p = -p^T g(x),$$

so in this case p is a downhill direction. In fact, p is the minimizer of the quadratic model $q(p)$.

If $H(x)$ fails to be positive definite, then the situation is not as nice. If the matrix $H(x)$ has a negative eigenvalue, we may walk uphill. Even worse, if the matrix $H(x)$ is singular and if $g(x)$ does not lie in its range, we have no solution to the linear system, so the search direction is not defined.

We can also get into trouble if $H(x)$ is close to singular, since in that case it is difficult to get a good solution to the linear system using floating-point arithmetic, so the computed direction may be almost orthogonal to $g(x)$.

POINTER 9.2. Matrix Inverses.

To find the Newton direction or an approximation to it, we solve a linear system of equations. Recall from Chapter 5 that although the solution to the linear system $\boldsymbol{H}(\boldsymbol{x})\boldsymbol{p} = -\boldsymbol{g}(\boldsymbol{x})$ is defined mathematically by $\boldsymbol{p} = -[\boldsymbol{H}(\boldsymbol{x})]^{-1}\boldsymbol{g}(\boldsymbol{x})$, computing the matrix inverse is generally more expensive and less accurate than using a matrix decomposition to solve the system.

So to run the basic Newton method safely, we need the Hessian $\boldsymbol{H}(\boldsymbol{x})$ to be sufficiently positive definite everywhere we need to evaluate it. Later, in Section 9.2.2, we'll study fixes for the cases in which this condition fails to hold, but for now, we'll just study the basic Newton algorithm, in which we step from $\boldsymbol{x}$ to $\boldsymbol{x} - \boldsymbol{H}(\boldsymbol{x})^{-1}\boldsymbol{g}(\boldsymbol{x})$.

9.2.1 How Well Does Newton's Method Work?

When it is good, it is very, very good! For example, if we let $\boldsymbol{e}^{(k)} = \boldsymbol{x}^{(k)} - \boldsymbol{x}_{opt}$ be the **error at iteration** k and assume that

- $\boldsymbol{H}(\boldsymbol{x})$ is **Lipschitz continuous**, so that there is a positive scalar λ such that

  $$\|\boldsymbol{H}(\boldsymbol{x}) - \boldsymbol{H}(\boldsymbol{y})\| \leq \lambda \|\boldsymbol{x} - \boldsymbol{y}\|$$

  for all points $\boldsymbol{x}, \boldsymbol{y}$ in a neighborhood of $\boldsymbol{x}_{opt}$,

- $\boldsymbol{x}^{(0)}$ is sufficiently close to $\boldsymbol{x}_{opt}$,

- $\boldsymbol{H}(\boldsymbol{x}_{opt})$ is positive definite,

then [50, p. 46] there exists a constant c such that

$$\|\boldsymbol{e}^{(k+1)}\| \leq c\|\boldsymbol{e}^{(k)}\|^2, \qquad k = 0, 1, \ldots.$$

This rate of convergence is called **quadratic convergence**, and it is remarkably fast. If, for example, $c = 1$ and we have an error of 10^{-1} at some iteration, then two iterations later the error is at most 10^{-4}. After four iterations it is at most 10^{-16}, as many figures as we carry in double precision arithmetic!

Newton's quadratic rate of convergence is nice, but Newton's method is not an ideal method:

- It requires the computation of $\boldsymbol{H}(\boldsymbol{x})$ at each iteration.

- It requires the solution of a linear system involving $\boldsymbol{H}(\boldsymbol{x})$.

- It can fail if $\boldsymbol{H}(\boldsymbol{x})$ fails to be positive definite.

- The convergence result only applies when we start "close enough" to the solution, and Newton's method may misbehave badly enough that we never get that close.

So we would like to modify Newton's method to make it cheaper and more widely applicable without sacrificing its fast convergence. A quadratic convergence rate is great, but we can settle for a **superlinear convergence rate**. We say that a sequence of errors $\boldsymbol{e}^{(k)}$ converges to zero with **rate** r and rate constant $c < \infty$ if

$$\lim_{k\to\infty} \frac{\|\boldsymbol{e}^{(k+1)}\|}{\|\boldsymbol{e}^{(k)}\|^r} = c.$$

(If $r = 2$, then this is a quadratic rate of convergence. If $r = 1$, then we need $c < 1$.)

The important point is that we can get a **superlinear convergence rate** (convergence with rate $1 < r < 2$) without walking exactly in the Newton direction. In fact, for the same class of functions included in the convergence theorem above, we obtain superlinear convergence if and only if [111, p. 304]

$$\lim_{k\to\infty} \frac{\|\boldsymbol{p}^{(k)}+\boldsymbol{H}(\boldsymbol{x}^{(k)})^{-1}\boldsymbol{g}(\boldsymbol{x}^{(k)})\|}{\|\boldsymbol{p}^{(k)}\|} = 0.$$

This means that if we choose a vector $\boldsymbol{p}^{(k)}$ that is a good approximation to the solution to the Newton equation (9.3), we can still achieve fast convergence. Therefore, we are free to safeguard Newton's method, or to substitute an approximate Newton direction, and still expect good results. This leads us to methods such as quasi-Newton methods (Section 9.5) and those in the next section.

9.2.2 Making Newton's Method Safe: Modified Newton Methods

We want to modify Newton's method whenever we are not sure that the direction it generates is downhill. If the Hessian is positive definite, we know the direction is downhill, although if $\boldsymbol{H} \equiv \boldsymbol{H}(\boldsymbol{x})$ is nearly singular, we may have some computational difficulties.

If the Hessian is semidefinite or indefinite, we might or might not get a downhill direction. So our strategy is as follows:

- We'll use the Hessian matrix whenever it is positive definite and not close to singular, because it leads to quadratic convergence.

- We'll choose a matrix $\widehat{\boldsymbol{E}}$ and replace $\boldsymbol{H}$ by $\widehat{\boldsymbol{H}} = \boldsymbol{H} + \widehat{\boldsymbol{E}}$ whenever $\boldsymbol{H}$ is close to singular or fails to be positive definite.

We want $\widehat{\boldsymbol{H}}$ to satisfy the following conditions:

- $\widehat{\boldsymbol{H}}$ is symmetric positive definite.

- $\widehat{\boldsymbol{H}}$ is not too close to singular; specifically, its condition number should be moderate in size, so that the smallest eigenvalue is not too close to zero, relative to its largest eigenvalue.

We consider two ways to modify $\boldsymbol{H}$ to satisfy these conditions: the Levenberg–Marquardt method and a modified Cholesky strategy.

The Levenberg–Marquardt Method

This strategy was actually proposed for least squares problems, but it works here, too. We choose $\widehat{\boldsymbol{E}} = \gamma \boldsymbol{I}$ for a fixed scalar γ and replace $\boldsymbol{H}$ with

$$\widehat{\boldsymbol{H}} = \boldsymbol{H} + \gamma \boldsymbol{I}.$$

This shifts every eigenvalue up by γ. The parameter γ is usually chosen by trial and error: seek a γ so that $\widehat{\boldsymbol{H}}$ is positive definite and $\|\boldsymbol{p}^{(k)}\| \leq h^{(k)}$, where $\{h^{(k)}\}$ is a given sequence of numbers. The Gerschgorin circle theorem (see Pointer 5.4) can be used to choose γ to ensure that $\widehat{\boldsymbol{H}}$ is positive definite; alternatively, we can choose a γ that makes $\boldsymbol{p}^T \boldsymbol{g}/(\|\boldsymbol{p}\| \|\boldsymbol{g}\|)$ small enough. If $h^{(k)}$ is small enough, then we indeed take a downhill step for $\alpha = 1$, so we can avoid the expense of a linesearch to find a suitable stepsize α.

Modified Cholesky Strategy

If $\boldsymbol{H}$ is positive definite, then it has a Cholesky decomposition

$$\boldsymbol{H} = \boldsymbol{L}\boldsymbol{D}\boldsymbol{L}^T.$$

(See Section 5.2.) But if $\boldsymbol{H}$ has zero or negative eigenvalues, then this factorization may fail. The idea behind the modified Cholesky class of algorithms is to diagnose and fix the failure when it occurs, thus generating a correction matrix $\widehat{\boldsymbol{E}}$ added to $\boldsymbol{H}$. We choose to make $\widehat{\boldsymbol{E}}$ diagonal. While factoring, if our pivot element $d_{ii} \leq 0$ (or, more practically, if $d_{ii} \leq \delta$ for some small parameter δ), we increase d_{ii} so that it is sufficiently large. This changes the factored matrix from $\boldsymbol{H}$ to $\widehat{\boldsymbol{H}}$. If modification is needed, we try to keep $\|\boldsymbol{H} - \widehat{\boldsymbol{H}}\|$ small so that we have an almost-Newton direction.

Our resulting modified Newton Algorithm is built upon the Basic Minimization Algorithm, Algorithm 9.1. We find a search direction $\boldsymbol{p}^{(k)}$ as indicated in Algorithm 9.2.

Algorithm 9.2 Modified Cholesky Strategy for Finding a Search Direction $\boldsymbol{p}$

Given $\boldsymbol{x}$ and a small parameter ϵ.
Calculate $\boldsymbol{g}(\boldsymbol{x})$ and $\boldsymbol{H} = \boldsymbol{H}(\boldsymbol{x})$.
Factor $\boldsymbol{H} + \widehat{\boldsymbol{E}} = \boldsymbol{L}\widehat{\boldsymbol{D}}\boldsymbol{L}^T$ using a modified Cholesky strategy.
if $\|\boldsymbol{g}(\boldsymbol{x})\| \leq \epsilon$ and $\widehat{\boldsymbol{E}} = 0$ **then**
  Halt with an approximate solution.
else
  if $\|\boldsymbol{g}(\boldsymbol{x})\| > \epsilon$ **then**
    Solve $\boldsymbol{L}\widehat{\boldsymbol{D}}\boldsymbol{L}^T \boldsymbol{p} = -\boldsymbol{g}(\boldsymbol{x})$ to get a downhill direction.
  else
    Get a direction of negative curvature. (Comment: The details of this are different for each algorithm to modify $\boldsymbol{L}$, but the cost is $O(n^2)$.)
  end
end

The main remaining issue for our algorithm is determining how long a step to take in the direction $\boldsymbol{p}^{(k)}$, and we consider two alternatives in the next two sections.

9.3 Descent Directions and Backtracking Linesearches

If we have a descent direction $\boldsymbol{p}^{(k)}$, then we know by Taylor series that a small enough step in this direction decreases f. But in order to make fast progress, we want to take large steps—in fact, the quadratic convergence rate of Newton's method depends on taking steps of size $\alpha_k = 1$. How do we choose α_k to make good progress but guarantee that f does not increase?

One way is to use a **backtracking linesearch**. We try a stepsize of 1. If we are not satisfied with that step, then we try successively smaller ones: perhaps $1, 1/2, 1/4, \ldots$, until success. There are better ways to choose the sequence of stepsizes, though. Let

$$F(\alpha) = f(\boldsymbol{x} + \alpha \boldsymbol{p}),$$

so that

$$F'(\alpha) = \boldsymbol{p}^T \boldsymbol{g}(\boldsymbol{x} + \alpha \boldsymbol{p}).$$

Then if the stepsize $\alpha = 1$ fails, we have values $F(0)$, $F'(0)$, and $F(1)$. This information allows us to build a model of F using quadratic interpolation, and we can use the minimizer of the quadratic to predict the minimizer of F, as in Figure 9.4. If the trial is unsuccessful, we can use the resulting function value to update the interpolant. So we have Algorithm 9.3, backtracking linesearch. This algorithm does not depend on $\boldsymbol{p}$ being the Newton direction, but if it is not, then we may need an initial **bracketing** phase that finds a good upper bound on α by testing successively larger values.

Algorithm 9.3 Backtracking Linesearch

Choose $\alpha = 1$ (to give the full Newton step).
while α is not good enough,
  Use recently obtained function values and derivatives of F to find a simple function that interpolates these values.
  Let the new $\alpha_{new} \in [0, \alpha]$ be the minimizer of the simple function and set $\alpha = \alpha_{new}$.
end

So far our situation is as follows:

- We have a downhill direction $\boldsymbol{p}$, so we know that for very small α, $F(\alpha) < F(0)$.
- If $\boldsymbol{p}$ is the Newton direction, then we predict that $\alpha = 1$ is the minimizer.
- We really can't perform an exact linesearch, one that determines the value of α that exactly minimizes $f(\boldsymbol{x} + \alpha \boldsymbol{p})$. We can do this for quadratic functions, since in that case a formula for α can be derived, but in general, exact linesearch is impossible and is only interesting because a lot of theorems demand it.

How do we decide that a candidate α in Algorithm 9.3 is good enough? Various conditions have been proposed. Two commonly used ones are the **Goldstein–Armijo condition** and the **Wolfe(1968)–Powell(1976) condition**. The Wolfe–Powell condition demands that

$$|\boldsymbol{p}^T \boldsymbol{g}(\boldsymbol{x} + \alpha \boldsymbol{p})| \leq \eta |\boldsymbol{p}^T \boldsymbol{g}(\boldsymbol{x})|,$$

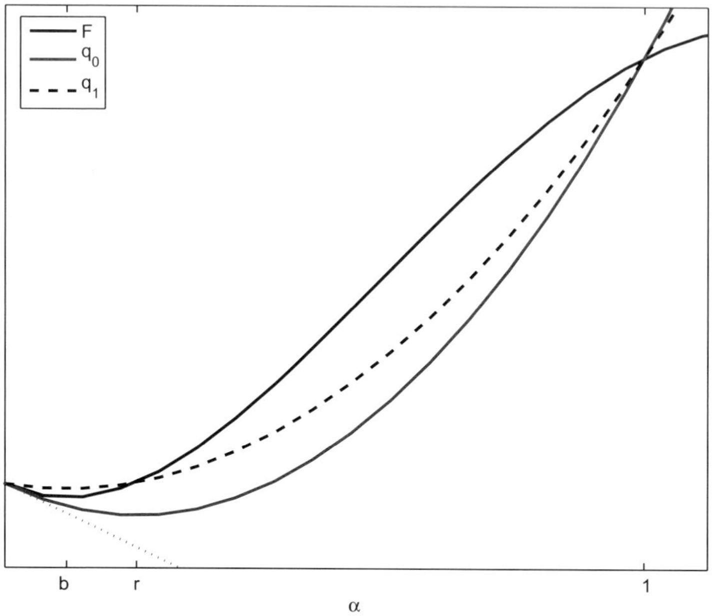

Figure 9.4. *Backtracking linesearch. We first try a value $\alpha = 1$ but find in this example that the function value is too high. Then we find the minimizer $\alpha_0 = r$ of the (red) quadratic polynomial q_0 that matches the data we have gathered: $F(0)$, $F'(0)$, and $F(1)$. ($F'(0)$ is the slope of the dotted line.) Since the function value $F(\alpha_0)$ is still too high, we find the minimizer $\alpha_1 = b$ of the (blue dashed) polynomial q_1 that matches the most recent data: $F(0)$, $F(1)$, and $F(\alpha_0)$. The function value $F(\alpha_1)$ is less than $F(0)$, so we might accept it and terminate the linesearch.*

where $\boldsymbol{\eta}$ is a fixed constant in the interval $[0,1)$. The parameter η controls how close we need to be to the exact linesearch, since $\boldsymbol{g}(\boldsymbol{x}+\alpha\boldsymbol{p}) = \boldsymbol{0}$ if the search is exact. The Goldstein–Armijo condition is similar but does not involve gradient evaluation:

$$f(\boldsymbol{x}+\alpha\boldsymbol{p}) \le f(\boldsymbol{x}) + \eta\alpha\boldsymbol{p}^T\boldsymbol{g}(\boldsymbol{x}).$$

Taking $\eta = 1$ in this condition says that the decrease in f should be at least as big as that predicted from Taylor series, and taking η close to zero requires only a small decrease in f.

Such conditions are built into good linesearch algorithms. It can be shown that points acceptable under these conditions exist as long as the minimizer is finite, and when using them we are assured of success. In particular, suppose we use these conditions in our linesearch to minimize a function f for which

- f is continuously differentiable and bounded below,
- $\boldsymbol{g}$ is Lipschitz continuous, so that there exists a constant L such that, for all $\boldsymbol{x}$ and $\boldsymbol{y}$,

$$\|\boldsymbol{g}(\boldsymbol{x}) - \boldsymbol{g}(\boldsymbol{y})\| \le L\|\boldsymbol{x} - \boldsymbol{y}\|.$$

Also suppose that the directions $\boldsymbol{p}^{(k)}$ are downhill and bounded away from orthogonality to $\boldsymbol{g}^{(k)}$:

$$\boldsymbol{g}^{(k)T}\boldsymbol{p}^{(k)} \leq -\delta\|\boldsymbol{g}^{(k)}\|\,\|\boldsymbol{p}^{(k)}\|, \quad k=0,1,\ldots,$$

for some fixed $\delta > 0$. Then either $\boldsymbol{g}^{(k)} = \boldsymbol{0}$ for some k or $\boldsymbol{g}^{(k)} \to \boldsymbol{0}$ [50].

9.4 Trust Regions

Trust regions are a popular alternative to linesearches. In these methods, we choose $\boldsymbol{p}$ and α simultaneously. The idea is again based on the quadratic model (9.2):

$$f(\boldsymbol{x}+\boldsymbol{p}) \approx q(\boldsymbol{p}) = f(\boldsymbol{x}) + \boldsymbol{p}^T\boldsymbol{g} + \frac{1}{2}\boldsymbol{p}^T\boldsymbol{H}\boldsymbol{p}.$$

But we should only trust the model when $\|\boldsymbol{p}\| < h$ for some small scalar h. So we step to the point $\boldsymbol{x}_{new} = \boldsymbol{x} + \boldsymbol{p}_{opt}$, where $\boldsymbol{p}_{opt}$ solves

$$\min_{\|\boldsymbol{p}\| \leq h} q(\boldsymbol{p}). \tag{9.4}$$

If the solution $\boldsymbol{p}_{opt}$ satisfies $\|\boldsymbol{p}_{opt}\| < h$, then solving (9.4) yields a step of size $\alpha = 1$ in the Newton direction, but it gives a different direction otherwise.

The shape of the **trust region** $\{\boldsymbol{p} : \|\boldsymbol{p}\| \leq h\}$ depends on the norm we choose, and so does the algorithm for computing $\boldsymbol{p}_{opt}$. If we choose the infinity norm, for example, then our problem becomes

$$\min_{|p_i| \leq h} q(\boldsymbol{p}).$$

This is a **quadratic programming problem with bound constraints**. If we instead use the 2-norm, we show in the next challenge that the problem is closely related to an algorithm we have seen before.

CHALLENGE 9.4.

Show, using a Lagrange multiplier, that if we define the region by $\|\boldsymbol{p}\|_2 = h$, then the solution corresponds to the direction in the Levenberg–Marquardt algorithm.

Hint: Write the constraint as $(1/2)(\boldsymbol{p}^T\boldsymbol{p} - h^2) = 0$.

In order to use the trust region method we need to decide how to determine h, how to adapt h as $\boldsymbol{x}^{(k)}$ gets closer to the solution, and how to solve the minimization problem for $\boldsymbol{p}_{opt}$. The parameter h determines the size of the region in which our model q is known to be a good approximation to f; for $\hat{h}\boldsymbol{p}$ in this region, we expect that

$$r \equiv \frac{f(\boldsymbol{x}+\hat{h}\boldsymbol{p}) - f(\boldsymbol{x})}{q(\hat{h}\boldsymbol{p}) - q(\boldsymbol{0})} \approx 1.$$

Therefore, often a heuristic rule is used for modifying h based on the solution to (9.4): For example,

- If r is too small ($< 1/4$), then reduce h by a factor of 4 and reject the step.

POINTER 9.3. Automatic Differentiation.

Automatic differentiation programs take a function `f` written in MATLAB or some other high-level language and produce a function that evaluates the derivatives of `f`. Automatic differentiation is an old idea.

- A **forward** (bottom-up) algorithm was proposed in the 1970s. To evaluate the gradient of a function f, this method requires $O(n)$ space and $O(n)$ times the time required for the evaluation of f.
- A **backward** (top-down) algorithm was proposed in the 1980s. To evaluate the gradient, this method can take $O(n^2)$ space but only requires 2 times the amount of time needed for evaluating f.
- Reliable software was first developed in the 1990s, and one package is found in [27].

- If r is large ($> 3/4$), then increase h by a factor of 2.

There is a pitfall in using trust region methods: if the problem is poorly scaled, then the trust region may remain very small and we may never be able to take large steps. For example, the function

$$f(\boldsymbol{x}) = f_1(x_1) + f_1(10000x_2),$$

where f_1 is a well-behaved function, is poorly scaled.

A trust region is a good substitute for a linesearch in that it, too, ensures success of our optimization algorithm: If [50, Thm. 5.1.1]

- the set $S_0 = \{\boldsymbol{x} : f(\boldsymbol{x}) \leq f(\boldsymbol{x}^{(0)})\}$ is bounded,
- f has two continuous derivatives on $\mathcal{S}_0$,

then the sequence $\{\boldsymbol{x}^{(k)}\}$ has an accumulation point $\boldsymbol{x}_{opt}$ that satisfies the first- and second-order necessary conditions for optimality.

9.5 Alternatives to Newton's Method

We now know how to recognize a solution and compute a solution using Newton's method. We have included safeguards in case the Hessian fails to be positive definite, and we have considered a linesearch or trust region to guarantee convergence. The resulting algorithm converges rather rapidly, but each iteration is quite expensive.

The most tedious and error-prone part of nonlinear optimization is writing functions that compute derivatives. An alternative is to let the computer do it, as discussed in Pointer 9.3. If that is not practical, then we use methods that require fewer derivatives. Often these algorithms have lower cost per iteration.

9.5.1 Methods that Require Only First Derivatives

Suppose we want to solve our problem

$$\min_{\boldsymbol{x}} f(\boldsymbol{x})$$

when $\boldsymbol{g}(\boldsymbol{x})$, but not $\boldsymbol{H}(\boldsymbol{x})$, can be computed.

One strategy is to construct a matrix $\boldsymbol{B}$ to approximate the Hessian matrix $\boldsymbol{H}$; alternatively, sometimes it is convenient to construct a matrix $\boldsymbol{C}$ to approximate $\boldsymbol{H}^{-1}$. We'll discuss two options:

- Approximate $\boldsymbol{H}(\boldsymbol{x})$ or $\boldsymbol{H}(\boldsymbol{x})^{-1}$ using **quasi-Newton methods** (also called **variable metric** methods).
- Estimate $\boldsymbol{H}(\boldsymbol{x})$ using finite differences, resulting in **discrete Newton** methods.

Note, though, that checking optimality is somewhat more problematic once $\boldsymbol{H}(\boldsymbol{x})$ is no longer available.

First Derivative Method 1: Quasi-Newton Methods

Recall that the Newton step is defined by

$$\boldsymbol{p} = -[\boldsymbol{H}(x)]^{-1}\boldsymbol{g}(x).$$

We define the quasi-Newton step as

$$\boldsymbol{p} = -[\boldsymbol{B}^{(k)}]^{-1}\boldsymbol{g}(x),$$

where

$$\boldsymbol{B}^{(k)} \approx \boldsymbol{H}(\boldsymbol{x}^{(k)})$$

and $\boldsymbol{B}$ is accumulated using free information! What information comes free? At step k, we know $\boldsymbol{g}(\boldsymbol{x}^{(k)})$ and we compute $\boldsymbol{g}^{(k+1)} \equiv \boldsymbol{g}(\boldsymbol{x}^{(k+1)})$, where $\boldsymbol{x}^{(k+1)} = \boldsymbol{x}^{(k)} + \boldsymbol{s}^{(k)}$. From Taylor series we know that

$$\boldsymbol{g}(\boldsymbol{x}+h\boldsymbol{s}) = \boldsymbol{g}(\boldsymbol{x}) + h\boldsymbol{H}(\boldsymbol{x})\boldsymbol{s} + O(h^2),$$

so the matrix $\boldsymbol{H}(\boldsymbol{x}^{(k)})$ satisfies

$$\boldsymbol{H}^{(k)}\boldsymbol{s}^{(k)} = \lim_{h\to 0} \frac{\boldsymbol{g}(\boldsymbol{x}^{(k)} + h\boldsymbol{s}^{(k)}) - \boldsymbol{g}(\boldsymbol{x}^{(k)})}{h}.$$

In fact, if f is quadratic, then

$$\boldsymbol{H}^{(k)}\boldsymbol{s}^{(k)} = \boldsymbol{g}(\boldsymbol{x}^{(k)} + \boldsymbol{s}^{(k)}) - \boldsymbol{g}(\boldsymbol{x}^{(k)}).$$

We'll ask the same property of our approximation $\boldsymbol{B}^{(k+1)}$ and call this the **secant condition**:

$$\boldsymbol{B}^{(k+1)}\boldsymbol{s}^{(k)} = \boldsymbol{g}^{(k+1)} - \boldsymbol{g}^{(k)}.$$

CHALLENGE 9.5.
Why do we demand that quasi-Newton matrices $B^{(k)}$ satisfy the secant condition?

Since we know how we want $B^{(k+1)}$ to behave in the direction $s^{(k)}$, and we have no new information in any other direction, we could require

$$B^{(k+1)}v = B^{(k)}v \quad \text{if} \quad v^T s^{(k)} = 0.$$

There is a unique matrix $B^{(k+1)}$ that satisfies the secant condition and these **no-change conditions**. It is computed by **Broyden's good method**:

$$B^{(k+1)} = B^{(k)} - (B^{(k)}s^{(k)} - y^{(k)})\frac{s^{(k)T}}{s^{(k)T}s^{(k)}},$$

where

$$s^{(k)} = x^{(k+1)} - x^{(k)},$$
$$y^{(k)} = g^{(k+1)} - g^{(k)}.$$

CHALLENGE 9.6.
Verify that Broyden's good method satisfies the secant condition and the no-change conditions.

For Broyden's good method, $B^{(k+1)}$ is formed from $B^{(k)}$ by adding a rank-1 matrix, but the matrix is not necessarily symmetric, even if $B^{(k)}$ is. This is undesirable since we know H is symmetric.

In order to regain symmetry, we need to sacrifice the no-change conditions. Instead, we formulate the problem in a **least change** sense:

$$\min_{B^{(k+1)}} \|B^{(k+1)} - B^{(k)}\|$$

subject to a symmetry condition and the secant condition $B^{(k+1)}s^{(k)} = y^{(k)}$. The solution depends on the choice of norm.

We can impose other constraints, too. Frequently, algorithms demand that $B^{(k+1)}$ be positive definite. If we have extra information about the structure of the Hessian (for example, knowing that H is sparse), then we might want B to have the same structure.

An alphabet soup of quasi-Newton algorithms have been proposed. The oldest is the DFP method (Davidon 1959, Fletcher–Powell 1963). In this method we accumulate an approximation C to H^{-1} as

$$C^{(k+1)} = C^{(k)} - \frac{C^{(k)}y^{(k)}y^{(k)T}C^{(k)}}{y^{(k)T}C^{(k)}y^{(k)}} + \frac{s^{(k)}s^{(k)T}}{y^{(k)T}s^{(k)}}.$$

DFP is one of the most popular quasi-Newton methods because it has many desirable properties. But the BFGS method (Broyden, Fletcher, Goldfarb, Shanno 1970) defined by

$$\boldsymbol{B}^{(k+1)} = \boldsymbol{B}^{(k)} - \frac{\boldsymbol{B}^{(k)}\boldsymbol{s}^{(k)}\boldsymbol{s}^{(k)T}\boldsymbol{B}^{(k)}}{\boldsymbol{s}^{(k)T}\boldsymbol{B}^{(k)}\boldsymbol{s}^{(k)}} + \frac{\boldsymbol{y}^{(k)}\boldsymbol{y}^{(k)T}}{\boldsymbol{y}^{(k)T}\boldsymbol{s}^{(k)}} \tag{9.5}$$

is the most successful method.

CHALLENGE 9.7.

Verify that the BFGS matrix satisfies the secant condition.

Algorithm 9.4 Quasi-Newton Algorithm

Initialize $\boldsymbol{x}^{(0)}$ and $\boldsymbol{C}^{(0)}$ (or $\boldsymbol{B}^{(0)}$).
Set $k = 0$.
while $\boldsymbol{x}^{(k)}$ is not a good enough solution,
  Compute a search direction $\boldsymbol{p}^{(k)}$ from $\boldsymbol{p}^{(k)} = -\boldsymbol{C}^{(k)}\boldsymbol{g}(\boldsymbol{x}^{(k)})$ (or solve $\boldsymbol{B}^{(k)}\boldsymbol{p}^{(k)} = -\boldsymbol{g}(\boldsymbol{x}^{(k)})$).
  Set $\boldsymbol{x}^{(k+1)} = \boldsymbol{x}^{(k)} + \alpha_k \boldsymbol{p}^{(k)}$, where α_k satisfies the Goldstein–Armijo or Wolfe–Powell linesearch conditions.
  Form the updated matrix $\boldsymbol{C}^{(k+1)}$ (or $\boldsymbol{B}^{(k+1)}$).
  Set $k = k + 1$.
end

The quasi-Newton algorithm, Algorithm 9.4, looks very similar to Newton's method, but now we need to initialize $\boldsymbol{B}^{(0)}$ (or $\boldsymbol{C}^{(0)}$) as well as $\boldsymbol{x}^{(0)}$. We take $\boldsymbol{B}^{(0)} = \boldsymbol{I}$, or a multiple of $\boldsymbol{I}$, or some better guess.

Quasi-Newton methods generally have an n-, $2n$-, or $(n+2)$-step quadratic convergence rate if the linesearch is exact. An n-step quadratic convergence rate, for example, means that there is a constant $c < \infty$ such that

$$\lim_{k\to\infty} \frac{\|\boldsymbol{x}^{(k+n)} - \boldsymbol{x}_{opt}\|}{\|\boldsymbol{x}^{(k)} - \boldsymbol{x}_{opt}\|^2} = c\,.$$

Weakening the linesearch to a Wolfe–Powell or Goldstein–Armijo search generally gives superlinear convergence; see [111, p. 356] for a typical result.

Some implementation issues remain:

- Near a stationary point, $\boldsymbol{H}^{-1}$ does not exist. How do we keep $\boldsymbol{C}$ from deteriorating?
- What happens if $\boldsymbol{H}$ is indefinite?

These questions concern **stability of the algorithm**. We are faced with a dilemma in trying to balance stability and efficiency:

- Updating $\boldsymbol{C}$ can be hazardous when $\boldsymbol{H}$ is close to singular.

- Updating $\boldsymbol{B}$ leaves the problem of solving a linear system at each iteration to determine the search direction.

The resolution comes by using an appropriate matrix decomposition. Note that (9.5) shows that the BFGS matrix $\boldsymbol{B}$ is changed by a rank-2 update at each iteration. The most stable way to implement the algorithm is to update a Cholesky decomposition of $\boldsymbol{B}$. If we have a Cholesky decomposition of $\boldsymbol{B}^{(k)}$ as $\boldsymbol{B}^{(k)} = \boldsymbol{L}^{(k)}\boldsymbol{D}^{(k)}\boldsymbol{L}^{(k)T}$, then we can obtain $\boldsymbol{B}^{(k+1)} = \boldsymbol{L}^{(k+1)}\boldsymbol{D}^{(k+1)}\boldsymbol{L}^{(k+1)^T}$, using techniques similar to those used in updating the QR decomposition in the case study of Chapter 7. This makes it easy to enforce symmetry and positive definiteness. The algorithms are $O(n^2)$, and details are given by Gill, Golub, Murray, and Saunders [58] and in some textbooks.

First Derivative Method 2: Finite Difference Newton Method

Evaluation of the matrix $\boldsymbol{H}(\boldsymbol{x})$ can be a major time sink in using Newton's method. One way to avoid evaluations of the entries

$$h_{ij} = \frac{\partial g_i(\boldsymbol{x})}{\partial x_j}$$

is to use Taylor series to approximate them by

$$h_{ij} \approx \frac{g_i(\boldsymbol{x} + \tau \boldsymbol{e}_j) - g_i(\boldsymbol{x})}{\tau}, i, j, \ldots, n,$$

where τ is a small number and $\boldsymbol{e}_j$ is the jth column of the identity matrix. We see, using Taylor series, that this approximation is accurate to $O(\tau)$. The cost is n extra gradient evaluations per iteration. Sometimes this is less than the cost of the Hessian evaluation, but sometimes it is more. The choice of τ is critical to the success of the method:

- If τ is large, the approximation is poor and we have large **truncation error**.
- If τ is small, then there is cancellation error in forming the numerator of the approximation, so we have large **rounding error**.

Usually we try to balance the two errors by choosing τ to make them approximately equal. If the problem is poorly scaled, we may need a different τ for each j. There are theorems that say that if we choose τ carefully enough, we can get superlinear convergence.

If you are considering using the finite difference Newton method, also consider the truncated Newton method of the next section, since it often gives comparable results at less cost.

9.5.2 Low-Storage First-Derivative Methods

Sometimes problems are too big to allow n^2 storage space for the Hessian matrix, so we consider three methods that avoid storage of a matrix:

- steepest descent,
- nonlinear conjugate gradient,
- truncated Newton.

Low-Storage Method 1: Steepest Descent

Let's return to that foggy mountain. If we repeatedly walk in the direction of **steepest descent** until we stop going downhill, we clearly are guaranteed to get to a local minimizer. The trouble is that the algorithm is terribly slow. For example, if we apply steepest descent to a quadratic function of n variables, then after many steps, the algorithm alternates between just two directions: those corresponding to the eigenvectors of the smallest and the largest eigenvalues of the Hessian matrix. In this case the convergence rate is only linear:

$$f(\boldsymbol{x}^{(k+1)}) - f(\boldsymbol{x}_{opt}) \le \left(\frac{\kappa - 1}{\kappa + 1}\right)^2 (f(\boldsymbol{x}^{(k)}) - f(\boldsymbol{x}_{opt})), \tag{9.6}$$

where κ, the condition number, is the ratio of the largest to the smallest eigenvalue of $\boldsymbol{H}$. See [111, p. 342] for proof. If steepest descent is applied to non-quadratic functions, using a good linesearch, then convergence is local and linear.

Instead of this method, consider using **nonlinear conjugate gradients**. It has the same advantages as steepest descent, requires only a few more vectors of storage, and gives a better convergence rate.

Low-Storage Method 2: Nonlinear Conjugate Gradient Methods

The (linear) conjugate gradient method [76] is a method for solving linear systems of equations $\boldsymbol{Ax} = \boldsymbol{b}$ when $\boldsymbol{A}$ is symmetric and positive definite. There are many ways to understand it, but here we can think of it as minimizing the function

$$\hat{f}(\boldsymbol{x}) = \frac{1}{2}\boldsymbol{x}^T\boldsymbol{A}\boldsymbol{x} - \boldsymbol{x}^T\boldsymbol{b},$$

which has gradient $\widehat{\boldsymbol{g}}(\boldsymbol{x}) = \boldsymbol{Ax} - \boldsymbol{b}$. So, by the first-order optimality conditions, a minimizer of $\hat{f}$ is a solution to the linear system $\boldsymbol{Ax} = \boldsymbol{b}$. We could use steepest descent, but we want something faster. The conjugate gradient method combines the concepts of descent and conjugate directions in order to improve on steepest descent. See Section 28.2 for a description of the algorithm. The nonlinear conjugate gradient algorithm, a generalization of this algorithm for minimizing arbitrary functions $f(\boldsymbol{x})$, is shown in Algorithm 9.5.

Algorithm 9.5 Nonlinear Conjugate Gradient Algorithm

Given $\boldsymbol{x}^{(0)}$, form $\boldsymbol{p}^{(0)} = -\boldsymbol{g}(\boldsymbol{x}^{(0)})$. Set $k = 0$.
while $\boldsymbol{x}^{(k)}$ is not a good enough solution,
  Use a linesearch to determine a parameter α_k and then set $\boldsymbol{x}^{(k+1)} = \boldsymbol{x}^{(k)} + \alpha_k \boldsymbol{p}^{(k)}$.
  Set $\boldsymbol{p}^{(k+1)} = -\boldsymbol{g}(\boldsymbol{x}^{(k+1)}) + \beta_{k+1}\boldsymbol{p}^{(k)}$, where β_{k+1} is a scalar parameter.
  Set $k = k + 1$.
end

If f is quadratic, then there is a formula for α_k.

CHALLENGE 9.8.

Let $f(x) = \frac{1}{2}x^T Hx - x^T b$, where H and b are constant, independent of x, and H is symmetric positive definite. Given vectors $x^{(0)}$ and $p^{(0)}$, find the value of the scalar α that minimizes $f(x^{(0)} + \alpha p^{(0)})$. This is the formula for the stepsize α_k in the linear conjugate gradient algorithm.

The parameter β_{k+1} has many definitions that are equivalent for quadratic functions but different when we minimize more general nonlinear functions:

$$\beta_{k+1} = \frac{g(x^{(k+1)})^T g(x^{(k+1)})}{g(x^{(k)})^T g(x^{(k)})} \quad \text{Fletcher–Reeves,}$$

$$\beta_{k+1} = \frac{y^{(k)T} g(x^{(k+1)})}{g(x^{(k)})^T g(x^{(k)})} \quad \text{Polak–Ribière,}$$

$$\beta_{k+1} = \frac{y^{(k)T} g(x^{(k+1)})}{y^{(k)T} p^{(k)}} \quad \text{Hestenes–Stiefel.}$$

Good theorems have been proven about convergence of Fletcher-Reeves, but Polak-Ribière generally performs better.

Note that this method stores no matrix. We only need to remember a few vectors at a time, so it can be used for problems in which there are thousands or millions of variables.

The convergence rate is linear, unless the function has special properties, but generally faster than steepest descent: for quadratics, the rate is

$$f(x^{(k+1)}) - f(x_{opt}) \leq \left(\frac{\sqrt{\kappa} - 1}{\sqrt{\kappa} + 1} \right)^2 (f(x^{(k)}) - f(x_{opt})),$$

where again κ is the condition number of H. Comparing this with the steepest descent bound (9.6) we see that now κ appears with a square-root, which gives a smaller bound.

One important property of conjugate gradients is that if f is quadratic, then the conjugate gradient algorithm generates the same iterates as BFGS or DFP started from the same initial point, with $B^{(0)} = I$. Just as in the linear case, preconditioning can improve the convergence rate of conjugate gradients; see Section 28.3.

Low-Storage Method 3: Truncated Newton Method (Newton-CG)

Again we return to the way the Hessian approximation is used. Newton's method determines the search direction by solving the linear system

$$H(x)p = -g(x).$$

We usually think of solving this by factoring $H(x)$ and then using forward and back substitution. But if n is large, this might be too expensive, and we might choose to use an iterative method, like **linear conjugate gradients** (see Section 28.2) to solve the linear system. If we do, how do we use the Hessian? All we need to do is to multiply a vector by it at each step of the algorithm.

Now Taylor series tells us that, if $\boldsymbol{v}$ is a vector of length 1, then

$$\boldsymbol{g}(\boldsymbol{x}+h\boldsymbol{v}) = \boldsymbol{g}(\boldsymbol{x}) + h\boldsymbol{H}(\boldsymbol{x})\boldsymbol{v} + O(h^2),$$

so

$$\boldsymbol{H}(\boldsymbol{x})\boldsymbol{v} = \frac{\boldsymbol{g}(\boldsymbol{x}+h\boldsymbol{v}) - \boldsymbol{g}(\boldsymbol{x})}{h} + O(h).$$

Therefore, we can get an $O(h)$ approximation of the product of the Hessian with an arbitrary vector by taking a finite difference approximation (the blue term in the equation above) to the change in the gradient in that direction. This is akin to the finite difference Newton method, but much neater, because we only evaluate the finite difference in directions in which we need it.

So we'll compute an approximation to the Newton direction by solving the linear system $\boldsymbol{H}(\boldsymbol{x})\boldsymbol{p} = -\boldsymbol{g}(\boldsymbol{x})$ using a few steps of the (linear) conjugate gradient method, computing approximate matrix-vector products by extra evaluations of the gradient, and stopping the conjugate gradient method early, before the exact solution is computed.

To obtain a superlinear convergence rate, the result at the end of Section 9.2.1 tells us that we need our direction to converge to the Newton direction as $k \to \infty$. We ensure this by

- taking enough iterations of conjugate gradient to get a small residual to the linear system,
- choosing h in the approximation carefully, so that the matrix-vector products are accurate enough.

To make sure we understand how this truncated Newton algorithm works, let's implement it.

CHALLENGE 9.9.

Suppose we are using the truncated Newton method to minimize a function $f(\boldsymbol{x})$. This means that we use Algorithm 9.1 with "Find a search direction" meaning "Solve the equation $\boldsymbol{H}(\boldsymbol{x}^{(k)})\,\boldsymbol{p} = -\boldsymbol{g}(\boldsymbol{x}^{(k)})$ using the conjugate gradient method." We evaluate $f(\boldsymbol{x})$ and $\boldsymbol{g}(\boldsymbol{x})$ by calling a function `[f,g] = myfnct(x)`. Write a MATLAB function which (approximately) computes the product $\boldsymbol{H}(\boldsymbol{x}^{(k)})\boldsymbol{v}$ for any input vector $\boldsymbol{v}$. Include enough documentation so that a reader would know how to use your function.

9.5.3 Methods that Require No Derivatives

These are methods of last resort, generally used when

- derivatives are not available, or
- derivatives do not exist.

An example of the first case is when function evaluation is performed by running a physical experiment or a numerical simulation. This is an area of very active research currently and all the methods we discuss are slow. We'll consider three classes of methods here and one more in the case study of Chapter 17.

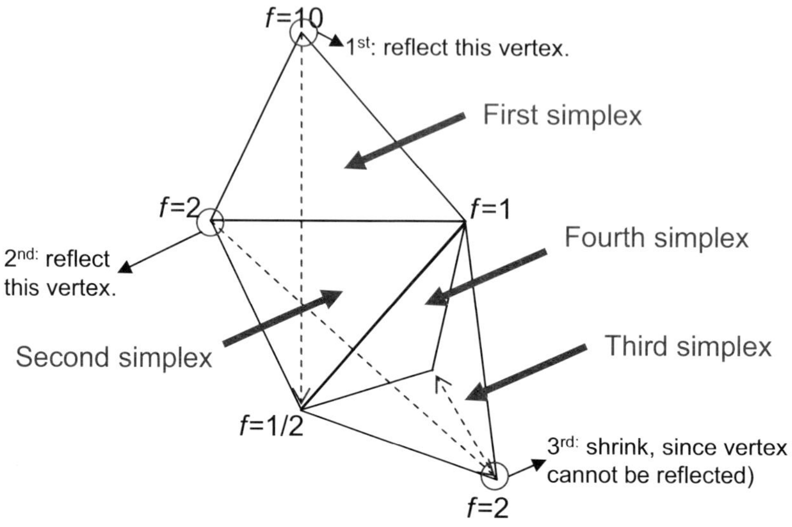

Figure 9.5. *A simplex-based method for function minimization*

No-Derivative Method 1: Finite Difference Methods

We could compute an approximate gradient using finite differencing on the function. This is not usually a good idea; for the same reason we rejected finite difference Newton methods, it is better to either use automatic differentiation methods to compute the gradient (See Pointer 9.3) or use pattern search methods to avoid the need for it.

No-Derivative Method 2: Simplex-Based Methods

The most popular of the simplex-based methods is the **Nelder–Mead algorithm**, and MATLAB has an implementation of this.[7] The idea behind simplex methods is to begin by evaluating the function at the vertices of a **simplex**. (In two dimensions, this is a triangle; in three, it is a tetrahedron, etc.) See Figure 9.5 for an example. We move one vertex of this simplex (usually the one with the largest function value), reflecting it around its current position, until we have enclosed the minimizer in the simplex. Then we shrink the size of the simplex to hone in on the minimizer.

Simplex-based algorithms have rather elaborate rules for determining when to reflect and when to shrink, and no algorithm that behaves well in practice has a good convergence proof. For that reason, it looks as if they will fade in popularity, being supplanted by **pattern search methods**.

[7]The terminology is confusing; these simplex-based methods are distinct from the more widely used simplex method for linear programming.

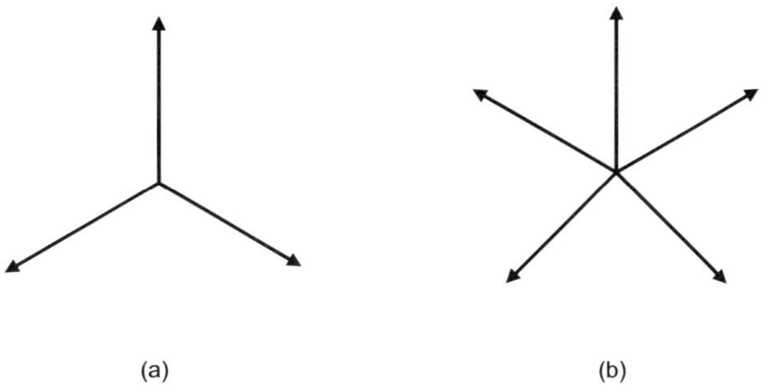

Figure 9.6. *Two positive spanning sets in two dimensions.*

No-Derivative Method 3: Pattern Search Methods

Suppose we are given an initial guess $\boldsymbol{x}$ for the solution to the minimization problem, and a set of at least $n+1$ directions $\boldsymbol{v}_i$, $i = 1, \ldots, N$, that form a **positive spanning set** for $\mathcal{R}^n$: this means that any vector in $\mathcal{R}^n$ can be expressed as a linear combination of these vectors, where the coefficients in the combination are nonnegative numbers. See Figure 9.6 for two examples of positive spanning sets in 2 dimensions.

At each step of a pattern search method, we do a linesearch in each of the directions (or perhaps just use a fixed stepsize) to obtain function values $f(\boldsymbol{x}+\alpha_i \boldsymbol{v}_i)$, and we replace $\boldsymbol{x}$ by the point with the smallest function value. We repeat this until convergence. This is a remarkably simple algorithm, but works well in practice and is provably convergent! Another desirable property is that it is easy to parallelize, and this is crucial to making a no-derivative algorithm effective when n is large.

9.6 Summary

We summarize in Figure 9.7 the process of choosing an appropriate algorithm for an unconstrained optimization problem. The next four challenges give us some experience with such choices.

CHALLENGE 9.10.

Explain how you would decide whether to choose the Nelder–Meade algorithm, the quasi-Newton algorithm, or the Newton algorithm for minimizing a function of many variables.

CHALLENGE 9.11.

You are asked to minimize a function of $n = 2000$ variables. Consider doing this by Newton's method, a quasi-Newton method, or pattern search. Give the main advantages and disadvantages of each. Which would you choose? Why?

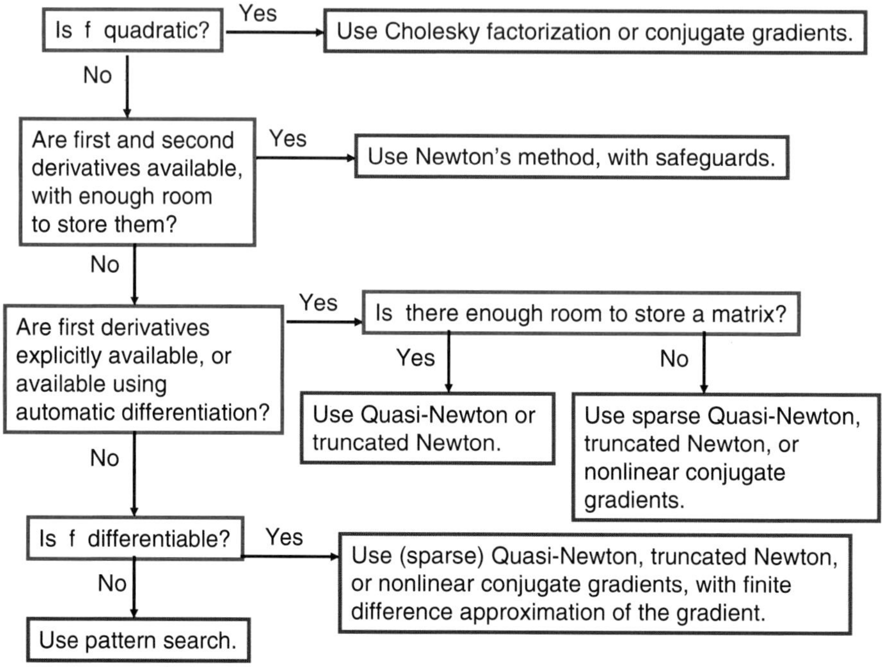

Figure 9.7. *Choosing an optimization algorithm. Note that derivatives might be available explicitly, or through automatic differentiation. The special case when f is the sum of squared terms is a* **nonlinear least squares problem***, and algorithms are discussed in the case study in Chapter* 13.

CHALLENGE 9.12.

Suppose that you have developed a model that uses an ordinary differential equation (ODE) to predict the amount of profit that you will receive on December 11 if you invest \$1000 today in various components of your business, and that profit depends on 5 parameters $x_1, \ldots, x_5$, so that

$$\begin{aligned} y'(t) &= f(t, y, \boldsymbol{x}), \\ y(0) &= 0, \\ y(1) &= \text{profit on December 11, using parameters } \boldsymbol{x}. \end{aligned}$$

You want to choose those 5 parameters in $\boldsymbol{x}$ to maximize $y(1)$ (which is a scalar value). (Then you will take the money and run.)

What numerical algorithms would you use to solve your problem and how would they pass information to each other? (Refer to Chapter 20 if you don't know methods for the numerical solution of ordinary differential equations.) Why did you choose these particular algorithms?

POINTER 9.4. Further Reading.

More information on unconstrained optimization algorithms can be found, for example, in textbooks by Nash and Sofer [111, Chap. 2, Chap. 10] and by Fletcher [50].

There are many modified Cholesky strategies in the literature; see, for example, [47].

Conn and Gould [28] give more information on trust regions.

Quasi-Newton methods could be the basis for a full course; there is a textbook by Dennis and Schnabel [40] and many other good references [22, 41].

Pattern search algorithms are discussed in [95].

CHALLENGE 9.13.

Fill in the following table, giving features of various algorithms for minimizing $f(\boldsymbol{x})$. The first line has been completed, as an example.

Method	Convergence rate	Storage	f evals/itn	$\boldsymbol{g}$ evals/itn	$\boldsymbol{H}$ evals/itn
Truncated Newton	> 1	$O(n)$	0	$\leq n+1$	0
Newton					
Quasi-Newton					
Steepest descent					
Conjugate gradients					

- Assume that all of these methods are convergent and that any linesearch is exact (i.e., the true optimal value of the stepsize parameter is used).
- Don't include the cost of the linesearch in the table entries. We are omitting this cost because it is the same, independent of method.
- f is the function, $\boldsymbol{g}$ is the gradient, and $\boldsymbol{H}$ is the Hessian matrix. "evals/itn" means the number of evaluations per iteration.
- The convergence rate should be "1" for linear, "> 1" for superlinear, or "2" for quadratic.
- Storage should be either $O(1)$, $O(n)$, or $O(n^2)$, where n is the number of variables (i.e., the dimension of $\boldsymbol{x}$).
- "Conjugate gradients" means the nonlinear conjugate gradient method, not the one for solving linear systems (minimizing quadratics).

Next we turn our attention to optimization problems with constraints.

Chapter 10

Numerical Methods for Constrained Optimization

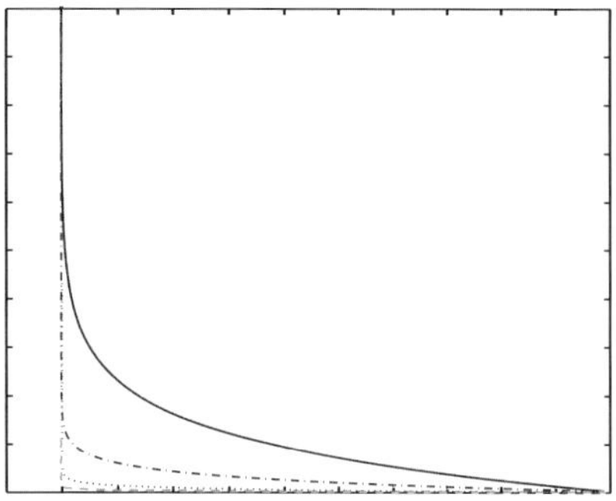

We define the **constrained optimization problem**:

Constrained Optimization Problem: Given a function $f : S \to \mathcal{R}$, find $\boldsymbol{x}_{opt}$ so that

$$f(\boldsymbol{x}_{opt}) = \min_{\boldsymbol{x} \in S} f(\boldsymbol{x}).$$

The set S is defined to be those points $\boldsymbol{x} \in \mathcal{R}^n$ that satisfy the constraints

$$\begin{aligned} c_i(\boldsymbol{x}) &= 0, \quad i = 1, \ldots, m_e, \\ c_i(\boldsymbol{x}) &\geq 0, \quad i = m_e + 1, \ldots, m, \end{aligned}$$

where $c_i : S \to R, \quad i = 1, \ldots, m.$

Often it is best to convert the problem to an unconstrained problem or a system of nonlinear equations, but sometimes it is best to treat the constraints explicitly, as in the **simplex algorithm for linear programming**. We'll concentrate in this chapter on the conversion approach, after first reviewing some fundamentals.

10.1 Fundamentals for Constrained Optimization

We assume that f and c_i are real-valued functions with two continuous derivatives on S. We say that $\boldsymbol{x}_{opt}$ is a solution to our problem if

- $\boldsymbol{x}_{opt}$ **is feasible**; i.e., $\boldsymbol{x}_{opt}$ satisfies all of the constraints.
- $\boldsymbol{x}_{opt}$ **is locally optimal**; i.e., for some $\epsilon > 0$, if $\|\boldsymbol{x} - \boldsymbol{x}_{opt}\| \leq \epsilon$, and if $\boldsymbol{x}$ satisfies the constraints, then $f(\boldsymbol{x}_{opt}) \leq f(\boldsymbol{x})$.

Note that unless both f and the feasible region are **convex**, local minimizers may exist and there is no guarantee of finding the global solution. (A region is convex if the line between any two points in the region lies entirely in the region.)

When solving any problem, it is important to be able to answer two questions:

- How do we recognize a solution?
- How sensitive is the solution to small changes in the data?

In this section we answer these questions for constrained optimization. First we derive the answer for a simple special case, in which all of the constraints are linear. Then we state the general results.

10.1.1 Optimality Conditions for Linear Constraints

For an unconstrained optimization problem, we measure the sensitivity of the function $f(\boldsymbol{x})$ to small changes in $\boldsymbol{x}$ through the gradient vector $\boldsymbol{g}(\boldsymbol{x})$, since if $\|h\boldsymbol{p}\|$ is small, then

$$f(\boldsymbol{x}+h\boldsymbol{p}) \approx f(x) + h\boldsymbol{p}^T g(x).$$

For constrained optimization, an equally interesting question is how sensitive the optimal function value is to small changes in the constraints.

To investigate this, we introduce **Lagrange multipliers**, which you may have studied in calculus. Suppose, to begin with a simple example, that the constraints are $\boldsymbol{c}(\boldsymbol{x}) = \boldsymbol{A}\boldsymbol{x} - \boldsymbol{b} = \boldsymbol{0}$, where $\boldsymbol{A}$ has m_e linearly independent rows. Define the **Lagrangian function** (used in Section 2.1) as

$$L(\boldsymbol{x}, \boldsymbol{\lambda}) = f(\boldsymbol{x}) - \boldsymbol{\lambda}^T \boldsymbol{c}(\boldsymbol{x}),$$

where $\boldsymbol{\lambda}$ is a vector containing the m_e Lagrange multipliers. The Lagrangian function has a saddle point, a minimizer with respect to $\boldsymbol{x}$ and a maximizer with respect to $\boldsymbol{\lambda}$, which is of interest to us. To find it, we set the partial derivatives of L to zero:

$$\begin{aligned} \boldsymbol{g}(\boldsymbol{x}) - \boldsymbol{A}^T \boldsymbol{\lambda} &= \boldsymbol{0}, \\ \boldsymbol{c}(\boldsymbol{x}) &= \boldsymbol{0}. \end{aligned}$$

(If you need to compute these derivatives component by component in order to be comfortable with these expressions, now is a good time to stop and do that exercise!) Note that differentiating with respect to the Lagrange multipliers just gives the constraints back again, while differentiating with respect to the original variables $\boldsymbol{x}$ gives the condition that the gradient of f should be a linear combination of the columns of the matrix $\boldsymbol{A}^T$, which are the constraint gradients.

Now suppose that at some point $\boldsymbol{x}$ satisfying the constraints $\boldsymbol{A}\boldsymbol{x} = \boldsymbol{b}$, the gradient of f is not a linear combination of the constraint gradients but has some additional component $-\boldsymbol{p}$. Then $\boldsymbol{g}(\boldsymbol{x})$ has some component outside of the range of $\boldsymbol{A}^T$, and therefore (by a theorem in linear algebra) that component is in the null space of $\boldsymbol{A}$:

$$\boldsymbol{A}\boldsymbol{p} = \boldsymbol{0}.$$

Since $\boldsymbol{p}^T \boldsymbol{g}(x) < 0$, if we take a small enough step h in the direction $\boldsymbol{p}$, we reduce the minimization function

$$f(\boldsymbol{x}+h\boldsymbol{p}) < f(\boldsymbol{x})$$

and still have a point that satisfies the constraints, since

$$\begin{aligned} \boldsymbol{c}(\boldsymbol{x}+h\boldsymbol{p}) &= \boldsymbol{A}(\boldsymbol{x}+h\boldsymbol{p})-\boldsymbol{b} \\ &= \boldsymbol{A}\boldsymbol{x}-\boldsymbol{b}+h\boldsymbol{A}\boldsymbol{p} \\ &= \boldsymbol{A}\boldsymbol{x}-\boldsymbol{b} \\ &= \boldsymbol{0}. \end{aligned}$$

Therefore, we have derived the **first-order optimality conditions** for a problem with linear equality constraints: the gradient of f must be a linear combination of the gradients of the constraints. Note that if there are no constraints, this condition forces the gradient to be zero, which is exactly the first-order optimality condition that we found for unconstrained optimization in Section 9.1.1.

We see that the Lagrange multipliers are useful in determining optimality, but they are not just an artificial tool. Suppose that we have a point $\widehat{\boldsymbol{x}}$ satisfying

$$\|\boldsymbol{x}_{opt}-\widehat{\boldsymbol{x}}\| \le \epsilon,$$

where $\boldsymbol{x}_{opt}$ satisfies the first-order optimality conditions and ϵ is small. Compute the vector $\boldsymbol{\delta}$ so that

$$\boldsymbol{A}\widehat{\boldsymbol{x}} = \boldsymbol{b}+\boldsymbol{\delta}.$$

Then Taylor series expansion tells us

$$\begin{aligned} f(\widehat{\boldsymbol{x}}) &= f(\boldsymbol{x}_{opt})+(\widehat{\boldsymbol{x}}-\boldsymbol{x}_{opt})^T\boldsymbol{g}(\boldsymbol{x}_{opt})+O(\epsilon^2) \\ &= f(\boldsymbol{x}_{opt})+(\widehat{\boldsymbol{x}}-\boldsymbol{x}_{opt})^T\boldsymbol{A}^T\boldsymbol{\lambda}_{opt}+O(\epsilon^2) \\ &= f(\boldsymbol{x}_{opt})+\boldsymbol{\delta}^T\boldsymbol{\lambda}_{opt}+O(\epsilon^2). \end{aligned}$$

This gives us a valuable insight: if we wiggle b_j by δ_j, then we wiggle f by $\delta_j(\boldsymbol{\lambda}_{opt})_j$. Therefore, $(\boldsymbol{\lambda}_{opt})_j$ is the change in f per unit change in b_j. Thus the jth Lagrange multiplier tells us the sensitivity of f to a small change in the constraint data b_j. For this reason, the Lagrange multiplier λ_j is sometimes called a **dual variable** or a **shadow price**.

So far we have only considered equality constraints. Suppose we have an additional linearly independent constraint $\boldsymbol{a}^T\boldsymbol{x}-b \ge 0$ and our current point $\tilde{\boldsymbol{x}}$ is a linear combination of the constraint gradients:

$$\boldsymbol{g}(\tilde{\boldsymbol{x}})-\boldsymbol{A}^T\boldsymbol{\lambda}-\lambda_{m_e+1}\boldsymbol{a}=\boldsymbol{0}.$$

Can we conclude that $\tilde{\boldsymbol{x}}$ is optimal? There are two cases to consider.

Case 1: Suppose that $\boldsymbol{a}^T\tilde{\boldsymbol{x}}-b=0$, so that the constraint is **active**. We need to remember that we are allowed to walk off the additional constraint if such a step reduces f. If we do choose to move off this constraint, we must still satisfy the remaining (equality) constraints, so we would need to find a vector $\boldsymbol{p}$ so that $\boldsymbol{A}\boldsymbol{p}=\boldsymbol{0}$ and $\boldsymbol{a}^T\boldsymbol{p}>0$. Suppose that we find such a vector. Then if the Lagrange multiplier for the inequality constraint is negative,

$$\begin{aligned} \boldsymbol{p}^T\boldsymbol{g}(\tilde{\boldsymbol{x}}) &= \boldsymbol{p}^T\boldsymbol{A}^T\boldsymbol{\lambda}+\lambda_{m_e+1}\boldsymbol{p}^T\boldsymbol{a} \\ &= \lambda_{m_e+1}\boldsymbol{p}^T\boldsymbol{a} \\ &< 0 \end{aligned}$$

and $\boldsymbol{p}$ is a descent direction that maintains feasibility. Therefore, for optimality, it is necessary that the Lagrange multiplier for the inequality constraint be nonnegative.

Case 2: Suppose that $\boldsymbol{a}^T\tilde{\boldsymbol{x}} - b > 0$ so that the constraint is not currently active. For optimality, the Lagrange multiplier λ_{m_e+1} must be zero, since we would obtain the same solution by solving the problem without that constraint.

So for a single linear inequality constraint, at an optimal point the Lagrange multiplier must be zero if the constraint is not active and nonnegative if the constraint is active. A similar result holds in general, and next we state these optimality conditions.

10.1.2 Optimality Conditions for the General Case

We now have an intuitive understanding of Lagrange multipliers for linear constraints. All of these ideas can be extended to general (differentiable) constraints by using the fact that

$$c_i(\boldsymbol{x}+h\boldsymbol{p}) = c_i(\boldsymbol{x}) + h\boldsymbol{a}_i^T(\boldsymbol{x})\boldsymbol{p} + O(h^2),$$

where $\boldsymbol{a}_i^T(\boldsymbol{x})$ is the gradient of the ith constraint at $\boldsymbol{x}$. In other words, locally (for small h) the constraints are almost linear. Rather than derive the results, we just summarize the conclusions.

Let the $m \times n$ matrix $\boldsymbol{A}(\boldsymbol{x})$ be defined by

$$a_{ij}(\boldsymbol{x}) = \frac{\partial c_i(\boldsymbol{x})}{\partial \boldsymbol{x}_j}.$$

We add one assumption, the **constraint qualification** that the gradients of the active constraints are linearly independent. The **Lagrangian function** is

$$L(\boldsymbol{x}) = f(\boldsymbol{x}) - \boldsymbol{\lambda}^T\boldsymbol{c}(\boldsymbol{x}),$$

where there is a **Lagrange multiplier** λ_i for each constraint.

If all of our constraints are inequalities and the constraint qualification holds, then the **first-order optimality conditions** for $\boldsymbol{x}$ to be a solution to the inequality-constrained problem are

$$\begin{aligned} \boldsymbol{A}^T(\boldsymbol{x})\boldsymbol{\lambda} &= \boldsymbol{g}(\boldsymbol{x}), \\ \boldsymbol{\lambda} &\geq \boldsymbol{0}, \\ \boldsymbol{c}(\boldsymbol{x}) &\geq \boldsymbol{0}, \\ \boldsymbol{\lambda}^T\boldsymbol{c}(\boldsymbol{x}) &= 0. \end{aligned}$$

Since $\boldsymbol{\lambda}$ and $\boldsymbol{c}(\boldsymbol{x})$ are nonnegative, the last condition is just a fancy way of saying that if the ith inequality constraint is not active, then the ith Lagrange multiplier must be zero. This condition is often called **complementarity**.

If the jth constraint is an equality constraint rather than an inequality, then we remove the nonnegativity constraint on its Lagrange multiplier and let λ_j be positive, negative, or zero for $j = 1, \ldots, m_e$. Other than that, the optimality conditions are unchanged.

CHALLENGE 10.1.
Let our minimization function be

$$f(\boldsymbol{x}) = x_1^2 + 4x_2^2 - x_1x_2 + 5x_1 + 3x_2 + 6.$$

For the following problems, write the optimality conditions for minimizing f subject to the given constraints. Solve each problem graphically and verify that the optimality conditions hold at the solution.

(a) Suppose that there are no constraints.

(b) The constraint is

$$x_1 + x_2 = 2.$$

(c) The constraint is

$$\boldsymbol{x} \geq \boldsymbol{0}.$$

(d) The constraints are

$$x_1^2 + x_2^2 \leq 1,$$
$$\boldsymbol{x} \geq \boldsymbol{0}.$$

A (necessary) **second-order condition for optimality** is that the matrix $\boldsymbol{Z}^T \nabla_{xx} L(x, \lambda) \boldsymbol{Z}$ is positive semidefinite, where ∇_{xx} denotes the matrix of second derivatives of L with respect to the $\boldsymbol{x}$ variables and where the columns of $\boldsymbol{Z}$ form a basis for the null space of the matrix whose rows are the gradients of the active constraints.

CHALLENGE 10.2.
Verify the second-order necessary condition for the problems in Challenge 10.1.

These optimality conditions are the basis for the **interior-point methods** discussed in Section 10.5. Before studying these methods, though, we consider some ways in which we might reduce a constrained problem to an unconstrained one so that we can apply the methods of Chapter 9.

10.2 Solving Problems with Bound Constraints

Most optimization software designed to solve the problems in Chapter 9 can also handle constraints of the form

$$\boldsymbol{\ell} \leq \boldsymbol{x} \leq \boldsymbol{u}.$$

If such software is not available, then variable transformation can be used to eliminate the constraints, and we give two examples of how this is done.

If the only constraints are that the variables be nonnegative, $\boldsymbol{x} \geq \mathbf{0}$, then we can replace the variables $\boldsymbol{x}$ with $x_j = e^{y_j}$ or $x_j = y_j^2$, for example, and minimize over $\boldsymbol{y}$ without constraint.

Similarly, if we have only upper and lower bounds on the variables, $-1 \leq x_j \leq 1$, for example, we might use the transformation $x_j = \cos(y_j)$.

The advantage of these variable transformations is the ability to use our unconstrained minimizers; the disadvantage is the possible introduction of multiple local minimizers $\boldsymbol{y}$ and the more complicated gradient expressions. For example, the transformation $x_j = y_j^2$ introduces 2^n minimizers $\boldsymbol{y}$ for each minimizer $\boldsymbol{x}$ (since $y_j^2 = (-y_j)^2$).

10.3 Solving Problems with Linear Equality Constraints: Feasible Directions

If the only constraints are linear equality constraints $\boldsymbol{Ax} = \boldsymbol{b}$, where $\boldsymbol{A}$ is an $m \times n$ matrix, then the **feasible direction formulation** can be very effective.

Note that usually $m < n$, since if $m = n$ and $\boldsymbol{A}$ has full rank, then there is only one point that satisfies the constraints, so there is nothing to optimize. In practice, though, problems are often presented in a form that contains redundant constraints, and $\boldsymbol{A}$ may fail to have full rank. Because of this it is important to use numerically stable variable transformations, such as those based on the QR decomposition, as discussed at the end of this section.

First we consider a simple example. Suppose our constraint is $x_1 + x_2 = 1$. Therefore $A = \begin{bmatrix} 1, & 1 \end{bmatrix}$ and $b = [1]$. Then all feasible points have the form

$$\boldsymbol{x} = \begin{bmatrix} 0 \\ 1 \end{bmatrix} + \alpha \begin{bmatrix} 1 \\ -1 \end{bmatrix}. \tag{10.1}$$

This formulation works because if $\boldsymbol{x}$ has this form, then

$$\boldsymbol{Ax} = \begin{bmatrix} 1, & 1 \end{bmatrix}\boldsymbol{x} = \begin{bmatrix} 1, & 1 \end{bmatrix}\begin{bmatrix} 0 \\ 1 \end{bmatrix} + \alpha\begin{bmatrix} 1, & 1 \end{bmatrix}\begin{bmatrix} 1 \\ -1 \end{bmatrix} = 1,$$

and all vectors $\boldsymbol{x}$ that satisfy the constraints have the form (10.1).

In general, if our constraints are $\boldsymbol{Ax} = \boldsymbol{b}$, to get feasible directions, we express $\boldsymbol{x}$ as

$$\boldsymbol{x} = \boldsymbol{x}_{good} + \boldsymbol{Zv},$$

where

- $\boldsymbol{x}_{good}$ is a particular solution to the equations $\boldsymbol{Ax} = \boldsymbol{b}$ (any one is fine),
- $\boldsymbol{v}$ is an arbitrary vector of dimension $(n - r) \times 1$, where r is the rank of the matrix $\boldsymbol{A}$,
- the columns of $\boldsymbol{Z}$ form a basis for the null space of $\boldsymbol{A}$, so that for all $\boldsymbol{v}$, $\boldsymbol{Ax} = \boldsymbol{A}(\boldsymbol{x}_{good}) + \boldsymbol{AZv} = \boldsymbol{Ax}_{good} = \boldsymbol{b}$.

(See Section 5.3.3 if these ideas need review.)

The null space of A defines the set of **feasible directions**, the directions in which we can step without violating a constraint. So we have succeeded in reformulating our constrained problem as an unconstrained one with a smaller number of variables:

$$\min_{v} f(x_{good} + Zv).$$

The most reliable way to implement this transformation when the constraints are $Ax = b$ is to use the QR decomposition of $A^T = QR$ to find Z; see Section 5.3. If A is full rank and we have $m < n$ equality constraints, then the last $n - m$ columns of the $n \times n$ matrix Q form an orthonormal basis for the null space. If we partition Q as $[Q_1, Z]$, with m and $n - m$ columns respectively, then a particular solution can be obtained by solving

$$R_1^T y_p = b,$$

where R_1 is the top $n \times n$ block of R, and then setting $x_{good} = Q_1 y_p$.

If A is not full rank, then we use the RR-QR decomposition (Section 5.4), and we also need to check that the constraints are consistent, so that a vector x_{good} that satisfies $Ax_{good} = b$ exists.

10.4 Barrier and Penalty Methods for General Constraints

Suppose our constraint is the inequality $x_1^2 + x_2^2 \leq 1$, so that $c_1(x) = 1 - x_1^2 - x_2^2$. Define the **barrier function** $-\log(1 - x_1^2 - x_2^2)$, and let

$$\begin{aligned} B_\mu(x) &= f(x) - \mu \log c_1(x) \\ &= f(x) - \mu \log(1 - x_1^2 - x_2^2), \end{aligned}$$

where $\mu > 0$ is a given **barrier parameter**. Suppose we minimize $B_\mu(x)$, starting from a point at which $1 - x_1^2 - x_2^2 > 0$. Then we never move to a point where $c_1(x) < 0$, because there is a barrier of infinite function values of B_μ that occur for values x satisfying $c_1(x) = 0$. We also notice, as illustrated in Figure 10.1, that when μ is sufficiently small, the barrier term is nearly zero for feasible points, so by minimizing $B_\mu(x)$ we approximately solve our original problem. In this reformulation of our constrained minimization problem, the barrier function erects a barrier to prevent the iteration from exiting the feasible region.

CHALLENGE 10.3.

Consider the problem

$$\min_{x} 5x_1^4 + x_1 x_2 + 6x_2^2$$

subject to the constraints $x \geq 0$ and $x_1 - 2x_2 = 4$. Formulate this problem as an unconstrained optimization problem using feasible directions to eliminate the equality constraint and a barrier function to eliminate the nonnegativity constraints.

Now define the **penalty function** $\max(0, (x_1^2 + x_2^2 - 1))^2$, and let

$$\begin{aligned} P_\mu(\boldsymbol{x}) &= f(\boldsymbol{x}) + \frac{1}{\mu} \max(0, -c_1(\boldsymbol{x}))^2 \\ &= f(\boldsymbol{x}) + \frac{1}{\mu} \max(0, (x_1^2 + x_2^2 - 1))^2 . \end{aligned}$$

If we minimize $P_\mu(\boldsymbol{x})$ for a sufficiently small value of the **penalty parameter** μ, then again, as illustrated in Figure 10.1, we have approximately solved our original problem. This formulation uses a penalty function to impose a penalty in the objective function when the constraint is violated.

In general, we can formulate a barrier method for a problem with m inequality constraints as minimizing

$$B_\mu(\boldsymbol{x}) = f(\boldsymbol{x}) - \mu \sum_{j=1}^{m} \log c_j(\boldsymbol{x}),$$

and a penalty method as minimizing

$$P_\mu(\boldsymbol{x}) = f(\boldsymbol{x}) + \frac{1}{\mu} \sum_{j=1}^{m} \max(0, -c_j(\boldsymbol{x}))^2.$$

Then we solve the minimization problem using our favorite unconstrained minimization method.

These examples illustrate the use of barrier and penalty functions. Functions other than the log and the square functions can be used, but these are the most common choices.

Minimizing B_μ or P_μ can be a difficult problem, because if μ is small and if $\boldsymbol{x}$ is near the boundary, then the gradient of the function is quite steep. We expect the solution to our original problem to occur on the boundary, so we cannot avoid this situation. Therefore, often we solve a sequence of barrier or penalty problems with a decreasing sequence of μ values, since larger values of μ give easier minimization problems with more gradual gradients. We decrease μ gradually, using the solution for one value as a good starting point for the next. This is an important computational strategy: we replace one hard problem (constrained minimization) by a sequence of easier (unconstrained) problems, each of which gives an approximate solution to the next. We see this again in Chapter 24 in **continuation methods** for solving systems of nonlinear equations.

Penalty methods are quite convenient since, unlike barrier methods, we do not need to start the iteration at a feasible point, and they can also be used for equality constraints. The disadvantage is that they cannot be used if the functions f or $\boldsymbol{c}$ are undefined outside the feasible region. An advantage of barrier methods is that even if they are stopped early, they yield a feasible point.

Penalty and barrier methods can be useful for simple optimization problems, but the methods discussed in the next section, based on barrier functions, are generally more reliable, since they give a prescription for choosing the sequence of barrier parameters in order to give fast convergence.

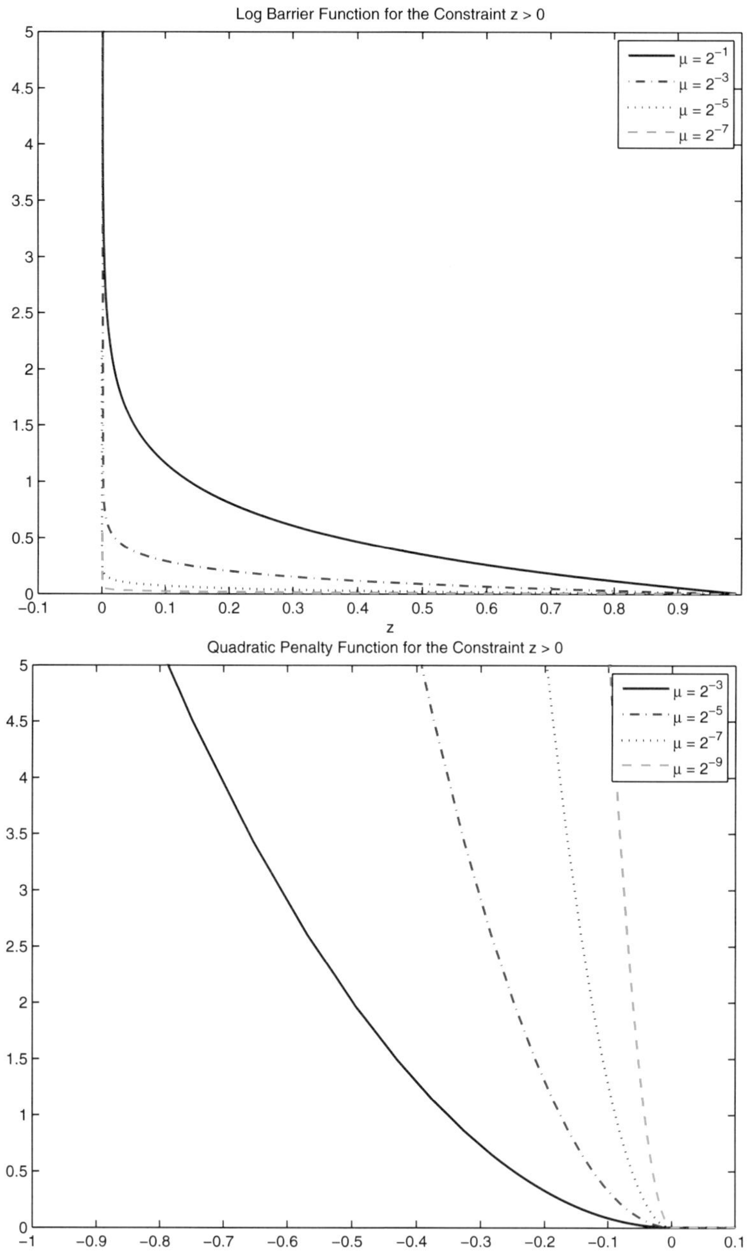

Figure 10.1. *Barrier and penalty functions for the constraint $z > 0$. As $\mu \to 0$, the log barrier (top) makes almost no change to function values of f at feasible points. The quadratic penalty (bottom) makes no change to f at feasible points, and as $\mu \to 0$ it is increasingly desirable to be feasible.*

POINTER 10.1. Linear Programming Problems.

If $f(x)$ is a linear function and if the constraints are linear, then we have a **linear programming problem**. There are two popular types of algorithm:

- **Simplex method** for linear programming. This was the most popular algorithm until the 1990s.
- **Interior-point methods**. These are generally faster on large problems and remain an active area of research.

Both of these methods are implemented in MATLAB's `linprog`.

10.5 Interior-Point Methods

Interior-point methods use log barrier functions in a systematic way in order to solve optimization problems. There are several variants, but we focus here, as an example, on **conic convex optimization** problems of the form

$$\inf_{\boldsymbol{x}} \langle \boldsymbol{c}, \boldsymbol{x} \rangle$$

subject to the constraints

$$\boldsymbol{A}\boldsymbol{x} = \boldsymbol{b}$$

and

$$\boldsymbol{x} \in K,$$

where $\langle \boldsymbol{c}, \boldsymbol{x} \rangle$ denotes inner product between $\boldsymbol{c}$ and $\boldsymbol{x}$ and $K \subset \mathcal{R}^n$ is a closed convex cone, meaning that

- **closed**: the limit point of any sequence of points in K also lies in K,
- **convex**: if $\boldsymbol{x}$ and $\boldsymbol{y}$ are two points in K, then the line segment connecting them also lies in K,
- **cone**: if $\boldsymbol{x} \in K$, then so is $\alpha\boldsymbol{x}$ for all nonnegative scalars α.

Many important classes of optimization problems can be expressed in the form of a conic convex optimization problem. Here are three examples:

- **Linear programming**: In this case we choose $K = \{\boldsymbol{x} : \boldsymbol{x} \geq \boldsymbol{0}\}$ to be the positive orthant, the set of vectors with nonnegative entries. The inner product is defined by $\langle \boldsymbol{c}, \boldsymbol{x} \rangle = \boldsymbol{c}^T \boldsymbol{x}$.
- **Second-order cone programming**: $K = \{\boldsymbol{x} : x_n \geq \|\boldsymbol{x}_{1:n-1}\|_2\}$ is the set of vectors whose last component is greater than or equal to the norm of the remaining components. The inner product is defined by $\langle \boldsymbol{c}, \boldsymbol{x} \rangle = \boldsymbol{c}^T \boldsymbol{x}$.
- **Semidefinite programming**: In this case K is the set of symmetric positive semidefinite matrices of dimension n, and the inner product $\langle \boldsymbol{c}, \boldsymbol{x} \rangle$ equals trace($\boldsymbol{c}\boldsymbol{x}$), where $\boldsymbol{c}$ and $\boldsymbol{x}$ are in K. The constraints become trace($\boldsymbol{a}_i \boldsymbol{x}$) $= b_i$, $i = 1, \ldots, m$, where $\boldsymbol{a}_i$ is a matrix.

POINTER 10.2. Min, Max, Inf, and Sup.

The mathematical notation inf (infinum) is used instead of min in conic optimization problems since it is possible that the function can be made arbitrarily small on the feasible set and thus no finite minimum exists. Similarly, sup (supremum) is used instead of max. This also covers the case when the cone is not closed and the optimal value is not achieved.

We associate with our conic optimization problem a **dual problem**

$$\sup_{\boldsymbol{w},\boldsymbol{s}} \langle \boldsymbol{w}, \boldsymbol{b} \rangle$$

subject to

$$\boldsymbol{A}^* \boldsymbol{w} + \boldsymbol{s} = \boldsymbol{c}$$

and

$$\boldsymbol{s} \in K^*,$$

where K^* denotes the set of points $\boldsymbol{y}$ for which $\langle \boldsymbol{x}, \boldsymbol{y} \rangle \geq 0$ for all $\boldsymbol{x} \in K$. The variables $\boldsymbol{w}$ are Lagrange multipliers for the constraints $\boldsymbol{A}\boldsymbol{x} = \boldsymbol{b}$, and the variables $\boldsymbol{s}$ are Lagrange multipliers for the cone constraint $\boldsymbol{x} \in K$. We assume the following:

- K has a nonempty interior (meaning that there is a point in K for which a small ball centered at the point is also in K).

- K contains no lines (meaning that there is no point in K at which we can walk infinitely far in one direction and also in the opposite direction and remain in K).

We associate with our two problems the following nonlinear system of equations defining the **primal-dual central path**:

$$\begin{aligned} \boldsymbol{A}\boldsymbol{x} &= \boldsymbol{b}, \\ \boldsymbol{A}^* \boldsymbol{w} + \boldsymbol{s} &= \boldsymbol{c}, \\ \mu F'(\boldsymbol{x}) + \boldsymbol{s} &= \boldsymbol{0}, \end{aligned}$$

where

- $F(\boldsymbol{x})$ is a barrier function for the cone K,

- $\boldsymbol{x} \in K$ and $\boldsymbol{s} \in K^*$,

- The barrier parameter μ is defined to be $\mu = \langle \boldsymbol{x}, \boldsymbol{s} \rangle / n$.

Some examples of barrier functions are given in Table 10.1, and we get some practice with them in the next two challenges.

CHALLENGE 10.4.

Use the barrier functions from Table 10.1 to write the system of equations defining the primal-dual central path for

(a) linear programming,

(b) semidefinite programming.

CHALLENGE 10.5.

(a) If $K = \{\boldsymbol{x} : \boldsymbol{x} \geq \boldsymbol{0}\}$, what is K^*?

(b) Consider the linear programming problem

$$\min_{\boldsymbol{x}} \boldsymbol{c}^T \boldsymbol{x}$$

subject to

$$\begin{aligned} \boldsymbol{A}\boldsymbol{x} &= \boldsymbol{b}, \\ \boldsymbol{x} &\geq \boldsymbol{0}. \end{aligned}$$

What is the dual problem?

(c) Show that for linear programming, the dual constraint $\boldsymbol{A}^*\boldsymbol{w} + \boldsymbol{s} = \boldsymbol{c}$ is equivalent to $\boldsymbol{A}^T\boldsymbol{w} \leq \boldsymbol{c}$.

(d) Assume that the $m \times n$ matrix $\boldsymbol{A}$ has rank m, so that the constraint qualification is satisfied. Write the first-order optimality conditions for the linear programming problem and the equations for the primal-dual central path. Try to convince yourself that as $\mu \to 0$, the solution to the primal-dual central path equations satisfies the first-order optimality conditions.

From the preceding challenge we conclude that for linear programming, if we solve the nonlinear system of equations corresponding to the primal-dual central path, for a sequence of values of μ decreasing to zero, the limit point of the solution sequence is a solution to our optimization problem. This is true in general for convex optimization problems and is the basis for one class of interior-point methods.

So our problem is reduced to solving the system of nonlinear equations. The reason that we use a sequence of μ values is related to the idea behind **continuation methods**: we want the solution when $\mu = 0$, but this is a difficult problem with steep gradients, and fast methods such as Newton's method may fail to converge unless we start very close to the optimal solution. Since the solution for one μ value is usually close to that for the next one, a method like Newton's method can be used quite effectively if μ is gradually decreased.

Interior-point methods have become the methods of choice for solving optimization problems. We discussed them only for conic convex optimization problems, but they have wide applicability. They provide a uniform framework, treating linear and nonlinear problems in a similar way. For some algorithms, it can be shown that the amount of work to

Table 10.1. *Barrier functions that can be used for various optimization problems in defining interior-point methods.*

Problem	Constraint	Barrier function
Linear programming	$\boldsymbol{x} \geq \boldsymbol{0}$	$F(\boldsymbol{x}) = -\sum_{i=1}^{n} \log(x_i)$
Second-order cone programming	$x_n \geq \|\boldsymbol{x}_{1:n-1}\|$	$F(\boldsymbol{x}) = -\log(x_n^2 - \|\boldsymbol{x}_{1:n-1}\|_2^2)$
Semidefinite programming	$\boldsymbol{X}$ pos. semi-def.	$F(\boldsymbol{X}) = -\log(\det(\boldsymbol{X}))$

find an approximate solution is $O(n^3)$ (or better). Algorithms like the traditional simplex method for linear programming have much worse bounds: in the worst case, they can visit every vertex of the feasible set, and the number of vertices can be an exponential function of n.

10.6 Summary

We have a variety of methods for constrained optimization problems:

- feasible direction methods for linear equality constraints.
- penalty and barrier methods if no better software is available.
- interior-point methods for convex optimization.

If the problem is not convex, then these methods can still be tried, but they may fail to converge or may converge to a point that is just a local minimizer rather than a global one.

CHALLENGE 10.6.

Consider the problem of minimizing the function

$$f(\boldsymbol{x}) = -(x_1 - 1/2)(x_2 - 1/2)$$

over the unit square $[0,1] \times [0,1]$. Convince yourself that there are two minimizers, at opposite corners of the square. Try to solve this problem using one of our algorithms. Find starting points so that both minimizers can be found. Then modify the function f so that one corner is only a local minimizer and again find starting points so that both local minimizers can be found. In nonconvex problems like this one, it is usually impossible to tell whether a global minimizer has been found.

In the next few chapters we try our algorithms on a variety of constrained and unconstrained optimization problems, emphasizing the fact that additional information about the problem can lead to a more efficient and reliable algorithm.

POINTER 10.3. Further Reading.

We have just touched the surface of the subject of constrained optimization. More detail can be found in the textbooks by Nash and Sofer [111] and Nocedal and Wright [115] and the notes of Nemirovski [114].

Chapter 11 / Case Study

Classified Information: The Data Clustering Problem

(coauthored by Nargess Memarsadeghi)

Many projects in engineering and science require the classification of data based on different criteria. For example,

- Designers classify automobile engine performance as acceptable or unacceptable based on a combination of efficiency, emissions, noise levels, and other criteria.
- Researchers routinely classify documents as "relevant to the current project" or "irrelevant."
- Genomic decoding divides RNA molecules into (coding) ones which translate into a protein and (noncoding) ones that do not.
- Pathologists identify cells as cancerous or benign.

We can classify data into different groups by **clustering** data that are close with respect to some distance measure. In this project, we investigate the design, use, and pitfalls of a popular clustering algorithm, the k-means algorithm, which solves an unconstrained optimization problem.

The Problem

For concreteness, we cluster the pixels in the image shown in Figure 11.1. Suppose our original image is of size $m \times p$, with the color for each of the mp pixels recorded by b bits. Then the total storage requirement is mpb bits. We choose k pixel values (colors) as **cluster centers** and map each pixel to one of these. This forms k clusters of pixels. This

Figure 11.1. *Use clustering algorithms to group the pixels of this image of Charlie (photographed by Timothy O'Leary).*

saves space (since we can store the cluster index for each pixel instead of the pixel value) and, in addition, filters out noise in the image.

The data for our sample problem is a 500×500 pixel image in jpeg format. For jpeg, the b bits for a pixel store $q = 3$ values (red, green, blue), each ranging between 0 and 255. We begin our investigation in Challenge 11.1 by seeing how clustering reduces data storage.

CHALLENGE 11.1.

Compare the number of bits required to store the original image and the image formed after clustering to k colors.

Note that it takes $\log_2 256 = 8$ bits to store a number that ranges between 0 and 255.

We can state our clustering problem this way. Given n data points $\boldsymbol{x}_i \in \mathcal{R}^q$, $i = 1,\ldots,n$, and given a value of k, we want to find k cluster centers $\boldsymbol{c}_j \in \mathcal{R}^q$, $j = 1,\ldots,k$, that are in some sense optimal and then assign each data point to a cluster. We assign $\boldsymbol{x}_i$ to cluster $\mathcal{C}_j$ if it is closer to that cluster's center than it is to any other center. (Break ties in an arbitrary way.) The distance from data point i to its cluster's center is thus

$$d_i = \min_{j=1,\ldots,k} \|\boldsymbol{x}_i - \boldsymbol{c}_j\|,$$

and we define the radius of cluster $\mathcal{C}_j$ as

$$r_j = \max_{i:\boldsymbol{x}_i \in \mathcal{C}_j} d_i .$$

For good clustering, we want each point to be close to one cluster's center. Therefore, we might want to minimize either

$$R = \sum_{j=1}^{k} r_j^{\ell}$$

or

$$D = \sum_{i=1}^{n} d_i^{\ell},$$

where $\ell = 1$ or 2. The variables in the minimization problem are the cluster centers.

Why the Problem Is Hard

In the next challenge, we consider some properties of the functions R and D.

CHALLENGE 11.2.

For this challenge, use the Euclidean norm with $q = 1$ and $\ell = 2$.

(a) If a function is **convex** and bounded below, then any local minimizer is a global minimizer. If not, then an algorithm for minimization might report a local minimizer rather than a global one. Consider the problem with $n = 2$ points and $k = 2$ clusters. Are D and R convex functions?

(b) Are D and R differentiable functions when $n = 2$ and $k = 2$?

(c) Derive a formula for the minimizer of D when $k = 1$ and n is arbitrary.

(d) Suppose we move one of our data points $\boldsymbol{x}_i$ very far away from the other points, making it an **outlier**. As that point moves further away from the others, what happens to the cluster centers determined by minimizing D or R?

From Challenge 11.2, we know that the minimization problem has some difficult properties, but let's try to compute the cluster centers using a standard optimization algorithm. To get the solution process started, we provide an array of k approximate centers. A common method for finding initial centers is to select k distinct points randomly among the data values, or perhaps to use k extreme values. When comparing algorithms, each should use the same initial centers. Challenge 11.3 investigates our clustering criteria and the behavior of optimization algorithms.

CHALLENGE 11.3.

Use your favorite optimization function (e.g., MATLAB's `fminunc`) to minimize R with $\ell = 2$ and the Euclidean norm. Use the data of Figure 11.1, and provide a function to evaluate R. Try $k = 3, 4, 5$. Also minimize D with the same parameters.

Write a function `map_to_cluster` that takes the data values and cluster centers as input and returns the cluster number for each data value and the counts of the number of data points assigned to each cluster. Use this function to generate the clustered image.

Algorithm 11.1 The k-Means Algorithm

Choose initial centers $\boldsymbol{c}_1, \ldots, \boldsymbol{c}_k$.
while the clusters are changing,
  for $i = 1, \ldots, n$,
    Assign each data point $\boldsymbol{x}_i$ to the cluster $\mathcal{C}_j$ whose center $\boldsymbol{c}_j$ is closest to it, breaking ties in an arbitrary way.
  end
  for $j = 1, \ldots, k$,
    Recompute the center $\boldsymbol{c}_j$ to be the mean (centroid) of the points in the cluster:

$$\boldsymbol{c}_j = \frac{1}{n_j} \sum_{i:\boldsymbol{x}_i \in \mathcal{C}_j} \boldsymbol{x}_i ,$$

    where n_j is the number of data points in $\mathcal{C}_j$.
  end
end

To keep the computation time reasonable, determine the cluster centers based on a sample of points in the image rather than using all 250,000 pixels. Choose the 1000 points in columns 210 and 211 in the sample image. Then experiment with other choices of points to study the algorithm's sensitivity to this choice.

(a) How does the number of variables increase with k?

(b) How does the running time increase with k?

(c) Evaluate the results of the various clusterings and justify the criteria that you choose to use in your evaluation. As one criterion, discuss how the clustered images look in comparison to the original image.

(d) How might a good value of k be determined experimentally?

The k-Means Algorithm

A general purpose minimization routine is a good tool to have, because it is useful for a wide variety of problems. But sometimes we can develop a better algorithm by taking advantage of special structure in the problem. Consider Algorithm 11.1, the k-**Means Algorithm**. It minimizes neither D nor R, but it iterates by clustering based on the current centers and then moving each center to the centroid of the points in the cluster.

In Challenge 11.4, we implement this algorithm.

CHALLENGE 11.4.

Implement the k-means algorithm and run it with the same data and values of k as Challenge 11.3. Compare its performance to that of the algorithm in Challenge 11.3.

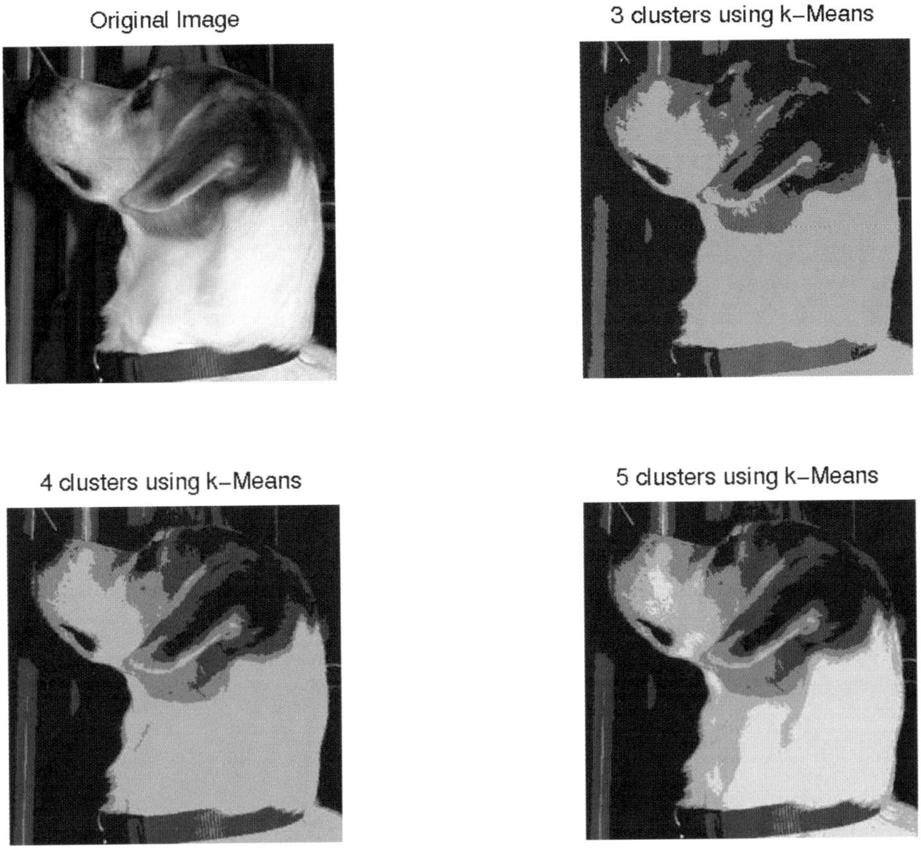

Figure 11.2. *The images resulting from k-means.*

Your implementation for this challenge should be rather general: write a function `mycluster` that takes as input the n data values, an initial guess for the k cluster centers, a convergence tolerance, and a maximum number of iterations. The output is assignments of each data point to a cluster, the set of k cluster centers, the number of data values in each cluster, and the radius of each cluster. Use another function to evaluate R or D, given the k final cluster centers. Compare the resulting clusters to those in Challenge 11.3.

The results of this challenge are shown in Figure 11.2.

Pitfalls in Data Clustering

This form of data clustering is quite useful, and the k-means algorithm is very successful in practice. Nevertheless, there are many pitfalls associated with its use. We investigate two

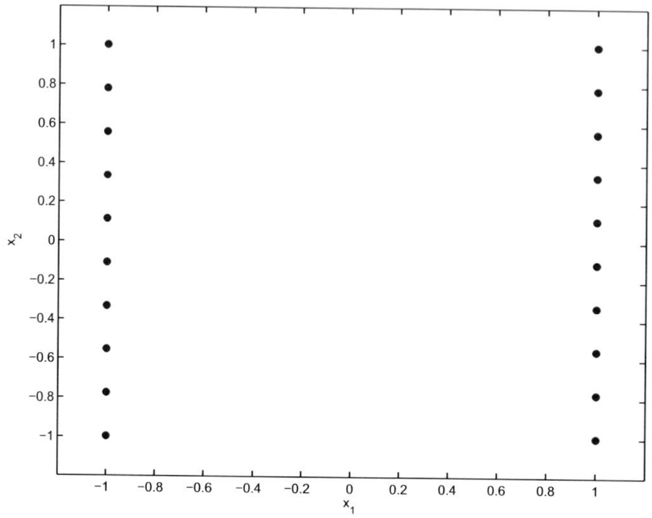

Figure 11.3. *This data set illustrates some of the pitfalls of clustering.*

of these, dependence of the answer on the starting data and on the number of clusters, in Challenge 11.5.

CHALLENGE 11.5.

Consider the data set of $n = 20$ data points with $q = 2$, shown in Figure 11.3:

$$(1, -1 + 2j/9),\ (-1, -1 + 2j/9),$$

for $j = 0, \ldots, 9$. Run the k-means algorithm with $k = 2, 3, 4$. Initialize the centers to the first k points in the list

$$(-1,-1), (1,1), (-1,1), (1,-1).$$

Display the clustered data. Discuss the effects of choosing the "wrong" value for k.

Then repeat the experiment, initializing the centers to

$$(0, -1 + 2j/(k-1)),$$

$j = 0, \ldots, k-1$. Note that although the answer is different, it is also a local minimizer of the (nonconvex) function R. Compare with the first set of answers and discuss the difficulty it illustrates with this kind of clustering.

Sensitivities of the clustering to the initial choice of centers and the number of clusters are serious pitfalls. As we see in Challenge 11.6, another serious pitfall arises from the sensitivity of the clustering to variable transformations.

POINTER 11.1. Further Reading.

The k-means algorithm used in Challenge 11.4 as well as other approaches to clustering are discussed, for example, in [52, 85, 84, 145].

Davidson [34] gives a nice discussion of the troubles we illustrate in Challenges 11.5 and 11.6, as well as many other pitfalls.

There are many implementations of variants of the k-means algorithm; see, for example, the software of Guan [44, 65].

CHALLENGE 11.6.

Consider the data set from Challenge 11.5, but multiply the second component of each data point by 100. Repeat the clustering experiments, applying the same transformation to the initial centers. Discuss why coordinate scaling is important in clustering algorithms.

Through our investigations, we see that despite its pitfalls, clustering is an important tool for data classification, noise reduction, and storage savings. Because of the nonconvexity of this problem, special purpose optimization algorithms, such as the ones discussed in this chapter, are preferred to the general purpose ones of Chapters 9 and 10.

Chapter 12 / Case Study

Achieving a Common Viewpoint: Yaw, Pitch, and Roll

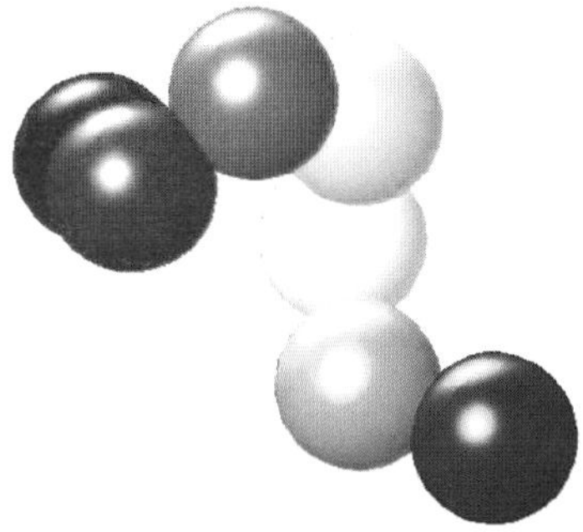

(coauthored by David A. Schug)

Tracking objects, controlling the navigation of a spacecraft, assessing the quality of machined parts, and identifying proteins seem to have little in common, but all of these problems (and many more problems in computer vision and computational geometry) share a core computational task: rotating and translating two objects so that they have a common coordinate system. In this case study, we study this deceptively simple optimization problem and its pitfalls.

"Life is about change; nothing ever stays the same." In particular, objects move, and tracking them is an essential ingredient in applications such as navigation and robot motion. Surprisingly, the same mathematical tools used in tracking are also used in the **absolute orientation problem** of comparing two objects, such as proteins or machine parts, to see if they have the same structure.

Consider molecule $\boldsymbol{A}$ in Figure 12.1, which we specify by the coordinates $\boldsymbol{a}_1,\ldots,\boldsymbol{a}_7$ of the centers of the seven spheres that represent some of its atoms, and the corresponding object $\boldsymbol{B}$, obtained by rotating $\boldsymbol{A}$. There are many ways to define 3D rotations, but in this case study, we specify the **yaw** ϕ, the **pitch** θ, and the **roll** ψ, as is common in flight control. In this coordinate system, the angles ϕ, θ, and ψ are called the **Euler angles**, and a rotation $\boldsymbol{Q}$ is defined by the product of three matrices

$$\boldsymbol{Q}(\phi,\theta,\psi) = \boldsymbol{Q}_{roll}\boldsymbol{Q}_{pitch}\boldsymbol{Q}_{yaw},$$

where

$$\boldsymbol{Q}_{roll} = \begin{bmatrix} 1 & 0 & 0 \\ 0 & \cos\psi & \sin\psi \\ 0 & -\sin\psi & \cos\psi \end{bmatrix},$$

$$\boldsymbol{Q}_{pitch} = \begin{bmatrix} \cos\theta & 0 & -\sin\theta \\ 0 & 1 & 0 \\ \sin\theta & 0 & \cos\theta \end{bmatrix},$$

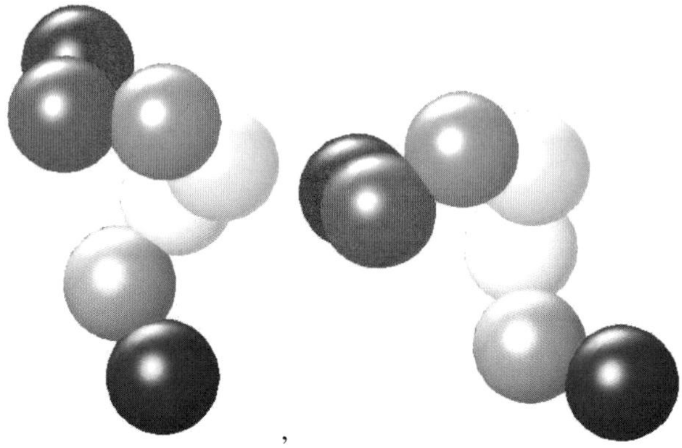

Figure 12.1. *How can we tell that molecule* **A** *(left) and molecule* **B** *(right) are the same?*

$$Q_{yaw} = \begin{bmatrix} \cos\phi & \sin\phi & 0 \\ -\sin\phi & \cos\phi & 0 \\ 0 & 0 & 1 \end{bmatrix}.$$

We impose the restrictions $-\pi < \phi < \pi$, $-\pi/2 < \theta < \pi/2$, and $-\pi < \psi < \pi$. Our first task is to develop some familiarity with this representation for rotation matrices.

CHALLENGE 12.1.

(a) Explain geometrically the effect of applying a rotation $\boldsymbol{Q}$ to a vector $[x,y,z]^T$ to create the vector $\boldsymbol{Q}[x,y,z]^T$.

(b) Show that if $\boldsymbol{Q}$ is any 3×3 orthogonal matrix (i.e., $\boldsymbol{Q}^T\boldsymbol{Q} = \boldsymbol{I}$), then $\boldsymbol{Q}$ can be expressed as $\boldsymbol{Q}_{roll}\boldsymbol{Q}_{pitch}\boldsymbol{Q}_{yaw}$ for some choice of angles ψ, θ, and ϕ.

Next, we need to determine the Euler angles (ϕ,θ,ψ) so that $\boldsymbol{Q}(\phi,\theta,\psi)$ rotates object $\boldsymbol{A}$ in Figure 12.1 to object $\boldsymbol{B}$. Let $\boldsymbol{A}$ be the $3\times n$ matrix ($n = 7$) whose columns are the coordinates of the first set of points: $\boldsymbol{A} = [\boldsymbol{a}_1,\ldots,\boldsymbol{a}_7]$. Define $\boldsymbol{B}$ similarly from the second set of points. Then we want to determine the three Euler angles so that

$$\boldsymbol{B} = \boldsymbol{Q}(\phi,\theta,\psi)\boldsymbol{A}.$$

Since life is about change and imperfection, we don't expect to get an exact equality, but we want to make the difference between $\boldsymbol{B}$ and $\boldsymbol{Q}(\phi,\theta,\psi)\boldsymbol{A}$ as small as possible. One reasonable way to measure this is by taking the sum of the squares of the differences in each component; then our task is to minimize

$$f(\phi,\theta,\psi) = \|\boldsymbol{B} - \boldsymbol{Q}(\phi,\theta,\psi)\boldsymbol{A}\|_F^2 \equiv \sum_{i=1}^{n} \|\boldsymbol{b}_i - \boldsymbol{Q}(\phi,\theta,\psi)\boldsymbol{a}_i\|_2^2.$$

Minimizing f is a **nonlinear least squares problem** with three variables, so let's experiment with solving the problem for various data sets. The square root of f/n is the **root mean squared distance** (RMSD) between the two objects. The factor $1/n$ applied to f forms the average (mean) of the squared distances between the corresponding points. RMSD provides us in Challenge 12.2 with a measure of how well our objects match.

CHALLENGE 12.2.

Use a nonlinear least squares solver (e.g., MATLAB's `lsqnonlin`) to find the Euler angles for the data sets $(\boldsymbol{A}, \boldsymbol{Q}(\phi,\theta,\psi)\boldsymbol{A})$ generated by taking the yaw $\phi = \pi/4$, roll $\psi = \pi/9$, and

$$\boldsymbol{A} = \begin{bmatrix} 0 & 0 & 1 & 1 & 0 & -1 & 0 \\ 0 & 1 & 1 & 0 & 0 & 1 & 2 \\ 0 & 1 & 2 & 3 & 4 & 4 & 4 \end{bmatrix}.$$

Let the pitch θ vary between $-\pi/2$ and $\pi/2$ in steps of $\pi/120$. Plot the computed Euler angles, and, in a separate plot, the Frobenius norm of the error in the computed $\boldsymbol{Q}$ and the RMSD in the computed positions. Discuss the time needed for solution and the accuracy obtained.

The problem we are considering is an old one, sometimes called the **orthogonal Procrustes problem**. In the next challenge we derive a better way to solve it.

CHALLENGE 12.3.

(a) Recall that the trace of a square matrix is the sum of its main diagonal entries, and that the trace of the sum of two matrices is the sum of the two traces. We need two additional facts about traces in order to derive our algorithm. Prove the following:

1. For any matrix $\boldsymbol{C}$, $\text{trace}(\boldsymbol{C}^T\boldsymbol{C}) = \|\boldsymbol{C}\|_F^2$.
2. For any matrix $\boldsymbol{D}$ for which the product $\boldsymbol{CD}$ is square, $\text{trace}(\boldsymbol{CD}) = \text{trace}(\boldsymbol{DC})$.

(b) Use the first fact to show that the $\boldsymbol{Q}$ that minimizes $\|\boldsymbol{B} - \boldsymbol{QA}\|_F^2$ over all choices of orthogonal $\boldsymbol{Q}$ also maximizes $\text{trace}(\boldsymbol{A}^T\boldsymbol{Q}^T\boldsymbol{B})$.

(c) Suppose that the singular value decomposition (SVD) of the $m \times m$ matrix $\boldsymbol{BA}^T$ is $\boldsymbol{U\Sigma V}^T$, where $\boldsymbol{U}$ and $\boldsymbol{V}$ are $m \times m$ and orthogonal, and $\boldsymbol{\Sigma}$ is diagonal with diagonal entries $\sigma_1 \geq \cdots \geq \sigma_m \geq 0$. Define $\boldsymbol{Z} = \boldsymbol{V}^T\boldsymbol{Q}^T\boldsymbol{U}$. Use these definitions and the second fact to show that

$$\text{trace}(\boldsymbol{A}^T\boldsymbol{Q}^T\boldsymbol{B}) = \text{trace}(\boldsymbol{Q}^T\boldsymbol{BA}^T) = \text{trace}(\boldsymbol{Z\Sigma}) \leq \sum_{i=1}^{m} \sigma_i.$$

(d) If $\boldsymbol{Z} = \boldsymbol{I}$, then

$$\text{trace}(\boldsymbol{Q}^T\boldsymbol{BA}^T) = \sum_{i=1}^{m} \sigma_i.$$

What choice of $\boldsymbol{Q}$ ensures that $\boldsymbol{Z} = \boldsymbol{I}$ and therefore ensures that the trace is maximized over all choices of $\boldsymbol{Q}$?

Challenge 12.3 shows that the optimal $\boldsymbol{Q}$ is $\boldsymbol{U}\boldsymbol{V}^T$, determined by computing an SVD of $\boldsymbol{B}\boldsymbol{A}^T$. This is much more efficient than solving the nonlinear least squares problem as we did in Challenge 12.2. Let's redo the computations.

CHALLENGE 12.4.

Use the SVD to find the Euler angles for the data in Challenge 12.2. Compare with your previous results.

So far we have assumed that the object has rotated with respect to the origin but has not translated. Now we consider a more general problem:

$$\boldsymbol{B} = \boldsymbol{Q}(\phi, \theta, \psi)\boldsymbol{A} + \boldsymbol{t}\boldsymbol{e}^T,$$

where the 3×1 vector $\boldsymbol{t}$ defines the translation and $\boldsymbol{e}$ is a column vector with n ones. How might we solve this problem?

One way is to solve a nonlinear least squares problem for $\boldsymbol{t}$ and the Euler angles. Here (as in the case study of Chapter 13), we could take advantage of the fact that given $\boldsymbol{t}$, it is easy to compute the optimal $\boldsymbol{Q}$, so we can express the problem as a function of just three variables: t_1, t_2, and t_3. It is interesting to implement this algorithm, but we'll just focus on a much more efficient approach.

The "easy" way arises from observing that the translation can be defined by the movement of the centroid of the points:

$$\boldsymbol{c}_A = \frac{1}{n}\sum_{j=1}^{n} \boldsymbol{a}_j,$$

$$\boldsymbol{c}_B = \frac{1}{n}\sum_{j=1}^{n} \boldsymbol{b}_j.$$

Luckily, the averaging in the centroid computations tends to reduce the effects of random errors, and Challenge 12.5 shows how $\boldsymbol{t}$ can be defined in terms of the centroids.

CHALLENGE 12.5.

Given a fixed rotation matrix $\boldsymbol{Q}$, show that the minimizer $\boldsymbol{t}_{opt}$ of $\|\boldsymbol{B} - \boldsymbol{Q}\boldsymbol{A} - \boldsymbol{t}\boldsymbol{e}^T\|_F$ satisfies

$$\boldsymbol{t}_{opt} = \boldsymbol{c}_B - \boldsymbol{Q}\boldsymbol{c}_A.$$

So we have an algorithm for solving our problem: we move both objects so that their centroids are at zero and then compute the resulting rotation $\boldsymbol{Q}$ using the SVD. Finally, we reconstruct the translation using the formula in Challenge 12.5. Let's see how this algorithm behaves.

CHALLENGE 12.6.

Implement this algorithm and try it on the data from Challenge 12.2 using $\theta = \pi/4$ and 20 randomly generated translations $\boldsymbol{t}$. Then repeat the experiment with 20 more translations, adding to each element of $\boldsymbol{A}$ a perturbation that is uniformly distributed between -10^{-3} and 10^{-3}, to see how sensitive the computation is to uncertainty in the measurements.

Through these computations (and further experimentation, if desired), you can see that the rotation matrix $\boldsymbol{Q}$ can almost always be computed quite accurately by the SVD algorithm; unfortunately, the Euler angles are not as well determined. In the next challenge, we study some degenerate cases.

CHALLENGE 12.7.

(a) Suppose that all of our points in $\boldsymbol{A}$ lie on a line. Is there more than one choice of $\boldsymbol{Q}$ that minimizes $\|\boldsymbol{B} - \boldsymbol{QA}\|$? Illustrate this with a numerical example.

(b) Use this insight to characterize the degenerate cases for which $\boldsymbol{Q}$ is not well determined.

(c) Suppose that our true data produces the angles $(\phi, \theta = \pi/2, \psi)$, but a small perturbation causes a small increase in the angle θ so that it is greater than $\pi/2$. Generate such an example, and observe that the computed angles are quite different. This jump in angle is called **gimbal lock**, a term borrowed from the locking of the mechanism that moves a stabilizing gyroscope in cases when the angle goes out of the range of motion of the device.

We can always choose a set of reference points for the object to make the matrix $\boldsymbol{Q}$ well determined, but unfortunately, this does not guarantee that the Euler angles are well determined.

One way to avoid this artificial ill-conditioning is to replace Euler angles by a better representation of the information in $\boldsymbol{Q}$. **Quaternions** are a common choice, and Pointer 12.1 gives references to more information on this subject.

In this case study we see that a rather complicated optimization problem reduces to a simple linear algebra problem, provided we use an appropriate matrix decomposition. We are not always this lucky, but when we are, we should certainly take advantage of this structure.

POINTER 12.1. Further Reading.

One important problem that we have ignored is that of getting a set of corresponding points from the two objects. This is treated, for example, in [146].

To help with Challenge 12.1, a nice demonstration of the parameters for yaw, pitch, and roll is found at [138]. Other rotation coordinate systems are described, for example, in [154].

The orthogonal Procrustes problem in Challenge 12.3 is considered in [64].

The use of quaternions instead of Euler angles is discussed, for example, in [96].

Chapter 13 / Case Study

Fitting Exponentials: An Interest in Rates

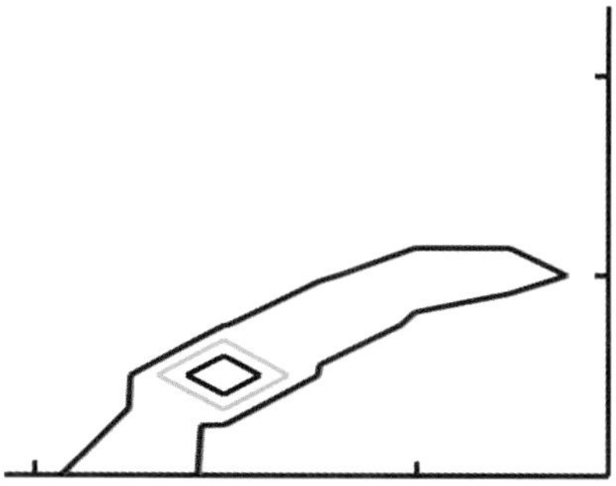

In this case study we investigate the problem of fitting a sum of exponential functions to data. This problem occurs in many real situations, but we will see that getting a good solution requires care.

Suppose we have two chemical reactions occurring simultaneously, with the amount y of a reactant changing due to both processes and behaving as a function of time t as

$$y(t) = x_1 e^{\alpha_1 t} + x_2 e^{\alpha_2 t},$$

where x_1, x_2, α_1, and α_2 are fixed parameters. The negative values α_1 and α_2 are **rate constants**; in time $-1/\alpha_1$, the first exponential term drops to $1/e$ of its value at $t = 0$. Often we can observe $y(t)$ fairly accurately, and we would like to determine the rate constants and the **amplitude constants** x_1 and x_2. This involves fitting the parameters of the sum of exponentials.

In this project we study efficient algorithms for solving this problem, but illustrate that for many data sets, the solution is not well determined.

How Sensitive Are the Amplitude Constants to Errors in the Data?

In this section, we investigate how sensitive the y function is to choices of parameters $\boldsymbol{x}$, assuming that we are given the $\boldsymbol{\alpha}$ parameters exactly.

Typically, we observe the function $y(t)$ for m fixed t values, perhaps $t_i = 0, \Delta t, 2\Delta t, \ldots, t_{final}$. For a given parameter set $\boldsymbol{\alpha}$ and $\boldsymbol{x}$, we can measure the goodness of the fit of model to data by calculating the residual

$$r_i = y(t_i) - y_e(t_i), \qquad i = 1, \ldots, m, \tag{13.1}$$

where $y_e(t) = x_1 e^{\alpha_1 t} + x_2 e^{\alpha_2 t}$ is the model prediction. Ideally, the residual vector $\boldsymbol{r} = \boldsymbol{0}$, but due to noise in the measurements, we never achieve this. Instead, we compute model parameters that make the residual as small as possible, and we often choose to measure size using the 2-norm: $\|\boldsymbol{r}\|^2 = \boldsymbol{r}^T \boldsymbol{r}$.

Note that if the parameters $\boldsymbol{\alpha}$ are given, then we find the $\boldsymbol{x}$ parameters by solving a **linear least squares problem**, since r_i is a linear function of x_1 and x_2. Thus, we minimize the norm of the residual, expressed as

$$r = y - Ax,$$

where $a_{ij} = e^{\alpha_j t_i}$, $j = 1,2$, $i = 1,\ldots,m$, and $y_i = y(t_i)$.

This problem can be easily solved using matrix decompositions such as the QR decomposition of A into the product of an orthogonal matrix times an upper-triangular one (see Section 5.3.3), or the singular value decomposition (SVD) (Section 5.6). We'll focus on the SVD since, although it is somewhat more expensive, it generally is less influenced by rounding error and also easily gives us a bound on the problem's sensitivity to small changes in the data.

The solution to Challenge 5.19 shows that the sensitivity of the parameters x to changes in the observations y depends on the condition number $\kappa(A)$. With these basic formulas in hand, we investigate this sensitivity in Challenge 13.1.

CHALLENGE 13.1.

Generate 100 problems with data $x_{true} = [0.5, 0.5]^T$, $\alpha = -[0.3, 0.4]$, and

$$y_{true}(j) = [e^{\alpha_1 t(j)}, e^{\alpha_2 t(j)}]x_{true},$$

where $t = [0, 0.01, \ldots, 6.00]$. Suppose there are errors in the measurements so that

$$y = y_{true} + \eta z,$$

where $\eta = 10^{-4}$ and the elements of the vector z are independent and uniformly distributed on the interval $[-1, 1]$. In one figure, plot the computed solutions $x^{(i)}$, $i = 1,\ldots,100$ obtained by your SVD algorithm assuming that α is known. In a second figure, plot the components $w^{(i)}$ of the solution in the coordinate system determined by the right singular vectors of A. Interpret these two plots using the results of Pointer 5.5. In particular, the points in the first figure are close to a straight line. What determines the line's direction? What determines the shape and size of the point cluster in the second figure? Verify your answers by repeating the experiment with $\alpha = -[0.3, 0.31]$, and also try varying η to be $\eta = 10^{-2}$ and $\eta = 10^{-6}$.

How Sensitive Is the Model to Changes in the Rate Constants?

Now we need to investigate the sensitivity to the nonlinear parameters α, where $-\alpha_1$ and $-\alpha_2$ are **rate constants**. In Challenge 13.2, we display how fast the function y changes as we vary these parameters, assuming that we compute the optimal x parameters using a linear least squares algorithm.

CHALLENGE 13.2.

Suppose that the chemical reactions are described by

$$y(t) = 0.5e^{-0.3t} + 0.5e^{-0.7t}.$$

Suppose that we observe $y(t)$ for $t \in [0, t_{final}]$, with 100 equally-spaced observations per second.

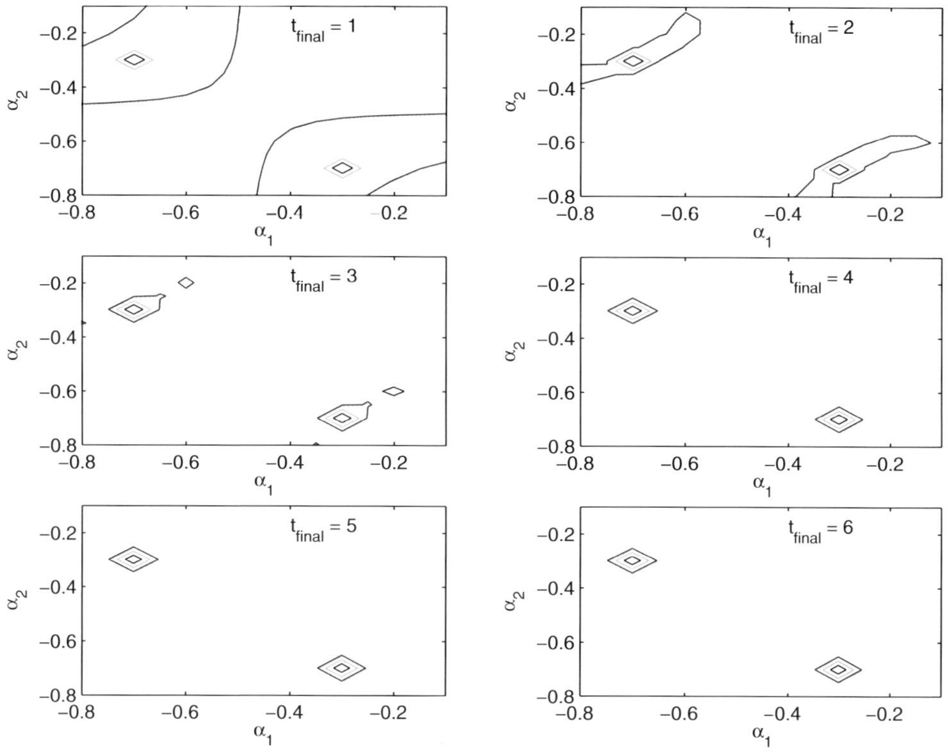

Figure 13.1. *Contour plots of the residual norm of the data fitting as a function of the estimates of $\boldsymbol{\alpha}$ for various values of t_{final}. The contours marked are* 10^{-2}, 10^{-6}, *and* 10^{-10}.

Compute the residual norm for various $\boldsymbol{\alpha}$ estimates, using the optimal values of x_1 and x_2 computed by the algorithm in Challenge 5.19(a) for each choice of $\boldsymbol{\alpha}$. Make 6 contour plots of the log of the residual norm as a function of $\boldsymbol{\alpha}$, letting the observation interval be $t_{final} = 1, 2, \ldots, 6$ seconds. Plot contours of -2, -6, and -10. How helpful is it to gather data for longer time intervals? How well determined are the $\boldsymbol{\alpha}$ parameters?

From the results of Challenge 13.2, shown in Figure 13.1, we learn that the parameters $\boldsymbol{\alpha}$ are not well determined; there is a broad range of $\boldsymbol{\alpha}$ values that lead to small residuals. This is an inherent limitation in the problem and we cannot change it. Nonetheless, we want to develop algorithms to compute approximate values of $\boldsymbol{\alpha}$ and $\boldsymbol{x}$ as efficiently as possible, and we next turn our attention to this computation.

Solving the Nonlinear Problem

If we are not given the parameters $\boldsymbol{\alpha}$, then minimizing the norm of the residual $\boldsymbol{r}$ defined in (13.1) is a **nonlinear least squares problem**. For our model problem, there are four parameters to be determined. We could solve this using standard optimization software, but it is more efficient to take advantage of the problem's least squares structure. In addition, since two parameters occur linearly, it is wise to take advantage of that structure, too. One very good way to do this is to use a **variable projection** algorithm. The reasoning is as follows: our residual vector is a function of all four parameters, but given the two $\boldsymbol{\alpha}$ parameters, the optimal values of the two $\boldsymbol{x}$ parameters are easy to determine by solving the linear least squares problem that we considered in Challenge 5.19. Therefore, we express our problem as a minimization problem with only two variables:

$$\min_{\boldsymbol{\alpha}} \|\boldsymbol{r}\|^2,$$

where the computation of $\boldsymbol{r}(\boldsymbol{\alpha})$ requires us to determine the $\boldsymbol{x}$ parameters by solving a linear least squares problem using, for instance, the SVD.

This is a very neat way to express our minimization problem, but we pay for that convenience when we evaluate the gradient of the function $f(\boldsymbol{\alpha}) = \boldsymbol{r}^T\boldsymbol{r}$. Since the gradient is quite complicated, we can choose either to use special purpose software to evaluate it (See Pointer 13.1) or use a minimizer that computes a difference approximation to it.

CHALLENGE 13.3.

(a) Use a nonlinear least squares algorithm to determine the sum of two exponential functions that approximates the dataset generated with $\boldsymbol{\alpha} = [-0.3, -0.4]$, $\boldsymbol{x} = [0.5, 0.5]^T$, and independent normally distributed error with mean zero and standard deviation $\eta = 10^{-4}$. Provide 601 values of $(t, y(t))$ with $t = 0, 0.01, \ldots, 6.0$. Experiment with the initial guesses

$$\boldsymbol{x}^{(0)} = \begin{bmatrix} 3 \\ 4 \end{bmatrix}, \quad \boldsymbol{\alpha}^{(0)} = \begin{bmatrix} -1, -2 \end{bmatrix}$$

and

$$\boldsymbol{x}^{(0)} = \begin{bmatrix} 3 \\ 4 \end{bmatrix}, \quad \boldsymbol{\alpha}^{(0)} = \begin{bmatrix} -5, -6 \end{bmatrix}.$$

Plot the residuals obtained from each solution. Repeat the experiment with $\boldsymbol{\alpha} = [-0.30, -0.31]$. How sensitive is the solution to the starting guess? (If no standard nonlinear least squares algorithm is available (such as `lsqnonlin` in MATLAB), use a general purpose minimization algorithm.)

(b) Repeat the runs of part (a), but use variable projection to reduce to two parameters, the two components of $\boldsymbol{\alpha}$. Discuss the results.

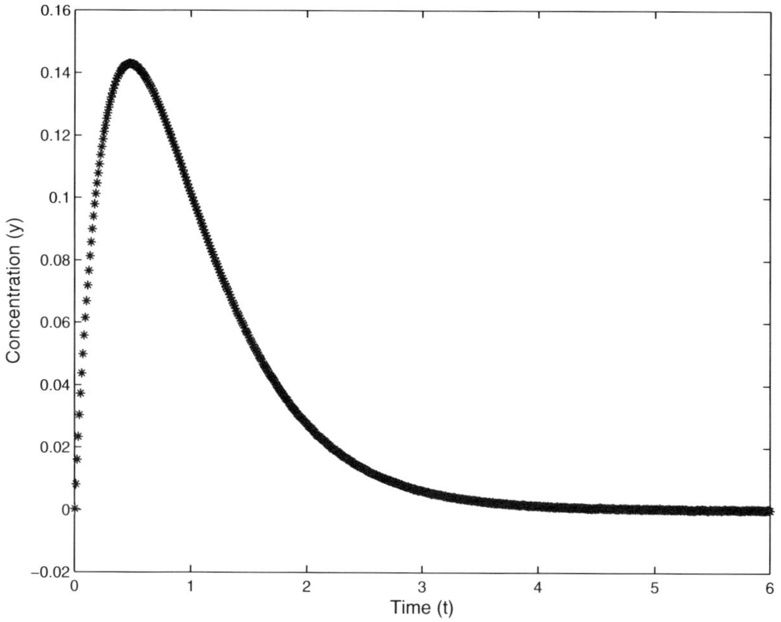

Figure 13.2. *Data for Challenge* 13.4. *Given these measurements of the species concentration (mg/ml) vs. time (sec) or drug concentration (mg/liter) vs. time (hours), find the rate constants.*

To finish our investigation of exponential fitting, let's try dealing with the data of Figure 13.2.

CHALLENGE 13.4.

Suppose that we gather data from a chemical reaction involving two processes; one process produces a species and the other depletes it. We have measured the concentration of the species as a function of time. (If you prefer, suppose that the data arises from measurements of the amount of drug in a patient's bloodstream while the drug is being absorbed from the intestine and excreted by the kidneys.) We display the data in Figure 13.2; it is available on the website. Suppose that your job (or even the patient's health) depends on determining the two rate constants and a measure of uncertainty in your estimates. Find the answer and document your computations and your reasoning.

In this case study we found efficient algorithms for solving data fitting problems that lead to very sensitive optimization problems. Like the image deblurring problem in the case study of Chapter 6, fitting rate constants is an ill-posed problem: small changes in the data can make very large changes in the solution. In the next two case studies we continue our study of such sensitive problems.

POINTER 13.1. Further Reading.

In the case study of Chapter 8, we also studied exponential fitting in order to determine directions of arrival of signals. That problem was somewhat better posed, since the data did not decay.

Fitting a sum of exponentials to data is necessary in many experimental systems including fluorescence of molecules [25], kinetics of voltage formation [73], studies of scintillators using x-ray excitation [42], drug metabolism, and predator-prey models. Often, though, the publication of a set of rate constants elicits a storm of letters to the editor, criticizing the methods used to derive them. It is important to do the fit carefully and to document the methods used.

See Chapter 5 for further information on perturbation theory, SVD, and numerical solution of linear least squares problems.

Looking at the contours of a function is a useful way to understand it. The MATLAB function `contour` is one way to construct such a plot.

The variable projection algorithm `Varpro`, which solves nonlinear least squares problems by eliminating the linear variables, was described by Gene Golub and Victor Pereyra. Linda Kauffman noticed that each iteration would run faster if certain negligible but expensive terms in the derivative computation were omitted. A recent review of the literature on the algorithm and its applications was written by Golub and Pereyra [63].

Although bad computational practices are often used in published papers involving fitting exponentials, the pitfalls are discussed quite lucidly in many sources. See, for example, the work of Shrager and Hendler [137] and the series of tutorials by Rust [129].

Chapter 14 / Case Study

Blind Deconvolution: Errors, Errors Everywhere

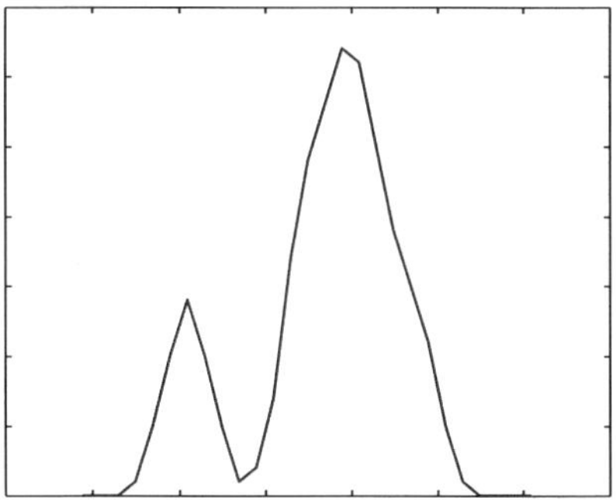

We focus in this case study on a class of methods that accounts for uncertainty in the model as well as the data. Our example concerns spectroscopy, where we try to reconstruct a true spectrum from an observed one. The problem we are considering is sometimes called **blind deconvolution**, since we are trying to unravel not only the spectrum but the function that caused the blurring. These problems also arise in image deblurring.

Consider the data of Figure 14.1, representing counts measured by a spectrometer. Suppose we have particles whose energy ranges from e_{lo} to e_{high}, and define some intermediate energy levels $e_{lo} = e_0 < e_1 < \cdots < e_{n_b-1} < e_{n_b} = e_{high}$. This creates n_b bins, where the count for the jth bin is the number of particles determined to have energies between e_{j-1} and e_j. Our spectrometer records n_b counts, one for each bin, and in the figure we have passed a curve through these counts.

Now ideally, the count in bin j is exactly the number of particles with energies in the range $[e_{j-1}, e_j]$. But some blurring occurs due to the measurement process, and a particle in that energy range might instead be included in the count for a different nearby bin. The probability that a particle with energy e is assigned to bin j is often modeled as a normal distribution with mean $(e_j + e_{j-1})/2$ and variance s_j^2.

We would like to determine the correct counts f_j and the correct blurring given the measured counts g_j, $j = 1, \ldots, n_b$ and estimates of the values s_j.

One model of this process is the matrix equation $(\boldsymbol{K}+\boldsymbol{E})\boldsymbol{f} = \boldsymbol{g}+\boldsymbol{r}$, where $\boldsymbol{E}$ accounts for errors in modeling the spectrometer's blur and $\boldsymbol{r}$ accounts for errors in counts. The matrix entry $k_{j\ell}$ is our estimate of the probability that a particle whose energy is in the interval $[e_{\ell-1}, e_\ell]$ is assigned to bin j ($j, \ell = 1, \ldots, n_b$). As in the case study of Chapter 6, this linear system of equations is an approximation to an integral equation of the first kind

$$\int_0^\infty K(e,\tilde{e}) f(\tilde{e}) d\tilde{e} = g(e),$$

and our information about the **kernel function** $K(e,\tilde{e})$ is incomplete.

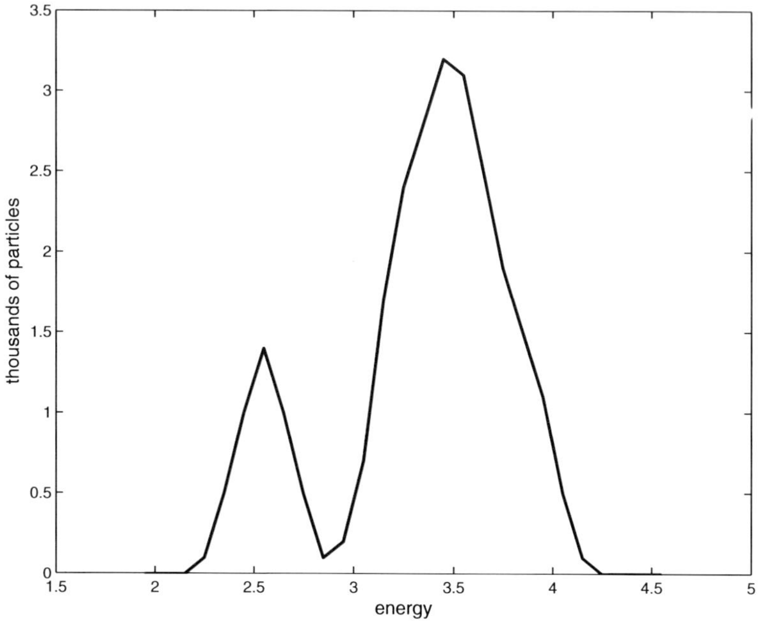

Figure 14.1. *(Simulated) data from a spectrometer. Given that there are particles with at most 5 different energy levels, determine these energies and the relative abundance of the particles.*

There are several sources of differences between the true spectrum $\boldsymbol{g}+\boldsymbol{r}$ and the recorded spectrum $\boldsymbol{g}$:

- We effectively assign energy $(e_j + e_{j-1})/2$ to all particles in bin j, and this is not correct.

- A count's value depends on the number of particles with the energies that it represents, but there is some blurring, so that it also depends on the number of particles with nearby energies.

- The values of the counts often have some error, since they are finite precision representations rather than the infinite precision "real ones." In our data, the counter incremented by 0.1 for every hundred particles detected.

We could assume that the main error is in our estimates of the variances s_j, and try to estimate the correct values of the variances and the correct counts. We take a different approach, appropriate even when the probabilities are not exactly normal and a correction needs to be applied to them. Our approach is useful whenever both $\boldsymbol{K}$ and $\boldsymbol{g}$ have errors whose relative variance is known or can be estimated. We experiment with two models:

- Model 1: Least squares. This model assumes that most of the error is in $\boldsymbol{g}$, so $\boldsymbol{E}$ is assumed to be zero.

- Model 2: Total least squares. This model assumes that there is significant error in both $\boldsymbol{g}$ and $\boldsymbol{K}$.

Model 1 was used in the case study of Chapter 6. As in that problem, the matrix $\boldsymbol{K}$ ($m \times n$) can be quite ill-conditioned, so small changes in the measured counts $\boldsymbol{g}$ make large changes in the resulting $\boldsymbol{f}$. Therefore, as in that case study, we add **regularization** to make the problem better conditioned, and we focus on the truncated SVD as a tool for doing this.

Our major algorithmic task is to figure out how to determine $\boldsymbol{f}$, $\boldsymbol{E}$, and $\boldsymbol{r}$ without much computational effort. In particular, we don't want to need to start over again if we change our mind about how many singular values to drop. In fact, we want to compute a singular value decomposition (SVD) only once.

Let's consider each case in turn.

Model 1: Least Squares ($E = 0$).

Define the SVD of the $m \times n$ matrix $\boldsymbol{K}$ to be

$$\boldsymbol{K} = \boldsymbol{U}\boldsymbol{\Sigma}\boldsymbol{V}^T,$$

where

- $\boldsymbol{U}$ has dimension $m \times m$ and $\boldsymbol{U}^T\boldsymbol{U} = \boldsymbol{I}$, the identity matrix.
- $\boldsymbol{\Sigma}$ has dimension $m \times n$, the only nonzeros are on the main diagonal, and they are nonnegative real numbers $\sigma_1 \geq \sigma_2 \geq \cdots \geq \sigma_n$,
- $\boldsymbol{V}$ has dimension $n \times n$ and $\boldsymbol{V}^T\boldsymbol{V} = \boldsymbol{I}$.

In Model 1, we determine $\boldsymbol{f}$ by solving the least squares problem

$$\min_{\boldsymbol{f}} \|\tilde{\boldsymbol{K}}_{\tilde{n}}\boldsymbol{f} - \boldsymbol{g}\|_2,$$

where

$$\tilde{\boldsymbol{K}}_{\tilde{n}} = \boldsymbol{U}\begin{bmatrix} \boldsymbol{\Sigma}_{\tilde{n}} & \boldsymbol{0} \\ \boldsymbol{0} & \boldsymbol{0} \end{bmatrix}\boldsymbol{V}^T.$$

We have already considered this problem in Section 5.6.2. There are $(n - \tilde{n})$ zero columns in the middle matrix, so $\boldsymbol{\Sigma}_{\tilde{n}}$ is a diagonal matrix with entries $\sigma_1, \ldots, \sigma_{\tilde{n}}$. The solution to the minimization problem is not unique if $\tilde{n} < n$, but the solution of minimal norm is found by taking

$$\boldsymbol{f}_{\tilde{n}} = \boldsymbol{V}\begin{bmatrix} \boldsymbol{\Sigma}_{\tilde{n}}^{-1} & \boldsymbol{0} \\ \boldsymbol{0} & \boldsymbol{0} \end{bmatrix}\boldsymbol{U}^T\boldsymbol{g}. \tag{14.1}$$

(Here, there are $(m - \tilde{n})$ zero columns in the matrix following $\boldsymbol{V}$.) Thus we can compute different estimates of the solution, for various values of $\tilde{n}$, without recomputing the SVD.

Which value of $\tilde{n}$ should we pick? One rule of thumb (called the **discrepancy principle**) is to choose the value of $\tilde{n}$ that makes the residual norm $\|\boldsymbol{g} - \boldsymbol{K}\boldsymbol{f}_{\tilde{n}}\|$ close to its expected value. If the errors in the data values $\boldsymbol{g}$ are independent and normally distributed with mean zero and variance δ^2, then this expected value is $\delta\sqrt{m}$. To estimate the value δ^2, other rules of thumb are used. For example, for many ill-posed problems, the left singular vectors corresponding to very small singular values are highly oscillatory, so they capture white noise in the measured data. Therefore, if we assume that these components are entirely due to noise, and if we believe that the noise has mean zero, we might estimate the variance by computing the variance of the last few components of $\widehat{g} = U^T g$ using the formula

$$\delta^2 \approx \frac{1}{m - \hat{m} + 1} \sum_{i=\hat{m}}^{m} \hat{g}_i^2$$

for some value of $\hat{m}$ close to m.

CHALLENGE 14.1.

Program the least squares algorithm and try it on the data of Figure 14.1 for various values of $\tilde{n}$. (The data can be found on the website.) The matrix K is 27×22, and we assume that the true counts for the first two and the last three bins are zero. Note how ill-conditioned the original matrix K is (by recording $\kappa(K) = \sigma_1/\sigma_n$).

Model 2: Total Least Squares (TLS)

If we allow both E and r to be nonzero, how can we solve the problem?

First we need a way to measure the size of these quantities. One reasonable way is to use the Frobenius norm of the errors:

$$\| [\ E \quad r\ ] \|_F^2 = \sum_{i=1}^{m} \sum_{j=1}^{n} e_{ij}^2 + \sum_{i=1}^{m} r_i^2 . \tag{14.2}$$

(If we expect the errors e_{ij}^2 to be much different in size than the errors r_i^2, then we might want to use weights for each term in this expression, but for this case study we just leave them equally weighted.)

Let's rewrite $Kf \approx g$ as

$$[\ K \quad g\ ] \begin{bmatrix} f \\ -1 \end{bmatrix} \approx \mathbf{0}.$$

Notice these facts:

- If this equation were exactly satisfied, then the columns of the matrix $[K, g]$ would be linearly dependent, so the rank of the matrix would be less than $n+1$. In this case there would be at least one singular value equal to zero and $[f^T, -1]^T$ would be a corresponding right singular vector.
- We need a matrix $[E, r]$ to add to $[K, g]$ to make

$$[\ K{+}E \quad g{+}r\ ] \begin{bmatrix} f \\ -1 \end{bmatrix} = \mathbf{0}. \tag{14.3}$$

- Among all such matrices $[\boldsymbol{E}, \boldsymbol{r}]$, we need the one with smallest Frobenius norm.

Finding this matrix is a well-studied problem.

CHALLENGE 14.2.

Suppose we have the SVD of $[\boldsymbol{K}, \boldsymbol{g}] = \widetilde{\boldsymbol{U}}\widetilde{\boldsymbol{\Sigma}}\widetilde{\boldsymbol{V}}^T$. Assume that $\boldsymbol{K}$ has rank n and that $\widetilde{v}_{nn} > \widetilde{v}_{n+1,n+1} \neq 0$. Show that the solution to

$$\min_{\boldsymbol{E},\boldsymbol{r}} \| \begin{bmatrix} \boldsymbol{E} & \boldsymbol{r} \end{bmatrix} \|_F,$$

subject to the constraint

$$\begin{bmatrix} \boldsymbol{K}+\boldsymbol{E} & \boldsymbol{g}+\boldsymbol{r} \end{bmatrix} \begin{bmatrix} \boldsymbol{f} \\ -1 \end{bmatrix} = \boldsymbol{0},$$

is

$$\begin{bmatrix} \boldsymbol{E} & \boldsymbol{r} \end{bmatrix} = -\widetilde{\sigma}_{n+1}\widetilde{\boldsymbol{u}}_{n+1}\widetilde{\boldsymbol{v}}_{n+1}^T,$$

with

$$\begin{bmatrix} \boldsymbol{f} \\ -1 \end{bmatrix} = -\frac{1}{\widetilde{v}_{n+1,n+1}}\widetilde{\boldsymbol{v}}_{n+1},$$

where $\widetilde{\boldsymbol{u}}_{n+1}$ is the $(n+1)$st column of $\widetilde{\boldsymbol{U}}$ and $\widetilde{\boldsymbol{v}}_{n+1}$ is the $(n+1)$st column of $\widetilde{\boldsymbol{V}}$.
Hint:
(a) First show that this solution satisfies the constraint and that the resulting $\| \begin{bmatrix} \boldsymbol{E} & \boldsymbol{r} \end{bmatrix} \|_F = \widetilde{\sigma}_{n+1}$.
(b) Show that $\|\widetilde{\boldsymbol{U}}^T\boldsymbol{A}\widetilde{\boldsymbol{V}}\|_F = \|\boldsymbol{A}\|_F$ for any matrix $\boldsymbol{A}$ of size $m \times (n+1)$.
(c) Then transform the problem to minimizing $\|[\widetilde{\boldsymbol{E}}, \widetilde{\boldsymbol{r}}]\|_F$ subject to $(\widetilde{\boldsymbol{\Sigma}} + \widetilde{\boldsymbol{E}})\widetilde{\boldsymbol{f}} = \boldsymbol{0}$ for some vectors $\widetilde{\boldsymbol{f}}$ and $\widetilde{\boldsymbol{r}}$ and matrix $\widetilde{\boldsymbol{E}}$. Solve the problem in this coordinate system, and show that there is no solution that gives a value of the minimization function smaller than $\widetilde{\sigma}_{n+1}$.

If we want to truncate our model at $\tilde{n} < n$, then the solution to the problem of minimizing (14.2) subject to (14.3) becomes

$$\boldsymbol{f} = -\frac{1}{\|\widetilde{\boldsymbol{V}}_{22}\|^2}\widetilde{\boldsymbol{V}}_{12}\widetilde{\boldsymbol{V}}_{22}^T,$$

where $\widetilde{\boldsymbol{V}}_{12}$ consists of rows 1 through n and columns $\tilde{n}+1$ through $n+1$ of $\widetilde{\boldsymbol{V}}$, and $\widetilde{\boldsymbol{V}}_{22}$ contains the last row of these columns of $\widetilde{\boldsymbol{V}}$.

CHALLENGE 14.3.

Write a MATLAB function to solve Model 2 using this truncated technique for various values of $\tilde{n}$. The input values should be $\boldsymbol{K}, \boldsymbol{g}$, and a range of $\tilde{n}$ values. Include appropriate documentation, and use your function to solve our problem.

Finally, we are ready to answer our original question.

POINTER 14.1. Further Reading.

This case study is related to the case study of Chapter 6, but previously the matrix $\boldsymbol{K}$ was assumed to be known exactly.

Instead of using truncated SVD for regularization, we might use Tikhonov regularization. This method is quite well studied for least squares problems [68], and for TLS some analysis appears in a paper by Golub, Hansen, and O'Leary [61].

The standard reference for TLS is the book by Van Huffel and Vandewalle [147]. Fierro, Golub, Hansen, and O'Leary [48] give further information on the truncated TLS algorithm used in Challenge 14.3. TLS is closely related to the **errors-in-variables** method from statistics.

CHALLENGE 14.4.

Write a brief summary of the results you obtained using Model 1 and Model 2 to solve the problem of Figure 14.1. Give your best estimate of the number of different peaks (energy levels) in the original data $\boldsymbol{f}_{true}$, the relative heights of the peaks, and the centers of the peaks. Make a convincing argument to justify your estimate and your choice of parameters (δ and $\tilde{n}$) for each method.

The total least squares formulation has given us a good tool for improving our solution estimate when we have errors in both the matrix and the right-hand side. In the next case study we add constraints to try to further improve the estimate.

Chapter 15 / Case Study

Blind Deconvolution: A Matter of Norm

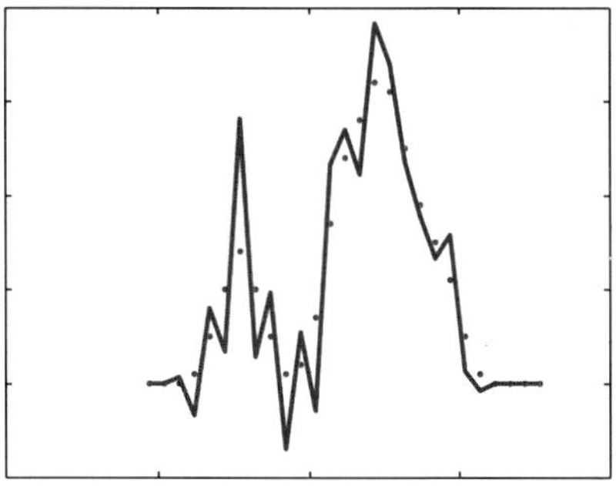

We continue the spectroscopy problem from the previous case study, trying to reconstruct a true spectrum from an observed one. Again we use blind deconvolution, but this time we impose some constraints on the error matrix $\boldsymbol{E}$, leading to a more difficult problem to solve but often a more useful reconstruction.

Recall that we have the counts of Figure 14.1, measured by a spectrometer, and we model our system by the matrix equation $(\boldsymbol{K}+\boldsymbol{E})\boldsymbol{f} = \boldsymbol{g}+\boldsymbol{r}$, where $\boldsymbol{E}$ accounts for errors in modeling the spectrometer's blur and $\boldsymbol{r}$ accounts for errors in counts. The matrix entry $k_{j\ell}$ is our estimate of the probability that a particle whose energy is in the interval $[e_{\ell-1}, e_\ell]$ is assigned to bin j ($j, \ell = 1, \ldots, n_b$).

We assume that there is significant error in both $\boldsymbol{g}$ and $\boldsymbol{K}$, but we note that in our data, the properties of each bin are the same, so that the rows of $\boldsymbol{K}$ have a pattern: for example, if the $m \times n$ matrix $\boldsymbol{K}$ were 5×5, then we would notice that

$$\boldsymbol{K} = \begin{bmatrix} k_5 & k_4 & k_3 & k_2 & k_1 \\ k_6 & k_5 & k_4 & k_3 & k_2 \\ k_7 & k_6 & k_5 & k_4 & k_3 \\ k_8 & k_7 & k_6 & k_5 & k_4 \\ k_9 & k_8 & k_7 & k_6 & k_5 \end{bmatrix},$$

so that there would be only $m+n-1 = 9$ distinct entries in $\boldsymbol{K}$. A matrix of this form is called a **Toeplitz matrix** and it is determined by the entries in its first row and column. Under this assumption, it might make sense to assume that the error matrix $\boldsymbol{E}$ also has this same structure, and therefore also depends on $m+n-1$ parameters instead of mn. We gather these parameters in a vector called $\widehat{\boldsymbol{e}}$.

Using the Euclidean Norm

In the case study of Chapter 14, we minimized (14.2) or, equivalently, we solved

$$\min_{\boldsymbol{E},\boldsymbol{f}} \frac{1}{2}\|\boldsymbol{E}\|_F^2 + \frac{1}{2}\|\boldsymbol{r}\|_2^2, \tag{15.1}$$

where

$$-\boldsymbol{r} = \boldsymbol{g} - (\boldsymbol{K}+\boldsymbol{E})\boldsymbol{f}. \tag{15.2}$$

With our new constraint that $\boldsymbol{E}$ be Toeplitz, our old solution is not feasible. So now we minimize the function, subject to the constraint that $\boldsymbol{E}$ be Toeplitz, over all choices of $\boldsymbol{f}$ and $\widehat{\boldsymbol{e}}$.

Our goal is to find an effective algorithm to solve this problem, and we do this in several steps. The first is to derive a useful alternative expression for the matrix-vector product $\boldsymbol{E}\boldsymbol{f}$.

CHALLENGE 15.1.

Show that if $\boldsymbol{E}$ is Toeplitz then $\boldsymbol{E}\boldsymbol{f}$ can be written as $\boldsymbol{F}\widehat{\boldsymbol{e}}$, where $\boldsymbol{F}$ is a matrix whose entries depend on the entries in the vector $\boldsymbol{f}$. In other words, find a matrix $\boldsymbol{F}$ so that $\boldsymbol{E}\boldsymbol{f} = \boldsymbol{F}\widehat{\boldsymbol{e}}$.

Let's use Newton's method (See Section 9.2) to solve our minimization problem. Recall that if we are minimizing some function $s(\boldsymbol{x})$, then the Newton direction is the solution $\boldsymbol{p}$ to the linear system

$$\boldsymbol{H}(\boldsymbol{x})\boldsymbol{p} = -\nabla s(\boldsymbol{x}),$$

where $\nabla s(\boldsymbol{x})$ is the gradient of s with respect to $\boldsymbol{x}$, and $\boldsymbol{H}(\boldsymbol{x})$ is the Hessian matrix, containing the second derivatives: $h_{ij}(\boldsymbol{x}) = \partial^2 s(\boldsymbol{x})/\partial x_i \partial x_j$. Let's derive a formula for $\boldsymbol{p}$.

CHALLENGE 15.2.

Derive the Newton direction for (15.1). To do this, use the definitions of $\boldsymbol{E}$ (in terms of $\widehat{\boldsymbol{e}}$) and $\boldsymbol{r}$ (equation (15.2)), and then differentiate the function in (15.1) with respect to $\widehat{\boldsymbol{e}}$ and $\boldsymbol{f}$.

Although the formula from Challenge 15.2 is mathematically correct, it is not the best computationally, and Challenge 15.3 provides a better alternative.

CHALLENGE 15.3.

Show that this Newton direction is approximately the same as the solution to the least squares problem

$$\min_{\Delta\widehat{\boldsymbol{e}},\Delta\boldsymbol{f}} \left\| \begin{bmatrix} \boldsymbol{F} & \boldsymbol{K}+\boldsymbol{E} \\ \boldsymbol{D} & \boldsymbol{0} \end{bmatrix} \begin{bmatrix} \Delta\widehat{\boldsymbol{e}} \\ \Delta\boldsymbol{f} \end{bmatrix} + \begin{bmatrix} -\boldsymbol{r} \\ \boldsymbol{D}\widehat{\boldsymbol{e}} \end{bmatrix} \right\|_2,$$

where $\boldsymbol{D}$ is a diagonal matrix of size $(m+n-1)\times(m+n-1)$ with entries equal to the square roots of $1,2,\ldots,n,\ldots,n,n-1,\ldots,1$. (In particular, the least squares solution is very close to the Newton direction if the model is good, so that $\|\boldsymbol{r}\|$ is small.)

If we were to solve problem (15.1) using our spectroscopy data, the solution would be quite contaminated by error. (That is why we truncated the SVD when solving the problem in the case study of Chapter 14.) Therefore, we make one further modification to obtain a useful model: we solve the problem

$$\min_{\boldsymbol{E}\boldsymbol{f}} \frac{1}{2}\|\boldsymbol{E}\|_F^2 + \frac{1}{2}\|\boldsymbol{r}\|_2^2 + \frac{1}{2}\lambda^2\|\boldsymbol{f}\|_2^2,$$

subject to the constraint that $\boldsymbol{E}$ be Toeplitz. The last term is a **Tikhonov regularization** term (as in the case study of Chapter 6), with a fixed parameter λ, added on to control the size of $\boldsymbol{f}$. In this case, the approximate Newton direction from Challenge 15.3 is computed from the solution to the least squares problem

$$\min_{\Delta\widehat{\boldsymbol{e}},\Delta\boldsymbol{f}} \left\| \begin{bmatrix} \boldsymbol{F} & \boldsymbol{K}+\boldsymbol{E} \\ \boldsymbol{D} & \boldsymbol{0} \\ \boldsymbol{0} & \lambda\boldsymbol{I} \end{bmatrix} \begin{bmatrix} \Delta\widehat{\boldsymbol{e}} \\ \Delta\boldsymbol{f} \end{bmatrix} + \begin{bmatrix} -\boldsymbol{r} \\ \boldsymbol{D}\widehat{\boldsymbol{e}} \\ \lambda\boldsymbol{f} \end{bmatrix} \right\|_2 .$$

Now we put these pieces together to solve our problem.

CHALLENGE 15.4.

Use a variant of Newton's method to solve our Toeplitz-constrained problem in a stable and efficient way, implementing it in a MATLAB function `[f,ehat,r,itn] = stls(K,g,lambda,tol)`. Use the least squares problem above to compute the approximate Newton direction. Start the iteration with $\widehat{\boldsymbol{e}} = \boldsymbol{0}$ and $\boldsymbol{f}$ equal to the least squares solution. Stop the iteration when the norm of the approximate Newton step is smaller than `tol`, and set `itn` to the number of iterations. Provide documentation for your function. Use it on the data from Figure 14.1, setting $\lambda = 0.06$ and `tol` $= 10^{-3}$. Plot the solution, and print the residual norm, the solution norm, and the number of iterations.

Using Other Norms

If the errors in our data are not normally distributed, then there are several reasonable alternatives to the choice of the Euclidean norm for the minimization function. For example, instead of minimizing

$$\frac{1}{2}\|\boldsymbol{E}\|_F^2 + \frac{1}{2}\|\boldsymbol{r}\|_2^2 = \frac{1}{2}\left\| \begin{bmatrix} \boldsymbol{r} \\ \boldsymbol{D}\widehat{\boldsymbol{e}} \end{bmatrix} \right\|_2^2,$$

we might instead minimize

$$\left\| \begin{bmatrix} \boldsymbol{r} \\ \boldsymbol{D}\widehat{\boldsymbol{e}} \end{bmatrix} \right\|_p, \tag{15.3}$$

where, if $p = 1$ the norm is defined as the sum of the absolute values of the components, and if $p = \infty$, the norm is the maximum of the absolute values of the components. (See Pointer 2.1.) Either of these choices has the effect of reducing the effects of outliers in our measurements.

To derive an algorithm to solve this problem for $p = 1$ or ∞, and to match our previous algorithm when $p = 2$, we reason this way. We need to satisfy the constraint

$$\boldsymbol{F}\widehat{\boldsymbol{e}} = \boldsymbol{E}\boldsymbol{f},$$

even after we replace $\boldsymbol{f}$ by $\boldsymbol{f}+\Delta\boldsymbol{f}$ and $\widehat{\boldsymbol{e}}$ by $\Delta\widehat{\boldsymbol{e}}$, so we require

$$(\boldsymbol{F}+\Delta\boldsymbol{F})(\widehat{\boldsymbol{e}}+\Delta\widehat{\boldsymbol{e}}) = (\boldsymbol{E}+\Delta\boldsymbol{E})(\boldsymbol{f}+\Delta\boldsymbol{f}),$$

where $\Delta\boldsymbol{E}$ is formed from $\Delta\widehat{\boldsymbol{e}}$ and $\Delta\boldsymbol{F}$ is formed from $\Delta\boldsymbol{f}$ so that $\Delta\boldsymbol{F}\widehat{\boldsymbol{e}} = \boldsymbol{E}\Delta\boldsymbol{f}$ and $\boldsymbol{F}\Delta\widehat{\boldsymbol{e}} = \Delta\boldsymbol{E}\boldsymbol{f}$. This means that

$$\Delta\boldsymbol{F}\Delta\widehat{\boldsymbol{e}} = \Delta\boldsymbol{E}\Delta\boldsymbol{f}.$$

Now let's examine the residual after we replace $\boldsymbol{f}$ by $\boldsymbol{f}+\Delta\boldsymbol{f}$ and $\widehat{\boldsymbol{e}}$ by $\widehat{\boldsymbol{e}}+\Delta\widehat{\boldsymbol{e}}$:

$$\begin{aligned}\boldsymbol{r}_{new} &= \boldsymbol{g} - (\boldsymbol{K}+\boldsymbol{E}+\Delta\boldsymbol{E})(\boldsymbol{f}+\Delta\boldsymbol{f})\\ &= \boldsymbol{g} - (\boldsymbol{K}+\boldsymbol{E})\boldsymbol{f} - \Delta\boldsymbol{E}\boldsymbol{f} - (\boldsymbol{K}+\boldsymbol{E})\Delta\boldsymbol{f} - \Delta\boldsymbol{E}\Delta\boldsymbol{f}.\end{aligned}$$

If both $\Delta\boldsymbol{f}$ and $\Delta\widehat{\boldsymbol{e}}$ are small, then the last term is negligible, and we can approximate

$$\boldsymbol{r}_{new} \approx \boldsymbol{r} - \boldsymbol{F}\Delta\widehat{\boldsymbol{e}} - (\boldsymbol{K}+\boldsymbol{E})\Delta\boldsymbol{f},$$

so that our minimization function (15.3) is approximated by

$$\left\| \begin{bmatrix} \boldsymbol{F} & \boldsymbol{K}+\boldsymbol{E} \\ \boldsymbol{D} & \boldsymbol{0} \end{bmatrix} \begin{bmatrix} \Delta\widehat{\boldsymbol{e}} \\ \Delta\boldsymbol{f} \end{bmatrix} + \begin{bmatrix} -\boldsymbol{r} \\ \boldsymbol{D}\widehat{\boldsymbol{e}} \end{bmatrix} \right\|_p .$$

So, to compute our step, we need to minimize a function of this form, and our next task is to develop an algorithm that does this.

CHALLENGE 15.5.

(a) Show that when $p=1$, minimizing

$$\left\| \begin{bmatrix} \boldsymbol{F} & \boldsymbol{K}+\boldsymbol{E} \\ \boldsymbol{D} & \boldsymbol{0} \\ \boldsymbol{0} & \lambda\boldsymbol{I} \end{bmatrix} \begin{bmatrix} \Delta\widehat{\boldsymbol{e}} \\ \Delta\boldsymbol{f} \end{bmatrix} + \begin{bmatrix} -\boldsymbol{r} \\ \boldsymbol{D}\widehat{\boldsymbol{e}} \\ \lambda\boldsymbol{f} \end{bmatrix} \right\|_p$$

over all choices of $\Delta\boldsymbol{f}$ and $\Delta\widehat{\boldsymbol{e}}$ is equivalent to solving the linear programming problem

$$\min_{\Delta\widehat{\boldsymbol{e}},\Delta\boldsymbol{f},\bar{\sigma}} \quad \bar{\sigma} = \sum_{i=1}^{m} \bar{\sigma}_{1i} + \sum_{i=1}^{q} \bar{\sigma}_{2i} + \sum_{i=1}^{n} \bar{\sigma}_{3i}$$

subject to

$$\begin{array}{rcccl} -\bar{\boldsymbol{\sigma}_1} & \le & \boldsymbol{F}\Delta\widehat{\boldsymbol{e}}+(\boldsymbol{K}+\boldsymbol{E})\Delta\boldsymbol{f}-\boldsymbol{r} & \le & \bar{\boldsymbol{\sigma}_1}, \\ -\bar{\boldsymbol{\sigma}_2} & \le & \boldsymbol{D}\Delta\widehat{\boldsymbol{e}}+\boldsymbol{D}\widehat{\boldsymbol{e}} & \le & \bar{\boldsymbol{\sigma}_2}, \\ -\bar{\boldsymbol{\sigma}_3} & \le & \lambda\Delta\boldsymbol{f}+\lambda\boldsymbol{f} & \le & \bar{\boldsymbol{\sigma}_3}, \end{array}$$

where $\boldsymbol{\sigma}_1 \in \bar{\mathbf{R}}^{m\times 1}$ and $\boldsymbol{\sigma}_2 \in \bar{\mathbf{R}}^{q\times 1}$, and $\boldsymbol{\sigma}_3 \in \bar{\mathbf{R}}^{n\times 1}$.

(b) Derive a similar linear programming problem that determines $\Delta\widehat{\boldsymbol{e}}$ and $\Delta\boldsymbol{f}$ when $p=\infty$.

Let's see how the choice of norm affects our computed solution.

CHALLENGE 15.6.

Use a variant of Newton's method to solve the problem when $p = 1$, and implement it in a MATLAB function `[f,ehat,r,itn] = stln1(K,g,lambda,tol)`. Use the solution to the linear programming problem as an approximate Newton direction. Start the iteration with $\widehat{\boldsymbol{e}} = \mathbf{0}$ and $\boldsymbol{f}$ equal to the vector of all ones. Stop the iteration when the norm of the approximate Newton step is smaller than `tol`, and set `itn` to the number of iterations. Use it on the data from Figure 14.1, setting $\lambda = 0.06$ and `tol` = 10^{-3}. Plot the solution, and print the residual norm, the solution norm, and the number of iterations.

Repeat for the case $p = \infty$.

Comparing Our Results

Recall that our goal is to reconstruct the spectrum of the particles fed into the spectrometer. Take some time now to compare the results we have obtained using various problem formulations.

CHALLENGE 15.7.

Results for various values of λ are shown in Figures 15.1 (`stls` from Challenge 15.4) and 15.2 (`stln` from Challenge 15.6). The red dots indicate the true spectrum. Compare these results with those in the case study of Chapter 14 by answering these two questions:

- How does the quality of results compare?
- How does the amount of work compare?

Our models in this case study use a variety of standard tools from matrix decompositions and optimization. Using these tools effectively, though, requires some cleverness in reformulating the problem and choosing methods appropriate to the data.

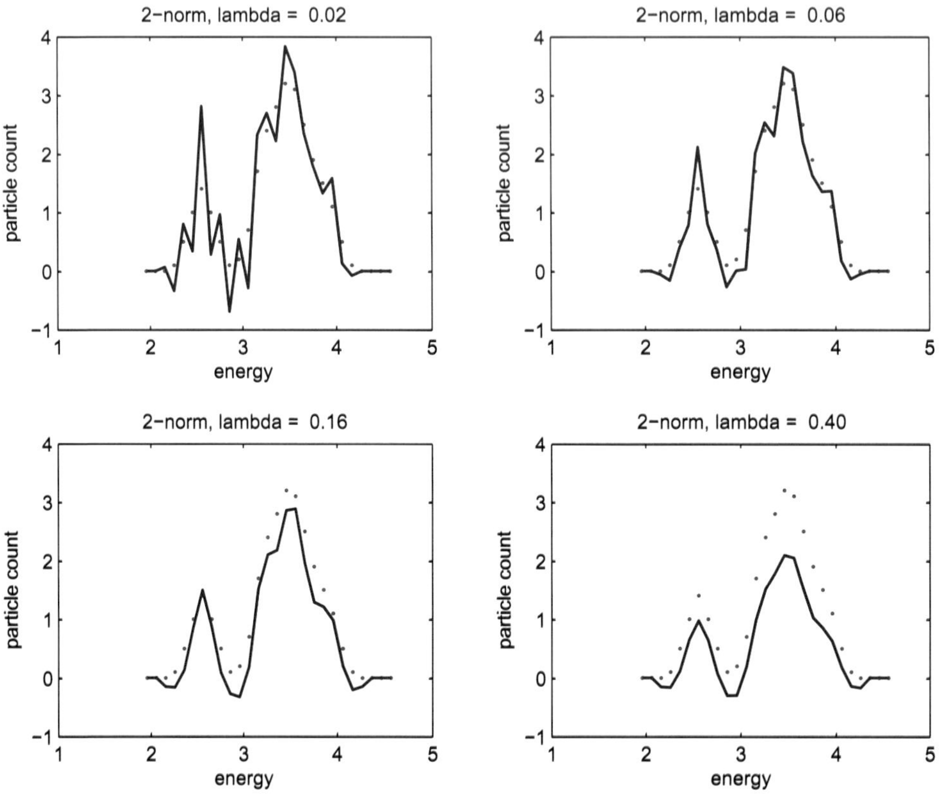

Figure 15.1. *Results from the structured total least squares algorithm for various values of* λ.

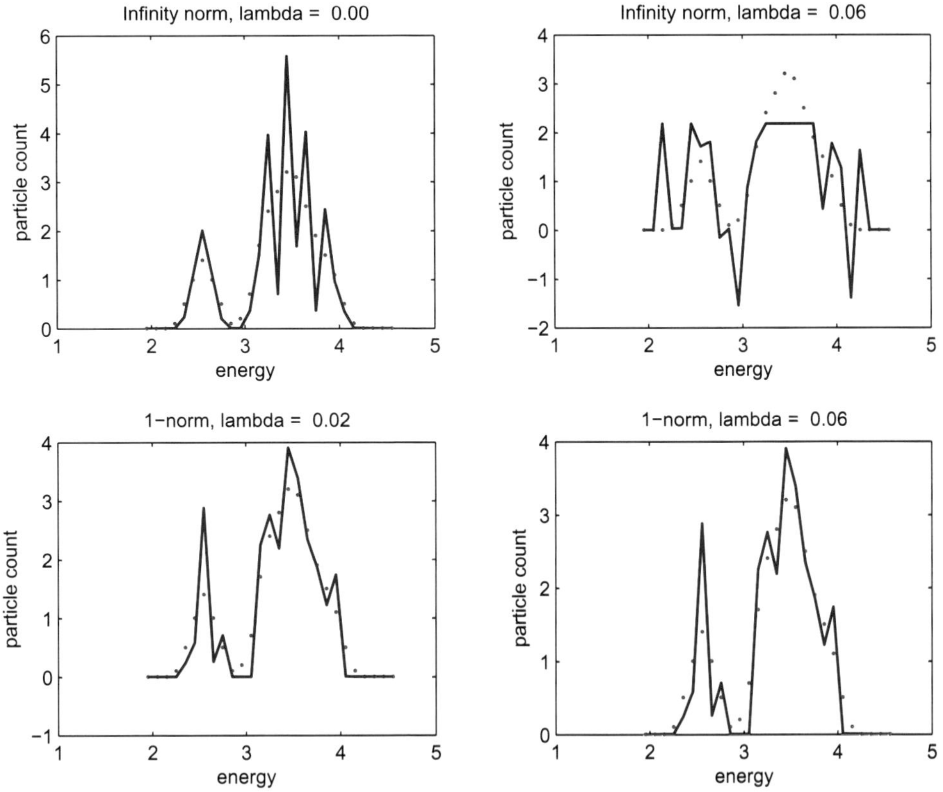

Figure 15.2. *Results from the structured total least norm algorithm, using the 1-norm and the ∞-norm, for various values of λ.*

POINTER 15.1. Further Reading.

MATLAB's `linprog` can be used to solve the linear programming problems.

Choices of vector norms are discussed in many elementary textbooks, but Ortega [119] gives a particularly nice discussion.

The use of regularization plus norm choice to solve our problem is discussed by Pruessner and O'Leary [126]. References to earlier work using regularization or norm choice can be found in that paper, too.

Unit IV

Monte Carlo Computations

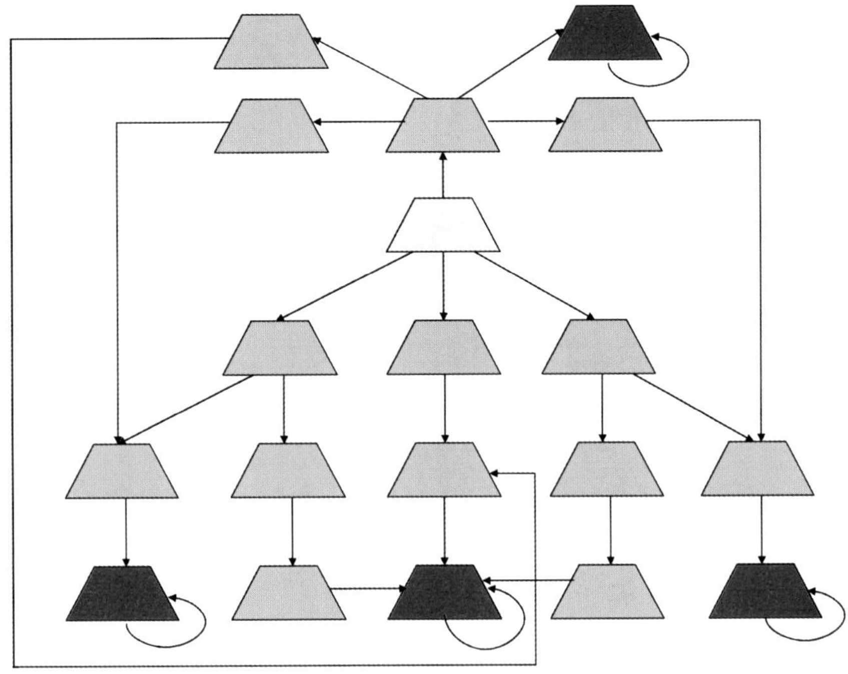

In a Monte Carlo method, the desired answer is a quantity in a stochastic model, and it is estimated by random sampling of the model. Applications of Monte Carlo methods range from estimating integrals, to minimizing difficult functions, to simulating complex systems. We have already seen its use in sensitivity analysis in Chapters 2, 12, and 13.

Our plan is to first discuss some basic statistical principles underlying Monte Carlo methods. Then we consider a variety of uses of these methods. Monte Carlo methods are also useful for minimizing nonconvex functions and for estimating distributions, and we illustrate these uses in Chapter 17. In Chapter 18, we develop Monte Carlo methods for estimating multidimensional integrals when conventional methods are too expensive. Finally, in Chapter 19, we develop and analyze a stochastic model for the spread of an epidemic.

BASICS: To understand this unit, the following background is helpful:

- **Statistics:** Past experience with random sampling and probability distributions is very helpful. See a basic statistics textbook such as [98] or a numerical book such as [71, Chap. 13].

MASTERY: After you have worked through this unit, you should be able to do the following:

- Write the probability density function for a given distribution.
- Distinguish between random numbers and pseudorandom numbers.
- Compute the mean and variance of a distribution.
- State the two properties that define a probability density function.
- State and use the central limit theorem.
- Explain why Monte Carlo algorithms are used for optimization problems and for counting problems.
- Explain or write a MATLAB algorithm to solve a problem like the traveling salesperson problem using the Metropolis algorithm (simulated annealing).
- Explain or write a MATLAB algorithm to implement the KRS algorithm for estimating the number of dimer arrangements on a lattice.
- Recognize when a product rule can be applied to a multidimensional integral.
- Write a MATLAB algorithm to estimate an integral using nested calls to a one-dimensional integrator.
- Write a MATLAB algorithm for Monte Carlo integration.
- Write a MATLAB algorithm for Monte Carlo integration using importance sampling.
- State and use the formulas for mean and variance of the estimates produced by Monte Carlo integration.

- Generate quasi-random numbers and explain their use.
- Set up Monte Carlo simulations of systems (such as epidemics).
- Analyze a Markov chain using its transition matrix to determine a stationary vector.

Chapter 16

Monte Carlo Principles

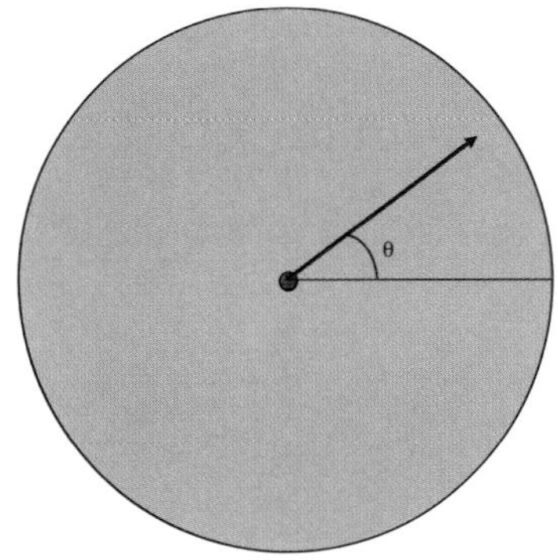

In a **Monte Carlo method**, the desired answer is a quantity in a **stochastic** (i.e., random) **model**, and it is estimated by random **sampling** of the model.

For example, suppose we have a cube, with the sides numbered 1 to 6. We might toss the cube 120 times, observe which side comes up on top each time, and study whether the sides each occur with frequency 1/6. These 120 random samples would not definitively answer the question of whether the frequencies are equal, since, for example, it is possible that one side would come up 120 times in a row, even if an infinitely long experiment would show that the frequencies are equal. Nevertheless, if we repeat the experiment many times, the proportion of tosses in which a given side comes up on top gives us a good estimate of the **probability** that the side would come up on top in a single toss.

As a second example, if we have a "black box" that takes a number between 0 and 1 as input and emits a number between 0 and 1 as output, we could feed the box m numbers and observe the m outputs of the box to develop insight into the hidden process.

In the second example, we could use random or non-random inputs to the box. There is an important difference between Monte Carlo methods, which estimate quantities using random samples for the inputs, and pseudo-Monte Carlo methods, which use input samples that might appear to be random but actually are more systematically chosen. We study random sampling in Section 16.1, and pseudo-random sampling in Section 16.4. In some sense, all practical computational methods involving randomization are **pseudo-Monte Carlo**, since "random" number generators implemented on machines are generally not truly random. We use the term Monte Carlo even for samples that are generated using **pseudorandom** numbers generated by a computer program.

In this chapter, we develop the basic statistical principles needed to understand random and pseudorandom numbers and their generation. It is important to remember that Monte Carlo methods are (at least in some sense) methods of last resort. They are generally quite expensive and only applied to problems that are too difficult to handle by **deterministic** (nonstochastic) methods.

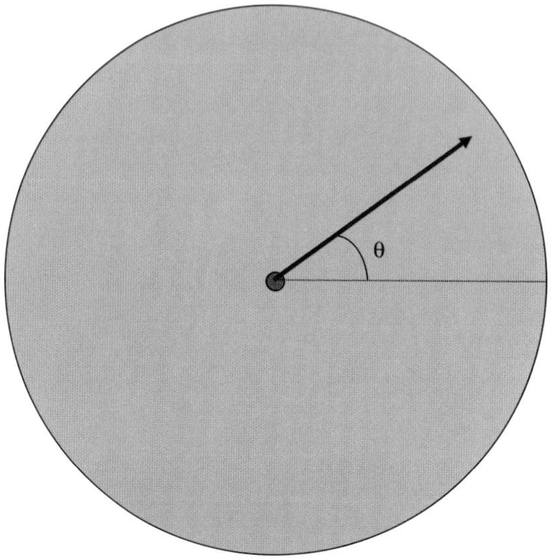

Figure 16.1. *A spinner that generates uniform random numbers* θ *in the interval* $[0, 2\pi)$.

16.1 Random Numbers and Their Generation

We generate random numbers by taking samples from a collection of numbers. For example, suppose we put n cards, numbered 1 to n, in a box and choose one at random. After we record the resulting number, we put the card back in the box. We have generated a random number that is **uniformly distributed** among the values $\{1, 2, \ldots, n\}$, since the probability $(1/n)$ of choosing one number is the same as that for any other number.

If we add k cards to the box, all numbered 1, then our distribution becomes **nonuniform**: the probability of choosing 1 is now $(k+1)/(n+k)$ and the probability of choosing any of the other numbers is $1/(n+k)$.

The most interesting random systems are those whose outcomes are numbers in some interval of the real line. For example:

- Make a spinner, as in Figure 16.1, by anchoring a needle at the center of a circle. Draw a radius line on the circle. Spin the needle, and measure the angle it forms with the radius line. You obtain random numbers that are **uniformly distributed** on the interval $[0, 2\pi)$.

- If, on average, a radioactive substance emits α-particles every μ seconds, then the time between two successive emissions has the **exponential distribution** with mean μ. (This is a special case of the **gamma distribution**.)

- The **normal distribution**, whose graph of probabilities is bell-shaped, is a good model in many situations:

  - The pattern formed by leaves that fall from a symmetric tree is approximately normal, so measuring the distance from the trunk gives a sample from a **univariate** normal distribution. There are many more leaves close to the trunk than far away. The two coordinates of a leaf, using the tree trunk as the origin, give a sample from a **bivariate** normal distribution.

- The IQ measure of intelligence was constructed so that the measures are normally distributed.
- Many physical characteristics of plants and animals (height, weight, etc.) are approximately normally distributed.
- The velocity distribution of molecules in a thermodynamic equilibrium (Maxwell–Boltzmann distribution) is normal [113], so measuring the velocity of a molecule would give a sample from a normal distribution.
- Measures of psychological variables such as reading ability, introversion, job satisfaction, and memory are approximately normal.
- Predicting probabilities in gambling was DeMoivre's (1667–1754) motivation in defining the normal distribution [98].

We can characterize a set of random samples by a set of numbers (**moments**) derived from them. The first two moments, the **mean** and the **variance**, are the most important. The **mean** or **average value** of **a set of samples** $\{x_i\}$, $i = 1, \ldots, n$, is

$$\mu_n \equiv \frac{1}{n} \sum_{i=1}^{n} x_i,$$

and the **variance** of the set of samples measures the variation of the samples from the mean value as[8]

$$\sigma_n^2 \equiv \frac{1}{n} \sum_{i=1}^{n} (x_i - \mu_n)^2 .$$

The mean and variance of the set of samples are estimates of the mean and variance of the distribution from which they were drawn, and we discuss these quantities in the next section. If the true mean μ of the distribution is known, then the variance of the distribution is estimated by

$$\sigma_n^2 \equiv \frac{1}{n} \sum_{i=1}^{n} (x_i - \mu)^2 .$$

CHALLENGE 16.1.

Compute the mean and variance of the following set of random samples: $\{1, 2, 5, 8, 6, 2\}$.

---

[8] Dividing by $n - 1$ is appropriate when the samples are from a normal distribution, since this gives the best unbiased estimator of the variance of the distribution; otherwise, we divide by n. MATLAB's function `std` computes both of these variants.

16.2 Properties of Probability Distributions

The particular system from which random samples are drawn is characterized by its nonnegative **probability density function** $f(x)$. The domain Ω of the function is the set of possible values that could be obtained by taking random samples of the function. The range of the function is a subset of the nonnegative numbers. The value

$$\int_{\alpha} f(x)\,\mathrm{d}x$$

is the **probability** that a sample from this distribution is in the set defined by α. (Replace the integral by a summation if the domain is discrete.) Because of these properties, we see that

$$\int_{\Omega} f(x)\,\mathrm{d}x = 1\,.$$

(In the discrete domain case, the sum of the probabilities is 1.)

For example, for our n cards in the box above, the domain is $\Omega = \{1,2,\ldots,n\}$ and $f(x) = 1/n$ for $x = 1,\ldots,n$. For our spinner, the domain is $\Omega = [0,2\pi)$ and $f(x) = 1/(2\pi)$.

The **mean** μ **of a distribution** (also known as the **expected value**) is the sum of all possible values, weighted by their probabilities. In other words, it is the expected value we would compute by averaging the values obtained for the sample mean if we took a large number of very large sets of samples. Similarly, the **variance** σ^2 **of the distribution** is estimated by the sample variances. The mean and variance of the distribution are defined by

$$\mu = \int_{\Omega} x f(x)\,\mathrm{d}x\,,$$
$$\sigma^2 = \int_{\Omega} (x-\mu)^2 f(x)\,\mathrm{d}x\,.$$

Again, for discrete distributions, we replace the integral by a summation. So for our n cards in the box,

$$\mu = \sum_{i=1}^{n} i\frac{1}{n} = \frac{n+1}{2},$$
$$\sigma^2 = \sum_{i=1}^{n}\left(i - \frac{n+1}{2}\right)^2 \frac{1}{n} = \frac{(n+1)(n-1)}{12}.$$

We give the distribution functions for three continuous distributions mentioned in our examples:

- The **uniform distribution** over the interval $[0,m]$ has probability density function

  $$f(x) = \frac{1}{m}.$$

  Its mean and variance are

  $$\mu = \int_0^m \frac{x}{m}\,\mathrm{d}x = \frac{m}{2},\quad \sigma^2 = \int_0^m \frac{1}{m}\left(x - \frac{m}{2}\right)^2 \mathrm{d}x = \frac{m^2}{12}.$$

- The **exponential distribution** with parameter μ has

$$f(x) = \frac{1}{\mu}e^{-x/\mu}$$

for $x \in [0,\infty)$. Its mean is μ and its variance is $\sigma^2 = \mu^2$.

- The **normal distribution** with parameters μ and σ has

$$f(x) = \frac{1}{\sqrt{2\pi\sigma^2}}e^{-(x-\mu)^2/(2\sigma^2)}$$

for $x \in (-\infty,\infty)$. Its mean is μ and its variance is σ^2.

A bit of practice makes the ideas of mean and variance more clear.

CHALLENGE 16.2.

Suppose we have a box with 8 cards numbered 1 through 8, and 2 cards with the number 10 written on them. Compute the mean and variance of the distribution corresponding to drawing a card from the box at random and recording the resulting number.

CHALLENGE 16.3.

Verify that $f(x) = 3x^2$ is a probability density function on the domain $[0,1]$. (In other words, verify that it is nonnegative and that its integral is equal to 1.) Find μ and σ^2.

16.3 The World Is Normal

The central importance of the normal distribution in statistics is due to two facts: the frequency of its occurrence in nature (See the examples in Section 16.1), and our ability to transform any set of observations into a set from a distribution that is approximately normal. This latter fact is summarized in the **central limit theorem**: Let $f(x)$ be any distribution with mean μ and (finite) variance σ^2. Take a random sample of size n from $f(x)$, and call the mean of the sample μ_n. Define the random variable y_n by

$$y_n = \sqrt{n}\,\frac{\mu_n - \mu}{\sigma}.$$

Then the distribution for y_n approaches the normal distribution with mean 0 and variance 1 as n increases.

Therefore, even if we know nothing about a distribution except its mean and variance, we can use samples from it to construct samples from a distribution that is near normal.

CHALLENGE 16.4.

Suppose the random number generator `y = randmy(n)` returns n random numbers from a distribution with mean 2 and variance 5. Using the central limit theorem, write a random number generator that uses 1000 numbers generated by `randmy` to produce a single random sample taken from an approximately normal distribution with mean 0 and variance 1.

16.4 Pseudorandom Numbers and Their Generation

In principle, we could generate random samples as discussed in Section 16.1. For example, when we wanted a uniform distribution, we could build a spinner and play with it for a while, writing down our list of samples.

In practice, when the random numbers are used in computer software, this does not work well:

- The samples are irreproducible; if someone else wanted to use the software, they would get a different result and not know whether it was because of their different sequence of random numbers or because of a bug.

- Computers run through enormous quantities of random numbers, and it is just not feasible to generate them manually.

- Sometimes we are very unlucky and generate a long string of random numbers that are not well-distributed.

Because of this, we usually use **pseudorandom numbers** in computer software. Pseudorandom numbers are generated by the computer using a **deterministic** (i.e., reproducible) procedure and appear to be random, in the sense that the mean, variance, and other properties of sequences of n samples match what we would expect to obtain from a random process. But the pseudorandom numbers actually cycle; i.e., if we ask for a long enough sequence, we see periodicity.

Pseudorandom number generators on computers usually use a **seed** (a number) to determine where in the cycle to begin. Thus, if other people want to reproduce our results, they simply use the same seed.

It is fairly cheap to generate pseudorandom numbers that appear to be uniformly distributed, and in that sense a pseudorandom number generator for the uniform distribution is easy to write, but it is very, very difficult to write a good one. Don't try to write your own unless you have a month to devote to it. Luckily, many good ones are readily available.

Samples from other distributions (for example, the normal or exponential distributions) are usually generated by taking two or more uniformly distributed pseudorandom numbers and manipulating them using functions such as `sin`, `cos`, and `exp`. This is much more costly, but let's consider a simple example that will be useful to us in the simulated annealing algorithm presented in Chapter 17.

POINTER 16.1. The Overhead of a Function Call.

Generating n random numbers in MATLAB is very slow if we do it one at a time:

```
for i = 1:n
  z = rand(1);
  (Comment: Use z in the calculations.)
end
```

This is because the overhead of starting up the `rand` function is quite high. We can improve speed (at the expense of storage) with this simple modification:

```
zrand = rand(n, 1);
for i = 1:n
  z = zrand(i);
  (Comment: Use z in the calculations.)
end
```

CHALLENGE 16.5.

Write MATLAB statements using the function `rand`, which generates a sample uniformly distributed in $[0,1]$, to generate a random number from the following distribution:

The probability that the number is 0 is 0.6.
The probability that the number is 1 is 0.4.

In other words, if $f(x)$ is the probability density function, it has domain $\{0,1\}$ with $f(0) = 0.6$ and $f(1) = 0.4$.

16.5 Mini Case Study: Testing Random Numbers

The program `demorand.m` from the website generates pseudorandom numbers using `rand` for the uniform distribution and `randn` for the normal distribution. It then plots histograms of these samples, showing the number of samples in each of 100 small intervals. This is a very simple test for random number generators. We expect that for the uniform distribution, the number of samples in each interval is approximately equal. For the normal distribution, we expect an approximately bell-shaped histogram. More rigorous tests of random numbers can be found in the references in Pointer 16.2.

The results of our simple test are shown in Figure 16.2. You can repeat the test with different seeds to judge the performance of `rand` and `randn`.

Now that we understand the basic principles behind pseudorandom numbers and their generation, we'll focus in the following case studies on how random numbers help us solve problems in computational science. We apply Monte Carlo methods to function minimization, discrete optimization, and counting in Chapter 17. Then in the case study of Chapter 18 we study their use in multidimensional integration, and we apply them to epidemic simulation in Chapter 19.

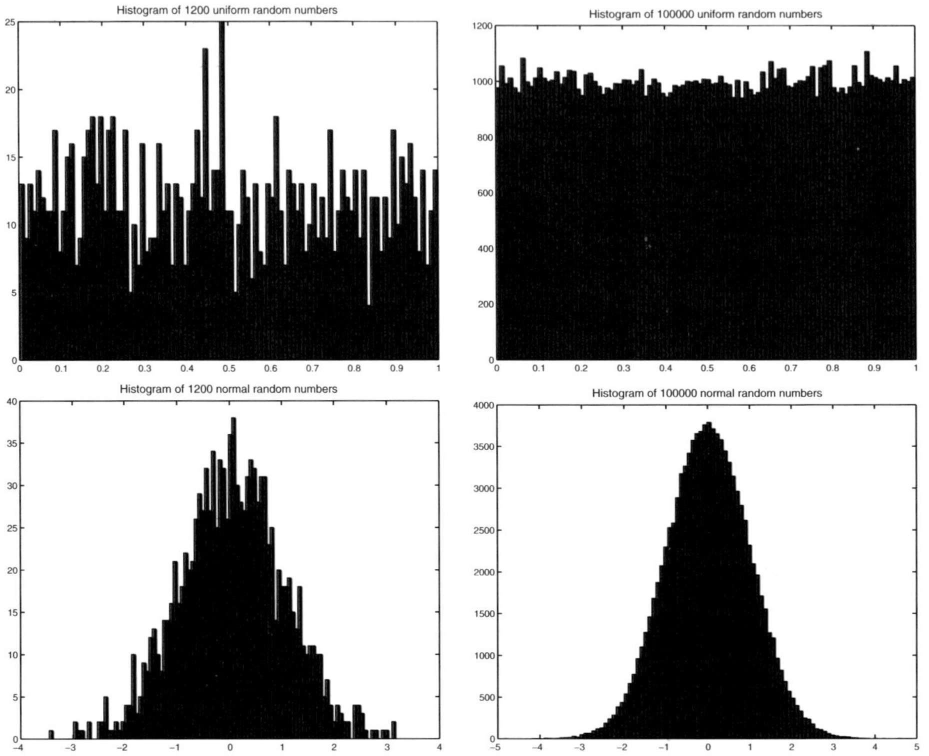

Figure 16.2. *Histograms, plotting the distribution of random samples from* `rand` *and* `randn`.

POINTER 16.2. Further Reading.

A whole course could be taught on how to generate pseudorandom numbers. The book by Knuth [91] is quite enlightening, talking about the generation of pseudorandom numbers as well as how to test whether a program is generating pseudorandom numbers that are a good approximation to random. See also [72]. The standard functions in many languages (including MATLAB) for generating pseudorandom normal and exponentially distributed sequences are adequate but not the best. With some searching and testing, you can find better ones when needed.

"True" random number generators, in contrast to pseudorandom number generators, can be found at [66]. They are based on atmospheric noise.

Chapter 17 / Case Study

Monte Carlo Minimization and Counting: One, Two, . . . , Too Many

(coauthored by Isabel Beichl and Francis Sullivan)

Given the choice between an electrical pump and a hand pump for getting water out of a well, most people would choose the electric. Yet the hand pump is quite reliable and can be used when other sources of power are not available. Similarly, Monte Carlo methods are remarkably versatile algorithms for solving difficult numerical problems when other methods are not practical. We focus in this case study on three uses of Monte Carlo methods: for function minimization, for discrete optimization, and for counting.

Function Minimization

A convex function $f(\boldsymbol{x})$ has a unique minimum that can be found using a variety of methods, including Newton's method and its variants, conjugate gradients, and (if derivatives are not available) pattern search algorithms. (See Chapter 9.) For functions that are nonconvex, such as that in Figure 17.1, these algorithms find a **local minimizer** such as x_1 but are not guaranteed to find the **global minimizer** x^*.

Minimization Using Monte Carlo Techniques

Monte Carlo methods provide a good means for generating starting points for optimization problems that are nonconvex. In its simplest form, a Monte Carlo method generates a random sample of points in the domain of the function. We use our favorite minimization algorithm starting from each of these points, and among the minimizers found, we report the best one. By increasing the number of Monte Carlo points, we increase the probability that we find the global minimizer.

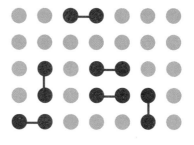

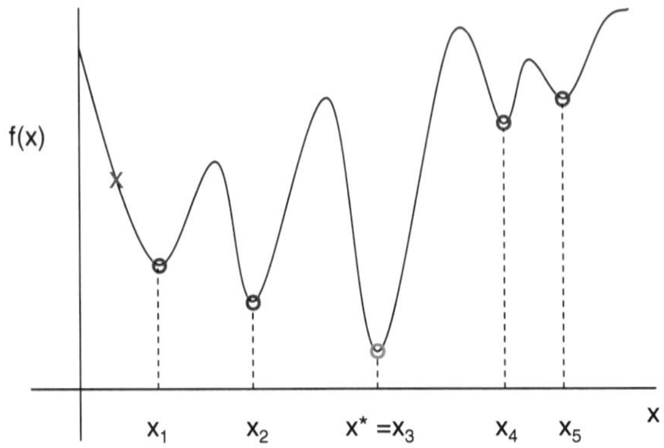

Figure 17.1. *A standard minimization algorithm might find the minimum $f(x_1)$ if we start at* ×*, but Monte Carlo minimization seeks the global minimum $f(x_3)$.*

This method can be improved by using any extra information we have about the function $f(\boldsymbol{x})$ that we are minimizing. For example, suppose we know a **Lipschitz constant** L for our function, so that for all $\boldsymbol{x}$ and $\boldsymbol{y}$ in the domain,

$$|f(\boldsymbol{x}) - f(\boldsymbol{y})| \leq L\|\boldsymbol{x} - \boldsymbol{y}\|.$$

To make the example specific, suppose that $L = 1$ and the domain is just the real line. If we know that $f(1) = 2$ and $f(4) = 0$, then using the Lipschitz inequality with $x = 1$, we obtain $|f(1) - f(y)| \leq (1)||1 - y||$, or $|2 - f(y)| \leq |1 - y|$. This says that $f(y)$ is bounded below by 0 for y between -1 and 3. Therefore, our global minimizer cannot lie here. This saves us the work of generating more points in this interval.

In the next challenge, we experiment with Monte Carlo minimization.

CHALLENGE 17.1.

Consider the function `myf.m` (on the website), with domain $0 \leq x \leq 7$.

(a) Write a function to generate 500 uniformly distributed starting points on the interval $[0, 7]$ and use your favorite minimizer (perhaps `fmincon`), starting from each one, to find a local minimum. Make a graph that shows which starting points result in which local minimizer. (There are many ways to do this; just make sure that your graph displays the information clearly.)

(b) $L = 98.5$ is a Lipschitz constant for the function `myf.m`. Try to speed up the minimization using this information. Document the changes to your function and compare the performance of the two methods.

(c) (Extra) Try Monte Carlo minimization on your favorite function of n variables for $n > 1$.

Minimization Using Simulated Annealing

Suppose we have a box of particles that have been allowed to slowly cool. Then we expect the **potential energy** of the system to be small. For example, if we make ice in a freezer,

POINTER 17.1. Simulated Annealing.

The simulated annealing algorithm has a long history. The basic idea first appeared in the literature in a paper coauthored by Nicholas Metropolis in 1953, so the algorithm sometimes bears his name.

---

the crystal structure that results is one that has a lower potential energy than most of the alternatives. If we drop the temperature too fast, then the system can easily get stuck in a configuration that has a higher potential energy, and we often see crack lines in ice cubes resulting from this. The physical process of slow cooling is called **annealing**, and with luck it can lead to something close to a perfect ice cube. The annealing process works because if the temperature is high, then each particle has a lot of kinetic energy and can easily move to positions that increase the potential energy of the system. This enables a system to avoid getting stuck in configurations that are local minimizers but not global ones. But as the temperature is decreased, it becomes less likely that a particle makes a move that gives a large increase in energy. This is illustrated in Figure 17.2. This enables a system to do the fine-tuning that produces an optimal final configuration.

This motivates an algorithm: if we want to minimize some function other than energy, can we simulate this annealing process? We need some artificial definition of temperature, a means of generating a new configuration and deciding whether to keep it, and also a cooling schedule for reducing the temperature. One way to implement the algorithm is shown in Algorithm 17.1. Notice that the accept/reject decision can be made in a way similar to that used in Challenge 16.5. Let's see how simulated annealing works.

Algorithm 17.1 Simulated Annealing

Given an initial point $\boldsymbol{x}$, initialize T to some large number and choose a number α between 0 and 1 and a positive number ϵ close to 0.

while $T > \epsilon$,

Generate a random move from the current $\boldsymbol{x}$ to a new $\widehat{\boldsymbol{x}}$.

If $t \equiv f(\widehat{\boldsymbol{x}}) - f(\boldsymbol{x}) < 0$, then our new point improves the function value and we accept the move, setting $\boldsymbol{x} = \widehat{\boldsymbol{x}}$.

If the function value did not improve, then we accept the move, setting $\boldsymbol{x} = \widehat{\boldsymbol{x}}$, with probability $e^{-t/T}$, where T is the current temperature. With probability $1 - e^{-t/T}$ we reject the move and leave $\boldsymbol{x}$ unchanged.

Decrease the temperature, replacing T by αT.

end

CHALLENGE 17.2.

Use the simulated annealing algorithm to minimize `myf.m`. Experiment with various choices of T, α, and ϵ. Describe your experiment and the conclusions you can draw about how to choose parameters to make the method as economical and reliable as possible.

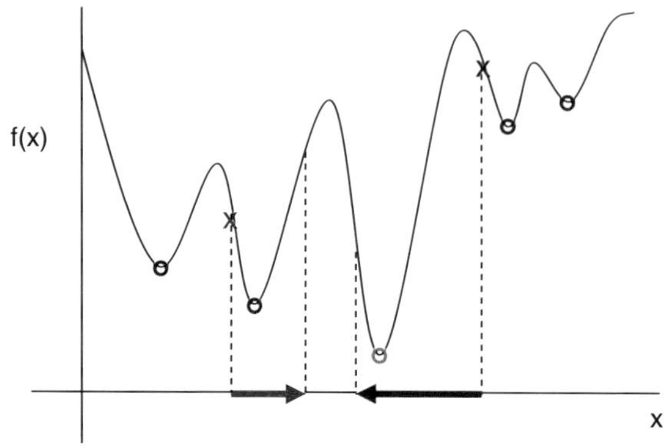

Figure 17.2. *In simulated annealing, we allow moves that increase the function value, such as that illustrated by the shorter arrow. At high temperature, such moves are likely to be accepted, while at low temperature they are likely to be rejected. At high temperature, we also might allow longer moves (horizontally), such as that illustrated by the left-pointing arrow, and these two features can enable us to escape from valleys that do not contain the global minimum.*

Note that although the simulated annealing algorithm is slow, it has two virtues:

- Compared to standard minimization algorithms, it gives a better probability of finding a global rather than a local minimizer.

- It does not require derivative values for f. In fact, it does not even require f to be differentiable.

The art of the method is determining the probability function and the temperature sequence appropriate to the specific problem.

Minimization of Discrete Functions

There are many optimization problems that are simple to state but difficult to solve. An example is the **traveling salesperson problem (TSP)**. This person needs to visit n cities exactly once and wants to minimize the total distance traveled and finish the trip at the starting point. To solve the problem, we need to deliver the permutation of the list of cities that corresponds to the shortest total distance traveled.

If, for definiteness, we specify the first city, then among the $(n-1)!$ permutations, we want to choose the best. This is an enormous number of possibilities for moderate values of n, and it is not practical to test each of them. Therefore, an approximate solution is often found either by generating random permutations and choosing the best (a Monte Carlo algorithm), or by using simulated annealing.

POINTER 17.2. Uses of Statistical Sampling.

Choosing a random sample of reasonable size and using it to estimate the exact counts of a much larger population has a long history in science, but it is even more widely used in commerce and public policy: Monte Carlo samples are the basis for opinion polls, census methodologies, traffic volume surveys, and for many other information estimates.

One step of a simulated annealing algorithm for TSP might look like this:

Start with an initial ordering of cities and an initial temperature T.
Randomly choose two cities.
if interchanging those cities decreases the length of the tour **then**
  interchange them!
else
  Interchange them with a probability that is an increasing function of temperature and a decreasing function of the change in tour length.
end
Decrease the temperature.

Let's experiment with Monte Carlo and simulated annealing solutions to TSP. One specific temperature sequence, the log cooling schedule, is discussed in the answer to the next challenge.

CHALLENGE 17.3.

(a) MATLAB provides a naïve Monte Carlo solution algorithm for the TSP. Given an ordering of the cities, it randomly generates a pair of cities and interchanges them if the interchange lowers the total distance. Experiment with the demonstration program `travel.m` and display the program using `type travel` to see how this works.

(b) Write a program for simulated annealing on the TSP problem and compare its performance with that of the algorithm of (a).

An Example of Monte Carlo Methods for Counting

Counting is so simple that we learned it as a young child. What could be so difficult about counting that would lead us to use Monte Carlo methods? It all depends on how many things we are asked to count!

Consider the lattice of Figure 17.3, and suppose it is the lattice of a crystal containing two types of molecules: **dimers**, which fill two adjacent sites, and **monomers**, which occupy only one site. Define $C(k)$ to be the number of distinct arrangements of k dimers

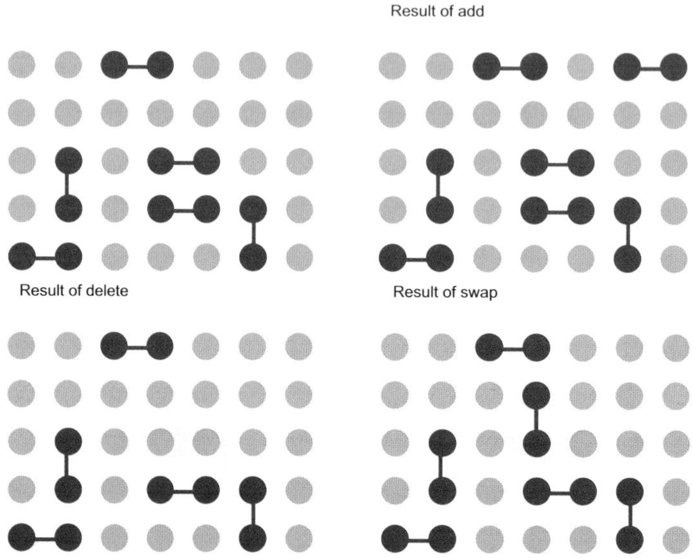

Figure 17.3. *Upper left: a lattice of* 6 *dimers (red) and* 23 *monomers (blue), used to illustrate the add, delete, and swap operations. Upper right: adding a dimer in row* 1*. Bottom left: deleting a dimer in row* 3*. Bottom right: swapping a horizontally oriented dimer in row three for a vertical one.*

on the lattice. (Clearly $C(0) = 1$ and $C(k)$ is nonzero only if k is an integer between 0 and half the number of sites in the lattice.) This function $C(k)$ gives the coefficients of the **generating function** for the problem.

For small lattices, we actually can count all the arrangements, and, in fact, there are analytic formulas for $C(k)$ for two-dimensional lattices. For many three-dimensional lattices, though, the function is not known exactly and computing it by counting all of the possibilities is too expensive. Therefore, Monte Carlo methods are used to estimate it.

The basic idea is to obtain a random arrangement of dimers, assume that all other sites are occupied by monomers, determine the number of dimers k, and add one to the count of occurrences of k dimers. We repeat this process as long as we can. We might find, for example, that we had 0 dimers in 5 of our trials and 3 dimers in 52 trials. Then, since we know that there is only 1 arrangement of 0 dimers on the lattice, our best guess at the number of arrangements of 3 dimers is $52/5 = 10.4$.

The difficulty is in randomly generating acceptable arrangements of dimers, ones in which the dimers do not overlap. In order to avoid generating mostly unacceptable arrangements, algorithms usually start with an acceptable arrangement and make a small change to it. The trouble with this is that the two arrangements are highly correlated, and therefore our estimate can have high variance. To fix this, we can make a large number of changes, enough so that the two arrangements are no longer correlated, before we record a count. One such algorithm is due to Kenyon, Randall, and Sinclair (KRS) and is given in Algorithm 17.2.

POINTER 17.3. Randomness vs. Correlation.

At each iteration of the KRS algorithm, either nothing happens or a dimer is added, deleted, or moved. If it has been long enough since our last recording, then the resulting configuration is added to the record. Why don't we record every configuration?

In order to get good estimates of the counts, the samples that we record must be (almost) uncorrelated. In other words, the probability of getting a particular sample must not depend on the value of the preceding sample.

It is **not** correct to say that we skip in order to get *random* samples. Some random samples are correlated and some are uncorrelated! For example, suppose we construct a string of numbers by starting with 0. To obtain the next entry in the string, we do this:

- If the latest entry is 0, the next entry is 0 with probability 0.8 and 1 with probability 0.2.
- If the latest entry is 1, the next entry is 0 with probability 0.3 and 1 with probability 0.7.

The entries in the string are *random*, because we don't make the same choices each time we generate a string. But the entries are *correlated*: for example, the probability of seeing the pattern 00 is much more likely than seeing the pattern 01.

Algorithm 17.2 KRS Algorithm

Start with a lattice of monomers.
repeat
  Choose an adjacent pair of sites.
  If both sites have monomers, add this dimer with some probability.
  If the sites are occupied by a single dimer, delete it with some probability.
  If one site is occupied with a dimer, swap that dimer for this one with some probability.
  If it has been long enough since our last recording, then add the resulting configuration to the record.
until enough steps are completed.

The add, delete, and swap moves for the KRS algorithm are illustrated in Figure 17.3. We'll use the KRS algorithm to estimate $C(k)$ for some simple lattices and compare it (on small lattices) with explicit counting. We see in solving the next challenge that although the problem is simple to state, the time for solution is quite large, so data structures must be carefully chosen to optimize efficiency.

CHALLENGE 17.4.

(a) Write a MATLAB function that computes the function $C(k)$ for an $n \times n$ lattice by explicit counting. Try to make it efficient.

(b) Implement the KRS algorithm in a MATLAB function and use it to estimate the function $C(k)$ for a 4×4 lattice. Try various choices of probabilities and updating intervals. Repeat

POINTER 17.4. Further Reading.

For background on probability, consult a standard textbook such as [109]. Simulated annealing is discussed in more detail by Kirkpatrick, Gelatt, and Vecchi [90].
More information on random counting algorithms and the estimation of generating functions can be found in papers by Kenyon, Randall, and Sinclair [89], Beichl and Sullivan [11], Caflisch [23], and Beichl, O'Leary, and Sullivan [10].

---

for a 6×6 lattice and then for a lattice as large as possible. Graph the function $C(k)$ for each lattice, and also report the sum of the $C(k)$ values for each lattice. Discuss the amount of work performed, the accuracy of your results, and how the work compares with that for explicit counting.

Chapter 18 / Case Study

Multidimensional Integration: Partition and Conquer

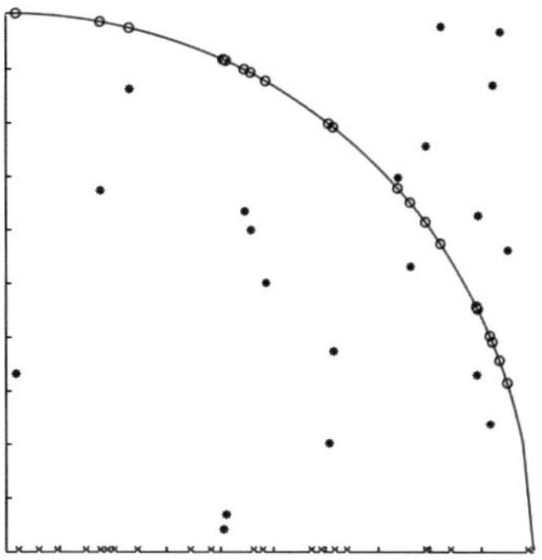

Numerical analysis textbooks provide excellent advice on approximating integrals over low-dimensional spaces, but many problems of interest are naturally posed in high dimensions and yield integrals over regions with a large number of variables. These problems arise, for example, in evaluating the failure rate of materials, the expected return on an investment, or the expected value of the energy of a system with a large number of particles. In this case study, we consider some algorithms appropriate for these problems, using a system of particles as a motivating example.

Understanding the behavior of particles subjected to forces is a basic theme in physics. The simplest system is a set of particles confined to motion along a line, but even this system presents computational challenges. For the **harmonic oscillator**, a particle is subjected to a force directed toward the origin and proportional to the distance between the particle and the origin. The resulting potential is $V(x) = -\frac{1}{2}\alpha x^2$, where α is a constant. This system is quite thoroughly understood, and quantities of interest can be computed in closed form. Alternatively, the Ginzburg-Landau anharmonic potential $-\frac{1}{2}\alpha x^2 + \frac{1}{4}\beta x^4$ (α and β constant) is related to solution of the Schrödinger equation, and quantities of interest are computed approximately. One method of obtaining such approximations is numerical integration, and that is our focus in this case study.

A system in thermodynamic equilibrium can be characterized by its **partition function**, an expression for the expected value of e raised to a power equal to the energy of the system divided by a normalization parameter kT. From the partition function, many quantities of interest can be computed (for example, the expected value of the energy, the **free energy**, the entropy, etc.), so it is a quite important function to compute.

Consider a chain of particles in which each particle interacts with its one or two neighbors in the chain. For this set of particles, the partition function is

$$Z_d(L) = \int_{-\infty}^{\infty} \rho_d(a, L)\, \mathrm{d}a, \tag{18.1}$$

where $L = 1/(\alpha T)$, T is the temperature, α is the Boltzmann constant, d is the number of particles, and ρ_d is

$$\rho_d(a,L) = \int_{-\infty}^{\infty}\int_{-\infty}^{\infty}\cdots\int_{-\infty}^{\infty} g(a,x_1,L)g(x_1,x_2,L)\cdots g(x_d,a,L)\,\mathrm{d}x_1\,\mathrm{d}x_2\ldots\mathrm{d}x_d,$$

where

$$g(x,y,L) = \frac{1}{\sqrt{2\pi\delta}}\exp\left(-\frac{1}{2\delta}(x-y)^2 - \frac{1}{2}\delta(V(x)+V(y))\right),$$

V is the potential, and $\delta = L/(d+1)$. Our goal in this chapter is to develop some algorithms for approximating $Z_d(L)$.

Numerical Integration Methods

Many methods produce excellent estimates of the value of the integral

$$I = \int_\Omega f(\mathbf{x})\,\mathrm{d}\mathbf{x}$$

when f is a smooth function of a single variable x and Ω is a finite interval $[a,b]$. For instance, we can partition the interval into small subintervals, construct a polynomial approximation to f in each subinterval by evaluating f at several points in the subinterval, approximate the integral over the subinterval by the integral of the polynomial, and sum these approximate values. This is the idea behind **Newton–Cotes** methods. If the function is slowly changing over some pieces of Ω, then we can reduce the error in the approximation by taking longer subintervals there and concentrating our work on regions where the function changes more rapidly.

For multidimensional integrals, however, the situation is much less satisfactory. The approach that is so successful in one dimension quickly becomes prohibitive. Suppose, for example, that $\mathbf{x} \in \mathcal{R}^{10}$ and Ω is the unit hypercube $[0,1]\times\cdots\times[0,1]$. Then, as an example, a polynomial of degree 2 in each variable would have terms of the form

$$x_1^{\square}x_2^{\square}x_3^{\square}x_4^{\square}x_5^{\square}x_6^{\square}x_7^{\square}x_8^{\square}x_9^{\square}x_{10}^{\square},$$

where the number in each box is 0, 1, or 2. So the polynomial has $3^{10} = 59{,}049$ coefficients, and we would need 59,049 function values to determine these. If we partition the domain by dividing each interval $[0,1]$ into 5 pieces, we make $5^{10} \approx 10$ million boxes, and we would need 59,049 function evaluations in each. Clearly, this method is expensive!

On the other hand, evaluating integrals this way is quite easy to program if we have a function `quad` that works for functions of a single variable, perhaps using a polynomial approximation method. Then we can use `quad` to compute

$$\int_0^1 h(x_1)\,\mathrm{d}x_1,$$

where

$$h(z) = \int_0^1\cdots\int_0^1 f(z,x_2,\ldots,x_{10})\,\mathrm{d}x_2\ldots\mathrm{d}x_{10},$$

POINTER 18.1. How to Use Nested Quadrature in MATLAB.

Suppose that we want to compute the volume of a half sphere with radius 1.

$$I = \int_{-1}^{1} \int_{-\sqrt{1-y^2}}^{\sqrt{1-y^2}} \sqrt{1 - x^2 - y^2}\, dx\, dy$$

We can accomplish this with nested calls to MATLAB's function quad using the following function definitions:

```
function I = nestedintegration()

I = quad(@ghat, -1, 1)


function z = ghat(y)

global ybar

% y is a vector of evaluation points from quad.

z = zeros(size(y));

for i = 1:length(y),
    ybar = y(i);

% Contribution to the integral is zero if (1-ybar^2) <= 0.

    if ((1-ybar^2)>0)
        z(i) = quad(@h, -sqrt(1-ybar^2), sqrt(1-ybar^2));
    end
end


function h = h(x)

global ybar

hsq = 1-x.^2-ybar^2;

% Set contributions from outside the domain to zero.

hsq = (hsq > 0).*hsq;
h = sqrt(hsq);
```

as long as we can evaluate $h(z)$ for any given value $z \in [0,1]$. But $h(z)$ is just an integration, so we can evaluate it using `quad`, too. We end up with 10 nested calls to `quad`. Again, this is very expensive but quite convenient! We illustrate this method in Pointer 18.1.

As an alternative, some functions $f(\boldsymbol{x})$ can be approximated quite well by a separable function $f(\boldsymbol{x}) \approx f_1(x_1) f_2(x_2) \dots f_{10}(x_{10})$. If so, we can approximate our integral by

$$I \approx \int_0^1 f_1(x_1)\,\mathrm{d}x_1 \dots \int_0^1 f_{10}(x_{10})\,\mathrm{d}x_{10}.$$

If this works, it is great, but we aren't often that lucky.

We need another option, since one of our methods is too expensive and the other is only special-purpose. If the function cannot be approximated well by a separable function, then the last resort is **Monte Carlo integration**.

Algorithm 18.1 Monte Carlo Volume Estimation

Let $\Gamma \subset \mathcal{R}^d$ be a set that contains Ω and has known volume J.
Generate n points $\{z^{(i)}\} \in \Gamma$ that are uniformly randomly distributed.
Let $\hat{n}$ be the number of these points that also lie in Ω, and estimate the volume of Ω by $\nu_n J$, where

$$\nu_n = \frac{\hat{n}}{n}.$$

Algorithm 18.2 Basic Monte Carlo Integration

Generate n points $\{z^{(i)}\} \in \Omega$ that are uniformly randomly distributed.
(For our 10-D hypercube example, this requires generating $10n$ random numbers, uniformly distributed in $[0,1]$.)
Then the average value of f in the region Ω is approximated by

$$\mu_n = \frac{1}{n} \sum_{i=1}^{n} f(z^{(i)}),$$

and therefore the value of the integral is

$$\int_\Omega f(\boldsymbol{x})\,\mathrm{d}\boldsymbol{x} = I \approx \mu_n \int_\Omega \mathrm{d}\boldsymbol{x}.$$

Monte Carlo Integration

There are two simple ways to use Monte Carlo integration. For the first, think of integration as computing the volume of some solid Ω, and embed that solid in a larger one Γ for which computing the volume is easy. (As a trivial example, we imbed the quarter circle Ω, illustrated in Figure 18.1, in a square Γ of side 0.5 in order to compute its area.) Then generate a sequence of random points in Γ, and determine the fraction ν of points that also

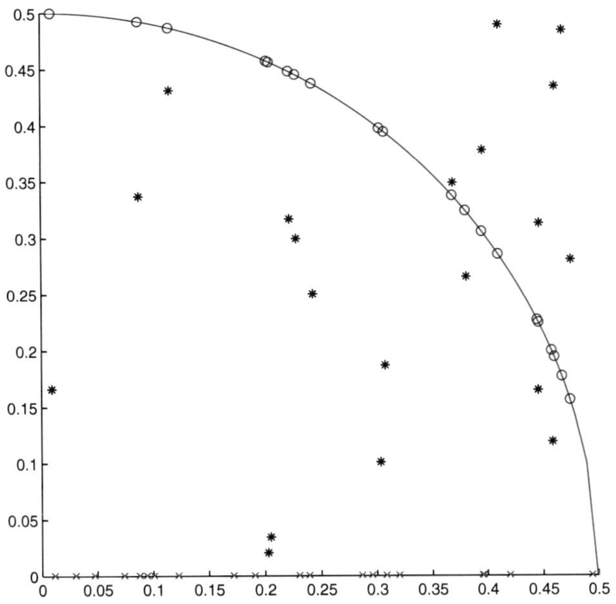

Figure 18.1. *Estimating the area* 0.19636 *of a quarter circle using Monte Carlo methods with* 20 *function evaluations. The volume estimator of Algorithm* 18.1 *with the random samples indicated in the figure gives* 13 *stars inside the circle and* 7 *outside, for an estimate of* $(13/20)(0.5)^2 = 0.163$. *The basic Monte Carlo algorithm of Algorithm* 18.2 *averages the* 20 *blue function values to give a somewhat better estimate,* 0.174. *Importance sampling in Algorithm* 18.3 *averages function values corresponding to the red x's to yield* 0.192.

lie in Ω. We can estimate the volume of Ω by ν times the volume of Γ. This gives us Algorithm 18.1.

Another viewpoint gives a somewhat better algorithm, one that obtains more information from function values. Again the idea is simple: according to the integral mean value theorem, the integral

$$I = \int_\Omega f(\boldsymbol{x})\, d\boldsymbol{x}$$

is equal to the average value of f on Ω times the volume of Ω, which is $\int_\Omega d\boldsymbol{x}$. This gives us a second method, Algorithm 18.2.

How good is the approximation I_n from Algorithm 18.2? The expected value of I_n is the true value of the integral, and this is very good! In fact, for large n, the estimates have an approximately normal distribution with variance σ^2/n, where

$$\sigma^2 = \int_\Omega (f(\boldsymbol{x}) - \mu_n)^2\, d\boldsymbol{x}.$$

Note that the variance is a constant that depends on the variation in f around its average value, but not on the dimension d of the integration domain Ω.

We illustrate the results of Algorithms 18.1 and 18.2 in Figure 18.1 and experiment with them in Challenge 18.1. We'll see how well these methods behave on a sample problem, finding the area of a quarter circle; but keep in mind that these methods are valuable for very high dimensional integrals, not the trivial one that we use in Challenge 18.1.

CHALLENGE 18.1.

Suppose we want to estimate the area of a quarter circle $\Omega = \{(x,y) : x^2 + y^2 \leq r^2, x \geq 0, y \geq 0\}$ with radius r. The area is the integral

$$I = \int_0^r \sqrt{r^2 - x^2}\,dx \equiv \int_0^r f(x)\,dx.$$

Let $r = 0.5$ and use two methods:

(a) Algorithm 18.1 with Ω equal to the quarter circle of radius r and Γ equal to the rectangle $[0,r] \times [0,r]$.

(b) Algorithm 18.2 applied to the integrand $f(x)$ and the interval $[0,r]$.

Compare the quality of the two estimates for 10, 100, 1000, 10000, and 100000 points by measuring the error and the convergence rate.

How many function evaluations does your favorite integration routine use to get an estimate of comparable quality?

CHALLENGE 18.2.

Write a MATLAB program to estimate the volume of the unit sphere $x_1^2 + x_2^2 + x_3^2 \leq 1$ using one of the methods for Monte Carlo integration.

Importance Sampling

The expected value of our estimate from either of the two methods is equal to the value we are looking for. But there is a nonzero variance to our estimate; we aren't likely to get the exact value of the integral. Most of the time, though, the value is close if n is big enough.

If we could reduce the variance of our estimate, then we could use a smaller value of n. This would mean less work, and it can be achieved by **importance sampling**.

Suppose that we want to estimate

$$I = \int_\Omega f(\boldsymbol{x})\,d\boldsymbol{x},$$

where Ω is a region in $\mathcal{R}^{10}$ with volume equal to one. Our Monte Carlo estimate of this integral involves taking uniformly distributed samples from Ω and taking the average value of $f(\boldsymbol{x})$ at these samples. We can improve on this by a good choice of a function $p(\boldsymbol{x})$ satisfying $p(\boldsymbol{x}) > 0$ for all $\boldsymbol{x} \in \Omega$, normalized so that

$$\int_\Omega p(\boldsymbol{x})\,d\boldsymbol{x} = 1.$$

POINTER 18.2. Facts about Importance Sampling.

Importance sampling is very good for decreasing the variance of the Monte Carlo estimates. In order to use it effectively,

- we need to be able to choose $p(x)$ appropriately;
- we need to be able to sample efficiently from the distribution with density $p(x)$.

Then

$$I = \int_{\Omega} \frac{f(\boldsymbol{x})}{p(\boldsymbol{x})} p(\boldsymbol{x}) \, \mathrm{d}\boldsymbol{x} .$$

Comparing this with our definition of expected value μ in Section 16.2, we see that this can be interpreted as the expected value of the function $f(\boldsymbol{x})/p(\boldsymbol{x})$ when the $\boldsymbol{x}$ are distributed as specified by $p(\boldsymbol{x})$. We can get a Monte Carlo estimate of this integral by taking samples from the distribution with probability density $p(\boldsymbol{x})$ and taking the average value of $f(\boldsymbol{x})/p(\boldsymbol{x})$ at these samples, and we call this **importance sampling**. One variant is defined in Algorithm 18.3 and illustrated in Figure 18.1.

Algorithm 18.3 Monte Carlo Integration by Importance Sampling

(Comment: Use this method only if $f(\boldsymbol{x})$ is nonnegative.)

Take a "few" samples of $f(\boldsymbol{x})$, and let $\hat{p}(\boldsymbol{x})$ be an approximation to $f(\boldsymbol{x})$ constructed from these samples. (For example, $\hat{p}(\boldsymbol{x})$ might be a piecewise constant approximation.)

Let $p(\boldsymbol{x}) = \hat{p}(\boldsymbol{x})/I_p$, where

$$I_p = \int_{\Omega} \hat{p}(\boldsymbol{x}) \, \mathrm{d}\boldsymbol{x}.$$

Generate points $z^{(i)} \in \Omega$, $i = 1, \dots, n$, distributed according to probability density function $p(\boldsymbol{x})$.

Then the average value of f/p in the region Ω is approximated by

$$\mu_n = \frac{1}{n} \sum_{i=1}^{n} \frac{f(z^{(i)})}{p(z^{(i)})},$$

and therefore the value of the integral is

$$I \approx \mu_n \int_{\Omega} p(\boldsymbol{x}) \, \mathrm{d}\boldsymbol{x} = \mu_n .$$

When is importance sampling better than basic Monte Carlo integration (Algorithm 18.2)? The variance of our estimate is proportional to

$$\sigma^2 = \int_\Omega \left(\frac{f(\boldsymbol{x})}{p(\boldsymbol{x})} - I \right)^2 p(\boldsymbol{x})\, d\boldsymbol{x}.$$

So if we choose p so that $f(\boldsymbol{x})/p(\boldsymbol{x})$ is close to constant, then σ is close to zero. (This probably won't be true if $f(\boldsymbol{x})$ changes sign on Ω.)

Intuitively, importance sampling works because in regions where $f(\boldsymbol{x})$ is big, $p(\boldsymbol{x})$ is also big, so there is a high probability that we sample from these regions that contribute most to the magnitude of I, even if the region itself is small.

The big unanswered question is how to get a good choice for $p(\boldsymbol{x})$. If f is nonnegative, then this can be done by sampling $f(\boldsymbol{x})$ at a few points and setting p to be piecewise constant, with value proportional to the sampled values. For our unit hypercube example, we could divide our domain into 3^{10} smaller cubes (with each side of length $1/3$) on which p is constant. If we evaluate our function f at the center of each of these, this gives a grid of 3^{10} values at meshpoints $\boldsymbol{w}_i$ with each component of $\boldsymbol{w}_i$ equal to $1/6$, $1/2$, or $5/6$. We set $p(\boldsymbol{x})$ in each cube to be the sampled value f at the center of the cube, divided by the sum of the sampled values. Then for our integration, among our n random values, we could choose the number of values in the ith cube to be $np(\boldsymbol{w}_i)$.

Let's repeat our experiment using importance sampling.

CHALLENGE 18.3.

Compute new estimates of the integral of Challenge 18.1 using Algorithm 18.3, taking 10 samples of $f(x)$ to determine the function $p(x)$. Compare with your previous results.

Using Quasi-Random Numbers

There are many uses of random numbers, and the properties that we need from them depend on the ultimate use that we intend. For example, simulation might require that the numbers be as independent of each other as possible, but in Monte Carlo integration, it is most important that the proportion of points in any region be proportional to the volume of that region. This property is actually better achieved by points that are correlated, and there have been many proposals for generating sequences of **quasi-random numbers** that guarantee a rather even distribution of points. For example, the van der Corput sequence generates the kth coordinate of the pth quasi-random number $\boldsymbol{w}_p$ in a very simple way. Let b_k be the kth prime number, so, for example, $b_1 = 2$, $b_2 = 3$, and $b_5 = 11$. Now write out the base-b_k representation of p,

$$p = \sum_i a_i b_k^i,$$

and set the coordinate to

$$w_{pk} = \sum_i a_i b_k^{-i-1}.$$

You might think that a regular mesh of points also has a uniform covering property, but it is easy to see (by drawing the picture) that large boxes are left with no samples at all if we choose a mesh. The van der Corput sequence, however, gives a sequence that rather uniformly covers the unit hypercube with samples, as we demonstrate in Challenge 18.4.

CHALLENGE 18.4.

Generate 500 pseudorandom points $\mathbf{v}_p$ in $\mathcal{R}^2$ and 500 quasi-random points $\mathbf{w}_p$ in $\mathcal{R}^6$ $(p = 1,\ldots,500)$. Plot the pseudorandom points. Then for the quasi-random points, make a plot of the first two coordinates, a separate plot of the third and fourth coordinates, and a final plot of the fifth and sixth coordinates. Discuss the desirability of using each of these four choices for "random" points.

How effective are quasi-random points in approximating integrals? For random points, the expected value of the error is proportional to $n^{-1/2}$ times the square root of the variance in f; for quasi-random points, the error is proportional to $V[f](\log n)^d n^{-1}$, where $V[f]$ is a measure of the variation of f, evaluated by integrating the absolute value of the dth partial derivative of f with respect to each of its variables, and adding on a boundary term. Therefore, if d is not too big and f is not too wild, then the result of Monte Carlo integration using quasi-random points probably has smaller error than using pseudorandom points.

CHALLENGE 18.5.

Compute new estimates of the integral of Challenge 18.1 using quasi-random numbers in Algorithm 18.2 instead of pseudorandom numbers. Compare the results.

Back to Our Partition Function

We now return to computing the partition function (18.1) that motivated our discussion.

CHALLENGE 18.6. (Extra)

(a) Let $L = 1$ and $\alpha = 1.38$ angstroms^2 g /sec^2 K and consider the harmonic oscillator potential $V(x) = -\frac{1}{2}\alpha x^2$. Determine finite integration limits for $a, x_1, \ldots, x_d$ so that the partition function $Z_d(L)$ in (18.1) can be approximated to 3 digit accuracy. (Do this by bounding the neglected part of the integral.)

(b) Use your favorite one-dimensional integration routine to estimate the partition function $Z_d(L)$. When you need a function value $\rho_d(a, L)$, use Monte Carlo integration to obtain it. Try $n = 100, 1000, 10000, 100000, 1000000$ (if possible) and $d = 1, 2, 4, 8, 16$.

POINTER 18.3. Further Reading.

The integration techniques we use here, as well as algorithms for generating quasi-random numbers, are discussed in [23]. Stratified sampling, also discussed there, is another tool for improving Monte Carlo estimates.

Beichl and Sullivan give a good introduction to importance sampling [11].

The partition function example is taken from [112], which also presents the divide-and-conquer approach to computing it. Monte Carlo integration is also used to compute integrals in rarefied gas dynamics [23], estimating the failure rate of materials [135], and quantum chromodynamics (QCD) [51].

For information on algorithms for approximating integrals in one-dimensional space, such as that used in MATLAB's `quad`, see a standard textbook such as [32, 148].

(c) Repeat the experiment using quasi-random points in the Monte Carlo integration. What can you say about the accuracy and convergence rate of your estimates?

Partition and Conquer

We noted earlier that if the function to be integrated is separable, then our problem reduces to a product of one-dimensional integrations. Although our partition function cannot be separated into $d+1$ factors, it can be partially separated. Notice that each variable x_i appears only in a pair of functions g, so it makes sense to accumulate

$$\hat{g}(x_{i-1}, x_{i+1}) = \int_{-\infty}^{\infty} g(x_{i-1}, x_i, L) g(x_i, x_{i+1}, L)\, \mathrm{d}x_i$$

for even values of i. If we repeat this trick on $\hat{g}$ for values of i that are multiples of 4, we reduce our problem further, and continuing, we can actually reduce the problem to a two-dimensional integration when d is a power of 2. This very nice observation is due to Nauenberg, Kuttner, and Furman [112]. Therefore, although the partition function provides a nice test problem for multidimensional integration, we should really compute it using this divide-and-conquer formulation.

Chapter 19 / Case Study

Models of Infection: Person to Person

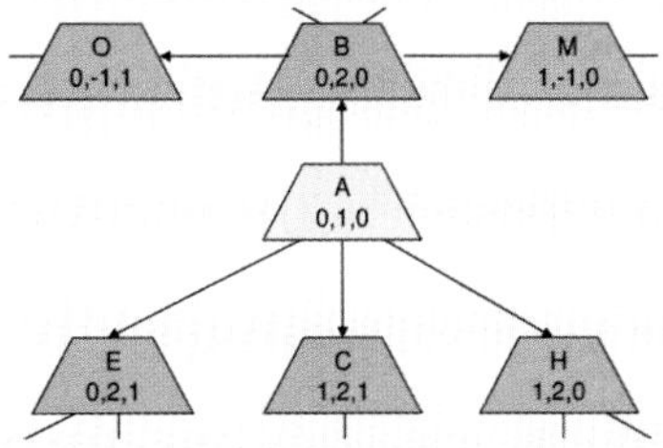

When faced with a spreading infection, public health workers would like to predict its path and severity in order to make decisions about vaccination strategies, quarantine policy, and use of public health resources. This is true whether dispersal of the pathogen is natural (as in SARS or the 1918 flu) or deliberate (for example, through terrorism). Effective mathematical models form a way to test the potential outcome of a public health policy and arrive at an effective response.

In this problem we focus on an overly simplified model of the spread of an infection and develop some tools that lend insight into its behavior.

To make our problem as easy as possible, we impose some rather artificial assumptions. Suppose that we have nm patients in a hospital ward and that their beds are arranged as n rows of m beds. Suppose that one of the patients, the one in the bed $\lceil m/2 \rceil$ in row $\lceil n/2 \rceil$, becomes infected and then can possibly infect patients in neighboring beds. How does this infection spread through the ward?

Insight through Monte Carlo Simulation

We'll need some parameters in our model. A patient, once infected, stays contagious for k days and then recovers, never to be infected again. During each day of infection, the probability that each susceptible neighbor (north, south, east or west) becomes infected is τ.

So there are three parts of the population to track. At day t, $I(t)$ is the proportion of the population that is infected and $S(t)$ is the proportion of the population that has never been infected. These quantities satisfy $0 \le I(t) \le 1$ and $0 \le S(t) \le 1$ for $t \ge 0$. The third part, $R(t)$, the proportion of the population that was once infected but has now recovered, can be derived from these two: $R(t) = 1 - I(t) - S(t)$.

We study this model by running a simulation. Each patient is in one of $k+2$ states: the patient has recovered from an infection, is susceptible to infection, or is in the ith day ($i = 1, \ldots, k$) of the k-day infection. It is convenient to use the integer values -1, 0, and

$1,\ldots,k$ to represent these different states. Each day, we update the status of each infected patient by incrementing the state of that patient (resetting it to -1 after k days), and for each susceptible neighbor, we generate a random number between 0 and 1; if that number is less than τ, then the neighbor's state is changed from 0 to 1, indicating infection. We continue this process until there are no infected patients; at that point, our model allows no possibility of any additional infections, so the epidemic ends.

Let's see how this model behaves.

CHALLENGE 19.1.

Run this model for $m = n = 10$, $k = 4$, $\tau = 0.2$ until there are no infected patients. Plot $I(t)$, $S(t)$, and $R(t)$ in a single graph.

If possible, display the epidemic as a movie. To do this, form a matrix of dimension $m \times n$, where the value of each entry corresponds to the state of the corresponding patient on a particular day. Using the `movie` command, we can display these matrices in sequence, day after day.

Since the model is stochastic, if we run it 10 times, we might get 10 different results, possibly ranging from no infections other than the original patient to infection of every patient, although these are both very low probability events. We need to investigate the variation in results, but first we'll add two complications.

The patients in our original model are immobile and only contact their four nearest neighbors. In most situations, population members move in more arbitrary ways. For example, epidemics jump from continent to continent by air or ship travel. In our hospital ward, let's assume that the nursing staff sometimes moves patients to other beds. For definiteness, we'll assume that each patient initiates a swap with probability δ. Then, at each time and for each patient we need to decide whether that patient initiates a swap. If so, we'll choose the bed indices for the second patient randomly, as $\lfloor r_2 n + 1 \rfloor, \lfloor r_2 m + 1 \rfloor$, where r_1 and r_2 are random samples from a uniform distribution on $[0,1]$.

CHALLENGE 19.2.

Modify your model to include mobility and run it for $\delta = 0.01$ until there are no infected patients. Display the results as in Challenge 19.1.

There are two major tools used to slow the spread of epidemics: quarantine, to isolate infected individuals, and vaccination, to protect susceptible individuals. To reduce the infections in our hospital model, infected individuals should be moved to a corner of the ward, with recovered individuals separating them from susceptible ones whenever possible. You can experiment with this quarantine strategy, but in the next problem we turn our attention to the use of vaccinations. For convenience, you might use the value -2 to indicate a vaccinated patient.

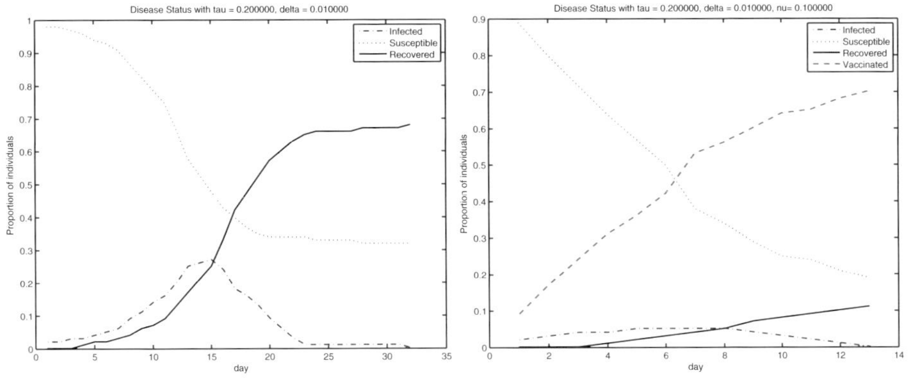

Figure 19.1. *Proportion of individuals (in one simulation) infected by day in a* 10×10 *grid of hospital beds, with infection rate* $\tau = 0.2$ *and mobility rate* $\delta = 0.01$ *and no vaccination (left) and with vaccination rate* $\nu = 0.1$ *(right).*

CHALLENGE 19.3.

Suppose that each day each susceptible individual has a probability ν of being vaccinated. Rerun your model with $\nu = 0.1$ until there are no infected patients. Display the results as in Challenge 19.1 and compare the results of the three models.

Some results from Challenges 19.2 and 19.3 are shown in Figure 19.1. Now we need to see how much variation is possible in the results if we run the model multiple times.

CHALLENGE 19.4.

Run the model of Challenge 19.3 1000 times, recording the number of individuals who become infected in each run. (Note that this is equal to the number of recovered individuals when the run is terminated.) Plot this data as a histogram (as in Figure 19.2), and compute the mean proportion of recovered individuals and the variance in this number. Try several different values of ν to see whether the variance changes.

From the results of Challenge 19.3, we can see that vaccinations can contain the spread of the epidemic. In Challenge 19.5, we take the role of a public health official trying to limit the spread of the epidemic.

CHALLENGE 19.5.

Develop a vaccination strategy that, on average, limits the epidemic to 20% of the population. Do this by using a nonlinear equation solver to solve the problem $R(\nu) - .2 = 0$, where $R(\nu)$ is the mean proportion of recovered individuals when we use a vaccination rate

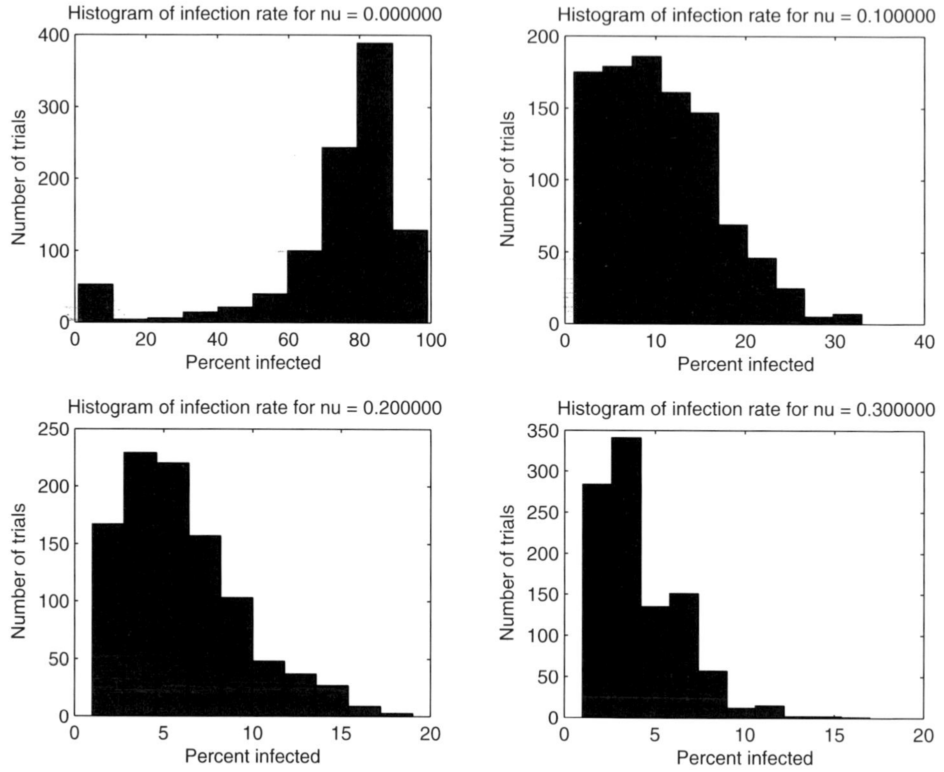

Figure 19.2. *Results of* 1000 *trials for a* 10×10 *grid of hospital beds, with infection rate* $\tau = 0.2$ *and vaccination rate* ν*, with* ν *varying.*

of ν. For each value of ν presented by the solver, you need to get a reliable estimate of R by running the model multiple times. Use the variance estimates from Challenge 19.4 to determine how many runs to use, and justify your choice.

A Markov Chain Model

The model we have developed has the **Markov property**: the status of each individual depends only on the status of the population on the previous day, and not on any older history. In fact, the system is a **Markov chain**. The **states** in the chain correspond to the possible statuses of the population; each state can be labeled $(d_1, \ldots, d_p)$, where there are p beds and d_i ranges from -2 to k, indicating that individual i $(i = 1, \ldots, p)$ is vaccinated, recovered, susceptible, or in day j $(1 \leq j \leq k)$ of the infection. There is an edge from one state to a second state if it is possible for the population to move from the first state to the second state the next day, and the weight on the edge is the probability of this happening. Figure 19.3 illustrates a Markov chain corresponding to three individuals in a single row of beds, with the middle one initially infected, a disease duration of $k = 2$ days, and no

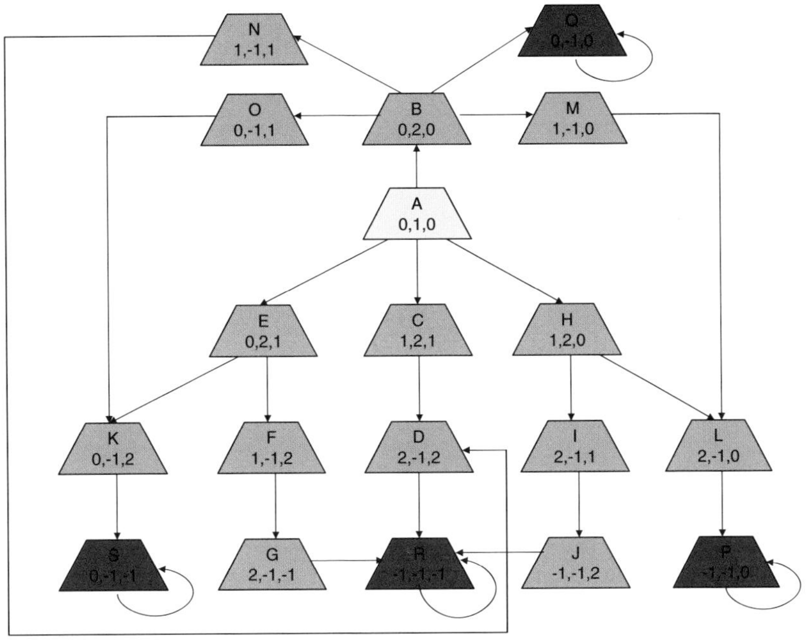

Figure 19.3. *A Markov chain that models three patients in a row of beds, with the middle patient infected and able to possibly infect two neighbors. The triple of numbers for each state gives the status of each of the three patients. The yellow state A is the state in which we start, and the red states are the possible outcomes when the infection runs its course, corresponding to* 1 *(state Q),* 2 *(states S and P), or* 3 *(state R) patients eventually infected.*

vaccination. (You will determine the edge weights in Challenge 19.6.) For this model, we are interested in three probabilities: the probability of terminating in

- state Q, corresponding to 33% of the population becoming infected,
- state R, corresponding to 100%,
- state P or state S, with the infection contained to 67% of the population.

CHALLENGE 19.6.

(a) Construct the **transition matrix** A corresponding to this Markov chain: element $a_{i,j}$ is the probability of transitioning to the ith state from the jth state and is therefore equal to the weight on that edge of the Markov chain in Figure 19.3.

(b) Let $\boldsymbol{e}_1$ be the first column of the identity matrix. If we begin in day 1 in the first state, then the vector $\boldsymbol{A}\boldsymbol{e}_1$ tells us the probabilities of being in each of the states on day 2. Prove this.

(c) Similarly, $A^2 e_1$ gives the probabilities for day 3. For efficiency, this should be computed as $A(Ae_1)$ rather than as $(A^2)e_1$. Explain why, by doing the operations counts.

(d) If we compute $z = A^j e_1$ for a large enough j, we have the (exact) probabilities of being in each state after the epidemic passes. Use this fact to compute the probabilities of having 1, 2, or 3 infected individuals, and compare these probabilities with the results of a Monte Carlo experiment performed as in Challenge 19.4 but using 3 individuals. How many Monte Carlo simulations does it take to get 2 digits of accuracy in the probabilities?

(e) In this simple problem, you can determine the three probabilities directly from Figure 19.3, by determining the probability of a transition from state A to states P, Q, R, and S. Show how these probabilities can be derived, giving the same answer as obtained by the Markov chain computation above.

In the preceding challenge, we estimated the probabilities of being in each state of a Markov chain at **steady state**, after a long time has passed. These estimates were formed by computing $A^j e_1$, where A is the transition matrix. The matrix A has several interesting properties, and it is worthwhile to take the time to verify them to see the relation between the steady state probabilities and the eigenproblem for A.

CHALLENGE 19.7.

Verify that the matrix A has the following properties:

- Its elements are all nonnegative, since they represent probabilities.
- The sum of the elements in every column is 1, since the transition probabilities out of a given state must sum to 1. We can write this as $e^T A = e^T$, where e is the vector with each component equal to 1.
- Therefore, e is a left eigenvector of the matrix A corresponding to eigenvalue $\lambda = 1$. So we can find the vector of steady state probabilities for a Markov chain by using any algorithm for solving the matrix eigenvalue problem. Any right-eigenvector z corresponding to $\lambda = 1$ is a steady state vector, since it satisfies $Az = z$.
- We can use the Gerschgorin theorem (See Pointer 5.4) to verify that no eigenvalue of A can be outside the unit circle.
- (More difficult) It turns out that for this problem, A has four eigenvalues equal to 1, and no other eigenvalue of A has magnitude 1. Using this fact, we can verify that the limit of the sequence $A^k e_1$ as $k \to \infty$ is a right eigenvector corresponding to the eigenvalue $\lambda = 1$. (You may assume that the matrix has a complete set of right eigenvectors $u_1, \ldots, u_n$ that form a basis for the space of $n \times 1$ vectors. Express e_1 as $\alpha_1 u_1 + \cdots + \alpha_n u_n$, where $\alpha_1, \ldots, \alpha_n$ are some coefficients, and then compute $A^k e_1$.)

Our Markov chain has an enormous number of states ($(k+3)^p$ for p patients), but many of these states provide more detail than we might need. For example, in Figure 19.3,

POINTER 19.1. Further Reading.

The 1918 flu epidemic killed more than 20 million people. Some investigators believe that it may have first taken hold in a US army base, but the disaster was worldwide, with millions killed in India alone. Travel of soldiers in World War I aided the spread of the infection. A book by Gina Kolata chronicles these events [94].

When doing Monte Carlo experiments, it is wise to use a high-quality pseudorandom number generator in order to get valid results. The classic reference for understanding such programs is Donald Knuth's book [92].

To go beyond the simple models investigated in this chapter, see, for example, the books by Britton [21] and by Hoppensteadt and Peskin [80] or the article by Zhuge [155].

Learn about Markov chain models and computing in the book by Stewart [142].

There are many approaches to aggregation of Markov chains. One starting point is an article by Marek [105].

---

if we are in state C, we always make the transition to state D, so these two states can be combined or **aggregated** without loss of information about infection totals. More importantly, states P and S represent different outcomes, but they are equivalent to us, since in each case 67% of the population becomes infected.

By aggregating states, we can reduce the size of the problem. Sometimes this can be done analytically, but when the model is too complicated (for instance, once mobility is added), we can do it by simulation, gathering data to determine the probability of transitions between aggregated states.

These simple Markov models can yield some insight into epidemics, but we have seen that the work of the Monte Carlo experiment, or the number of states in the original Markov chain, quickly grows with the size of the population. In the case study of Chapter 21 we investigate an alternative set of models.

Meanwhile, you might want to modify the models to explore more realistic variations. You also might consider how to model related systems, such as spread of a fungus in a tree farm, spread of contamination in a set of chicken coops, or spread of disease in a dormitory when the students also interact at school.

Unit V

Ordinary Differential Equations

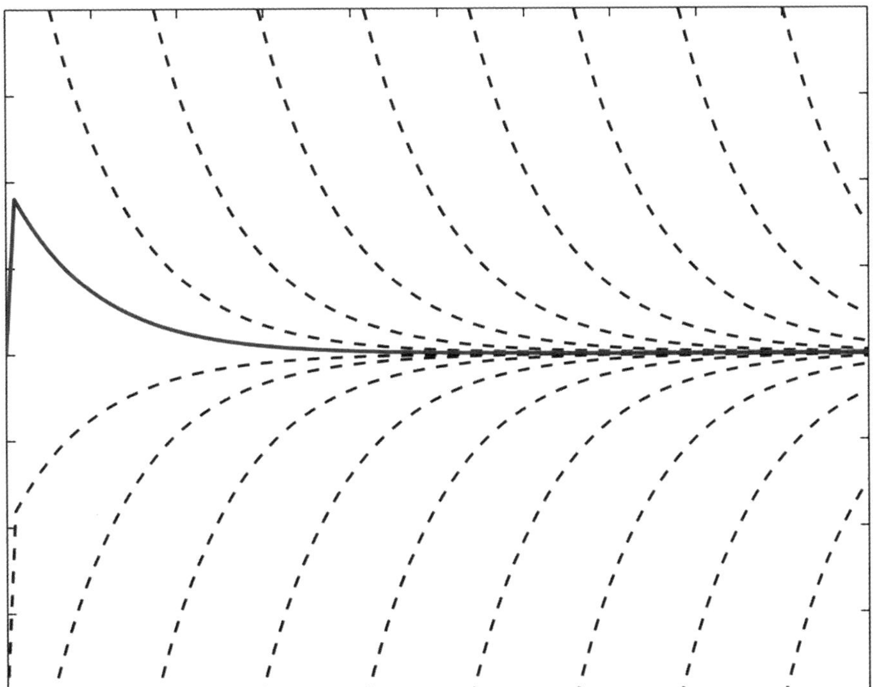

In this unit we study initial value problems (IVPs) and boundary value problems (BVPs) for ordinary differential equations (ODEs).

We first review basic theory and algorithms for ODEs in Section 20.1, and then discuss Hamiltonian systems. Then we discuss some basics and some numerical methods for differential-algebraic equations (DAEs) in Section 20.4. Section 20.5 focuses on boundary value problems and their solution using shooting methods and finite difference methods. In Chapter 21, we revisit the problem of modeling the spread of an epidemic, considered in Chapter 19; this time we model the problem using differential equations. More experience with stability and control of ODEs is given in Chapter 22, where we study a differential equation modeling motion of a robot arm, investigating its positioning using a nonlinear equation, and the energy required for positioning using optimization. Finite element methods for BVPs are discussed in Chapter 23.

BASICS: To understand this unit, the following background is helpful:

- The mean value theorem and Taylor series from calculus.
- Basic facts about differential equations. This information can be found in a basic textbook on ODEs or in some specialized numerical textbooks such as [75] or [99].
- Numerical solution of ordinary differential equations by Euler's method, discussed in basic scientific computing textbooks.

MASTERY: After you have worked through this unit, you should be able to do the following:

- Convert an ODE system to a "standard form" system.
- Use theorems to test whether there is a unique solution to an ODE.
- Determine whether an ODE is stable or stiff.
- Distinguish between local and global ODE error, estimate local error, and estimate a step length that keeps local error below a given tolerance.
- Derive the Euler method or the backward Euler method from Taylor series.
- Explain why the backward Euler method is sometimes more effective than the Euler method.
- Use Euler, backward Euler, or other Adams methods to solve differential equations.
- Plot ODE solutions vs. time, or make a phase plot of two components of a solution.
- Understand why predictor-corrector algorithms are used and be able to write one.
- Do order and stepsize control for an ODE method.
- Determine whether a numerical method for ODEs is stable.
- Know when to use the Gear methods instead of Adams or Runge–Kutta.

- Use a nonlinear equation solver to find a parameter in an ODE that satisfies a given condition.
- Define a Hamiltonian system and verify that a given system has a given Hamiltonian function.
- Incorporate a conservation principle into an ODE system by adding an extra variable.
- Put a DAE in standard form and use theorems to test whether a unique solution exists.
- Given an approximation to y' in terms of old function values, describe a numerical method for solving a DAE, using a finite difference formula to approximate y', and then using a nonlinear equation solver to compute y at the next timestep.
- Use theorems to test whether a unique solution to a BVP exists.
- Program a shooting method to solve a BVP.
- Formulate a delay differential equation.
- Use finite difference or finite elements to solve a BVP.
- Check Lyapunov stability of a system described by an ODE.
- Determine parameters in an ODE that minimize an energy function.

Chapter 20

Solution of Ordinary Differential Equations

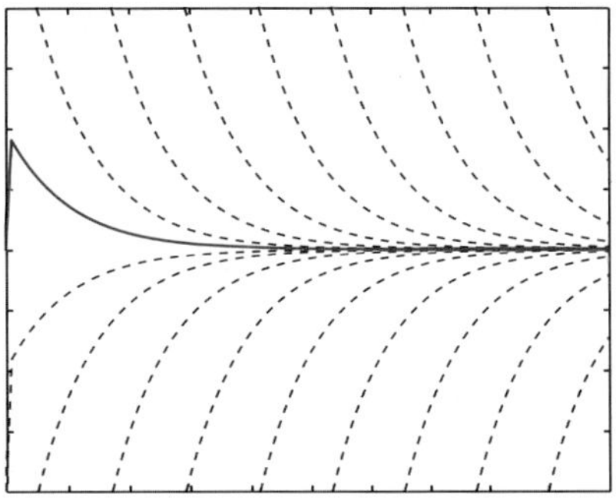

In this chapter we study some algorithms for solving various types of ordinary differential equation (ODE) problems. We'll consider problems in which data values are known at a single initial time (Section 20.1) as well as those for which values are known at two different points in time or space (Section 20.5). In between, we'll discuss problems that have algebraic constraints (Section 20.4).

The goal is to understand the characteristics of the problems and the methods well enough to choose an appropriate method and assess the results it returns.

The problems we consider are these:

Initial Value Problem (IVP) in standard form: Given a function $\boldsymbol{f} : [0,T] \times \mathcal{R}^m \to \mathcal{R}^m$, and an $m \times 1$ vector of initial values $\boldsymbol{y}_0$, find a function $\boldsymbol{y} : [0,T] \to \mathcal{R}^m$ satisfying

$$\boldsymbol{y}'(t) = \boldsymbol{f}(t, \boldsymbol{y}(t)), \tag{20.1}$$

$$\boldsymbol{y}(0) = \boldsymbol{y}_0 . \tag{20.2}$$

Higher-Order Initial Value Problem: Given a function $\boldsymbol{g} : [0,T] \times \mathcal{R}^{Km} \to \mathcal{R}^m$, and $m \times K$ initial values $\boldsymbol{u}_0$, find a function $\boldsymbol{u} : [0,T] \to \mathcal{R}^m$ satisfying

$$\boldsymbol{u}^{(K)}(t) = \boldsymbol{g}(t, \boldsymbol{u}(t), \boldsymbol{u}'(t), \ldots, \boldsymbol{u}^{(K-1)}(t)),$$

$$\boldsymbol{u}^{(j)}(0) = \boldsymbol{u}_0(:, j), \qquad j = 0, \ldots, K-1,$$

where $\boldsymbol{u}^{(j)}$ denotes the jth derivative of $\boldsymbol{u}$.

Boundary Value Problem (BVP): Given a function $\boldsymbol{f} : [0,T] \times \mathcal{R}^m \times \mathcal{R}^m \to \mathcal{R}^m$, and two $m \times 1$ vectors of values $\boldsymbol{u}_0$ and $\boldsymbol{u}_T$, find a function $\boldsymbol{u} : [0,T] \to \mathcal{R}^m$ satisfying

$$\boldsymbol{u}''(t) = \boldsymbol{f}(t, \boldsymbol{u}, \boldsymbol{u}'),$$

$$\boldsymbol{u}(0) = \boldsymbol{u}_0 ,$$

$$\boldsymbol{u}(T) = \boldsymbol{u}_T .$$

POINTER 20.1. Existence and Uniqueness of Solutions to IVPs for ODEs.

Before computing a numerical approximation to the solution to our ODE, it is important to know that such a solution exists! One condition guaranteeing existence and uniqueness of a continuous and differentiable solution on the interval $[0,a]$ is that the function $\boldsymbol{f}$ be **Lipschitz continuous**, meaning that there exists a constant L so that for all points t in $[0,a]$ and for all points $\boldsymbol{y}$ and $\widehat{\boldsymbol{y}}$ we have the bound

$$\|\boldsymbol{f}(t,\boldsymbol{y})-\boldsymbol{f}(t,\widehat{\boldsymbol{y}})\| \leq L\|\boldsymbol{y}-\widehat{\boldsymbol{y}}\|.$$

See standard textbooks such as [26, p. 251] for this and other existence results.

Differential-Algebraic Equation (DAE) in standard form: Given functions $\boldsymbol{M} : [0,T] \to \mathcal{R}^{m\times m}$, $\boldsymbol{A} : [0,T] \to \mathcal{R}^{m\times m}$, $\boldsymbol{f} : [0,T] \to \mathcal{R}^m$, and appropriate initial values $\boldsymbol{y}(0)$, find a function $\boldsymbol{y} : [0,T] \to \mathcal{R}^m$ satisfying

$$\boldsymbol{M}(t)\boldsymbol{y}'(t) = \boldsymbol{A}(t)\boldsymbol{y}(t) + \boldsymbol{f}(t).$$

20.1 Initial Value Problems for Ordinary Differential Equations

In this section, we review some standard ideas in the numerical solution of ODEs and discuss Hamiltonian systems.

20.1.1 Standard Form

We only work with initial value problems in **standard form**, as in (20.1), because that is what most software needs. Note that $\boldsymbol{y}'$ means the derivative with respect to t. In order to make sure there is no confusion between components of a vector and partial derivatives, in this chapter we denote the components of the vector $\boldsymbol{y}(t)$ by $y_{(j)}(t)$, so that writing our equations component by component yields

$$\begin{aligned} y'_{(1)}(t) &= f_1(t, y_{(1)}(t), \dots, y_{(m)}(t)), \\ &\vdots \\ y'_{(m)}(t) &= f_m(t, y_{(1)}(t), \dots, y_{(m)}(t)), \end{aligned}$$

with $y_{(1)}(0), \dots, y_{(m)}(0)$ given numbers.

It is not essential that we start at $t = 0$; any value is fine.

One famous system of ODEs is **Volterra's predator/prey model**. We can think of this as modeling a population of foxes, who eat rabbits, coexisting with a population of rabbits with an infinite food supply. If we denote the population of rabbits at time t by $r(t)$ and the population of foxes by $f(t)$, then the model is

$$\frac{\mathrm{d}r(t)}{\mathrm{d}t} = 2r(t) - \alpha r(t) f(t),$$

$$\frac{\mathrm{d}f(t)}{\mathrm{d}t} = -f(t) + \alpha r(t) f(t),$$

$$r(0) = r_0,$$
$$f(0) = f_0.$$

The parameter α is the **encounter factor**, with $\alpha = 0$ meaning no interaction between rabbits and foxes.

Let $y_{(1)}(t) = r(t)$ and $y_{(2)}(t) = f(t)$. Then we can write this ODE system in standard form:

$$y'_{(1)}(t) = 2y_{(1)}(t) - \alpha y_{(1)}(t) y_{(2)}(t),$$
$$y'_{(2)}(t) = -y_{(2)}(t) + \alpha y_{(1)}(t) y_{(2)}(t).$$

The solution is not unique until we specify α and the initial population of rabbits and foxes.

CHALLENGE 20.1.

(a) Use one of MATLAB's solvers (See Pointer 20.2) to explore the behavior of the Volterra model for various values of α. Suppose that initially the population of rabbits is 20 and the population of foxes is 10. Graph the solutions for $0 \le t \le 2$ for $\alpha = 10^{-2}, 10^{-1}$, and 1.

(b) Suppose we want to find a value of α so that the final population of rabbits is 4:

$$r_\alpha(2) - 4 = 0.$$

Note that the result of our experiment in (a) tells us that there is such a value of α between 0.01 and 1. Write a program that uses `ode45` along with the nonlinear equation solver `fzero` to find α.

(c) Repeat part (b), using the "Events" option in the solver to find α.

Higher-order initial value problems, those involving derivatives of order greater than first, can be converted to standard form. For example, suppose

$$u''(t) = g(t, u, u'),$$

where g is a given function. Let $y_{(1)}(t) = u(t)$ and $y_{(2)}(t) = u'(t)$. Then

$$u'(t) = y'_{(1)}(t) = y_{(2)}(t),$$
$$u''(t) = y'_{(2)}(t) = g(t, y_{(1)}, y_{(2)}),$$

and we have a system in standard form.

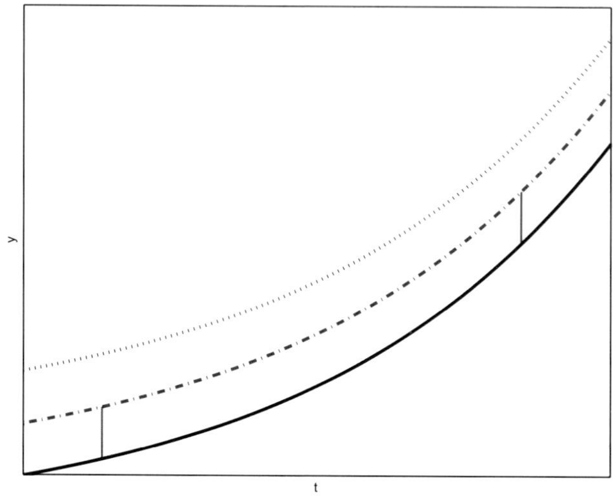

Figure 20.1. *Parallel solutions to an* ODE. *This graph shows three solutions to the same* ODE *but with different initial values* y_0. *This* ODE *is on the stability boundary.*

POINTER 20.2. ODE **Solvers in MATLAB.**

MATLAB has several solvers for differential equations, including:

`ode23`: Solves a nonstiff ODE using a low-order method.
`ode45`: Solves a nonstiff ODE using a medium-order method.
`ode23s`: Solves a stiff ODE using a low-order method.

The ODE solvers can do more complicated things, too; read the documentation carefully.

20.1.2 Solution Families and Stability

Given $\mathbf{y}'(t) = \mathbf{f}(t, \mathbf{y}(t))$, the **family of solutions** is the set of all functions $\mathbf{y}$ that satisfy this equation.

Let's consider three examples:

	ODE	Solution	Jacobian matrix	Illustration
Example 1.	$y'(t) = e^t$	$y(t) = c - e^t$	$J(t) = 0$	Figure 20.1
Example 2.	$y'(t) = -y(t)$	$y(t) = ce^{-t}$	$J(t) = -1$	Figure 20.2
Example 3.	$y'(t) = y(t)$	$y(t) = ce^t$	$J(t) = 1$	Figure 20.3

In the table, c is an arbitrary constant, determined by the initial conditions, and the **Jacobian matrix $\mathbf{J}$** is an $m \times m$ matrix with elements

$$J_{kj}(t, \mathbf{y}(t)) = \frac{\partial f_k(t, \mathbf{y})(t)}{\partial y_j}, \qquad k = 1, \ldots, m, \ j = 1, \ldots, m.$$

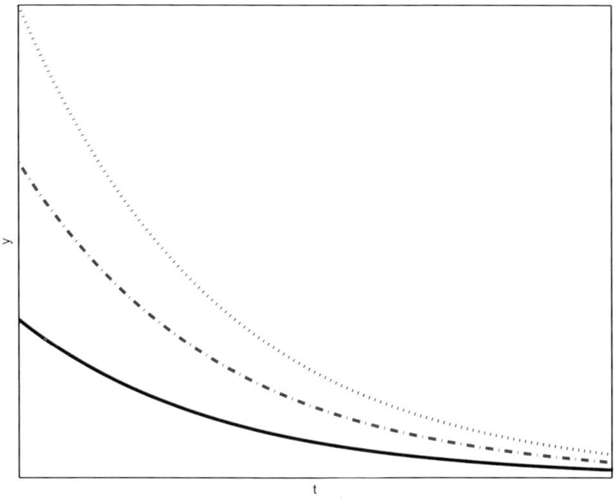

Figure 20.2. ODE *solutions that get closer to each other over time. The* ODE *is stable.*

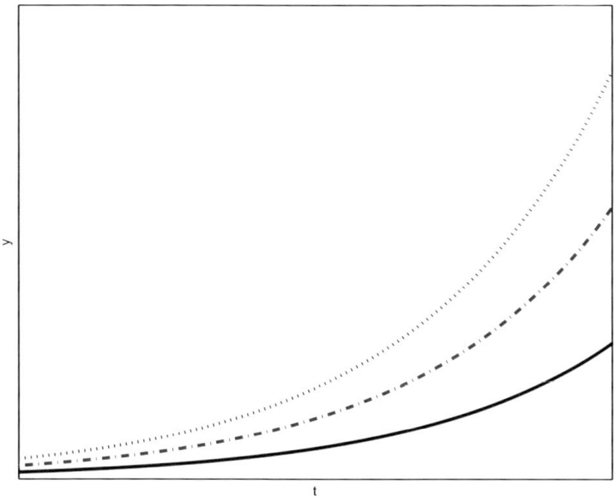

Figure 20.3. ODE *solutions that get further from each other. This* ODE *is unstable.*

For example, if

$$y'(t) = \begin{bmatrix} 2y_{(1)}(t) + 3y_{(2)}^2(t) \\ -6t + 7y_{(2)}(t) \end{bmatrix},$$

then

$$J(t,y) = \begin{bmatrix} 2 & 6y_{(2)}(t) \\ 0 & 7 \end{bmatrix}.$$

The Jacobian matrix of a single ODE is 1×1, hence scalar, as shown in the table above. In this case,

$$J(t,y) = f_y(t,y(t)) = \partial f(t,y)/\partial y.$$

We say that a single ODE is

- **stable** at a point $(\hat{t}, \hat{y}(\hat{t}))$ if $f_y(\hat{t}, \hat{y}(\hat{t})) < 0$,
- **unstable** at a point $(\hat{t}, \hat{y}(\hat{t}))$ if $f_y(\hat{t}, \hat{y}(\hat{t})) > 0$,
- **stiff** at a point $(\hat{t}, \hat{y}(\hat{t}))$ if $f_y(\hat{t}, \hat{y}(\hat{t})) << 0$.

A system of ODEs is

- **stable** at a point $(\hat{t}, \hat{\mathbf{y}}(\hat{t}))$ if the real parts of all the eigenvalues of the matrix $\mathbf{J}(\hat{t}, \hat{\mathbf{y}}(\hat{t}))$ are negative,
- **stiff** at a point $(\hat{t}, \hat{\mathbf{y}}(\hat{t}))$ if the real parts of more than one eigenvalue of $\mathbf{J}(\hat{t}, \hat{\mathbf{y}}(\hat{t}))$ are negative and wildly different. In this case, the solution to the IVP is determined by processes occurring on time scales that are radically different from each other, and this forces numerical methods to take small timesteps in order to follow the solution. Stiff problems are common, for example, in chemical reactions and weather prediction.

Thus, stability is determined by the Jacobian matrix, which can change over time, so a system can be stable for one time and unstable for another.

CHALLENGE 20.2.

Suppose we have a system of differential equations $\mathbf{y}'(t) = \mathbf{f}(t, \mathbf{y}(t))$ with 3 components and a Jacobian matrix $\mathbf{J}(t, \mathbf{y}(t))$ with eigenvalues $4 - t^2$, $-t - it$, and $-t + it$, where $i = \sqrt{-1}$. For what values of t is the equation stable?

To illustrate the effects of stability, consider

$$\mathbf{y}'(t) = \mathbf{A}\mathbf{y}(t) = \begin{bmatrix} a_{11} & a_{12} \\ a_{21} & a_{22} \end{bmatrix} \mathbf{y}(t).$$

The solution is

$$\mathbf{y}(t) = b_1 e^{\lambda_1 t} \mathbf{z}_1 + b_2 e^{\lambda_2 t} \mathbf{z}_2,$$

where λ_j and $\mathbf{z}_j$ are eigenvalues and eigenvectors of $\mathbf{A}$:

$$\mathbf{A}\mathbf{z}_1 = \lambda_1 \mathbf{z}_1,$$
$$\mathbf{A}\mathbf{z}_2 = \lambda_2 \mathbf{z}_2.$$

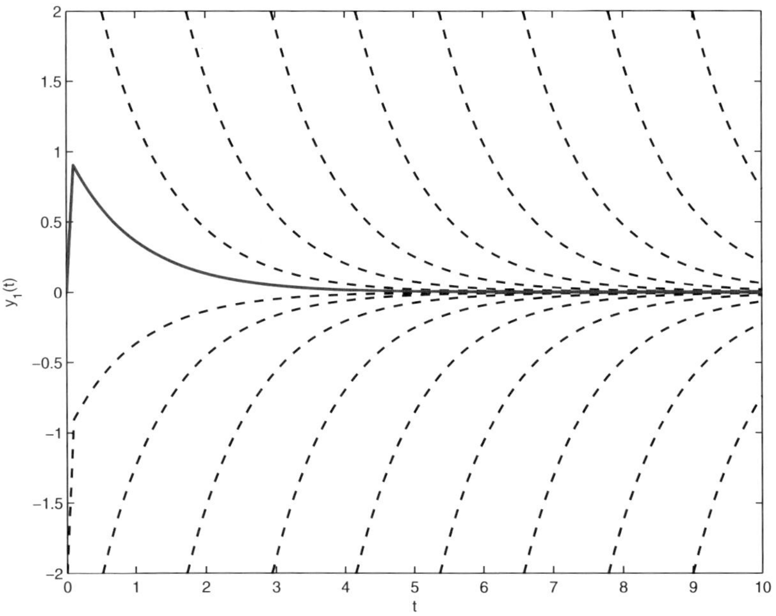

Figure 20.4. *Solution to the example stiff* ODE *problem. The red curve is the desired solution, and the blue dashed curves are in the same solution family; they solve the same* ODE *with different initial values.*

Let

$$\boldsymbol{A} = \begin{bmatrix} -1001 & 999 \\ 999 & -1001 \end{bmatrix}.$$

Then $\boldsymbol{y}'(t) = \boldsymbol{A}\boldsymbol{y}(t)$ has the Jacobian $\boldsymbol{J} = \boldsymbol{A}$ with eigenvalues $\lambda = -2, -2000$, so this system is stiff. Why does this trouble us? The solution is

$$\boldsymbol{y}(t) = b_1 e^{-2t} \begin{bmatrix} 1 \\ 1 \end{bmatrix} + b_2 e^{-2000t} \begin{bmatrix} -1 \\ 1 \end{bmatrix},$$

where b_1 and b_2 are constants determined from the initial conditions. If $y_{(1)}(0) = 0$ and $y_{(2)}(0) = 2$, then

$$b_1 = 1, \; b_2 = 1,$$

and the solution is shown in Figure 20.4. Notice that even after our solution is nearly constant, other solutions in the family are changing rapidly. This causes trouble for numerical methods since they introduce small perturbations in the solution and may jump from one member of the solution family to another, as we will see in Figure 20.7. We can cope with this difficulty of rapidly changing solutions by using an ODE solver designed to handle stiff systems.

Note that stability for an ODE is the same as conditioning. Consider, for example, the stable system $y'(t) = -y(t)$ with $y_0 = 1$. Then the solution is $y(t) = e^{-t}$. A small perturbation in initial condition to $y_0 = 1 + \epsilon$ changes the solution to $y(t) = (1 + \epsilon)e^{-t}$,

POINTER 20.3. Stability vs. Conditioning.

The terminology used here differs somewhat from that discussed in Chapter 1. Usually we say that a problem is well-conditioned, not stable, if small changes in data lead to small changes in the result. We use the term stability to refer to algorithms. Since the use of the term stability in ODEs is older than the term conditioning, the standard terminology for ODEs violates our convention.

Thus, stability of solution families is different from stability of numerical algorithms used to solve them. Note that an algorithm that solves a well-conditioned (stable) problem may or may not be numerically stable.

---

so the problem is well-conditioned. On the other hand, the unstable system $y'(t) = y(t)$ with $y_0 = 1$ has the exact solution $y(t) = e^t$. A small perturbation to $y_0 = 1 + \epsilon$ changes the solution to $y(t) = (1+\epsilon)e^t$. Thus an arbitrarily small change in input leads to a very different result for large t, and the problem is ill-conditioned. We'll consider other ways to measure stability in the case study of Chapter 22. Meanwhile, we investigate how a basic solution algorithm, Euler's method, works on stable and unstable problems.

20.2 Methods for Solving IVPs for ODEs

Now that we understand the characteristics of solutions to IVPs for ODEs, we introduce some methods for their numerical solution.

20.2.1 Euler's Method, Stability, and Error

We use Euler's method to illustrate the basic ideas behind numerical methods for ODEs. It is not a practical method because it is too slow, but it helps us understand the ideas. Three derivations of the method yield insight. Suppose that for some value t_n, we know the values $y_n = y(t_n)$ and $f_n = f(t_n, y_n)$. (Initially, $n = 0$, $t_0 = 0$, and we have the required data.) We use this information to step to a new point $t_{n+1} = t_n + h$.

Approach 1: Geometry

Euler's method can be derived from the left illustration in Figure 20.5. We have a function value y_n and a derivative value $f_n = f(t_n, y_n)$ at t_n. From these we obtain an approximation y_{n+1} by starting at y_n and moving along a line with a slope of f_n to $t = t_{n+1}$.

Approach 2: Polynomial interpolation

Euler's method can also be derived by determining the linear polynomial that matches the function value and derivative of y at t_n, and then evaluating it at $t = t_{n+1}$.

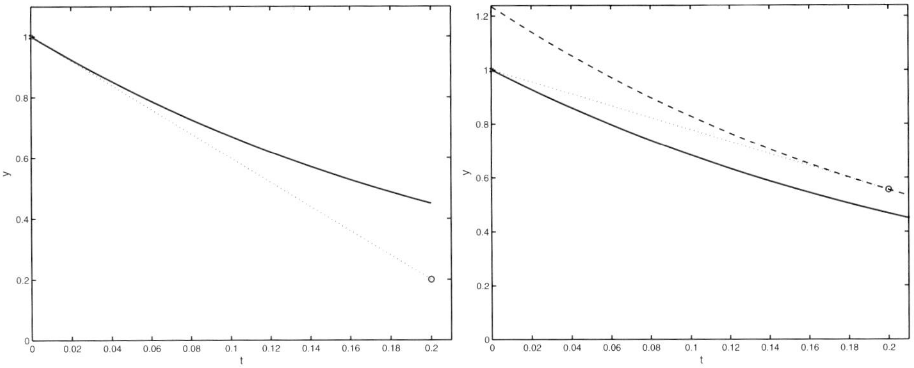

Figure 20.5. *Euler's method (left) and the backward Euler method (right) for solving an* ODE. *On the left, the blue curve is the true solution. An Euler step from $t = 0$, $y(0) = 1$ with stepsize $h = 0.2$ steps to the point marked with a circle. The red dotted line is tangent to the solution curve at $t = 0$, $y(0) = 1$, so its slope is $f(0, y(0))$. On the right, a backward Euler step from $t = 0$, $y(0) = 1$ with stepsize $h = 0.2$ steps to the point marked with a circle. The red dotted line passes through the point $t = 0$, $y(0) = 1$ and is tangent at $t = h$ to a solution curve (black dashed) for the same differential equation but with different initial condition.*

CHALLENGE 20.3.

Find the linear polynomial that passes through the point data (t_n, y_n) with slope f_n, and then evaluate it at t_{n+1}.

Approach 3: Taylor series

From Taylor series we know that if y is smooth enough, then

$$y(t+h) = y(t) + hy'(t) + \frac{h^2}{2}y''(\xi)$$

for some point $t \le \xi \le t+h$. Since

$$y'(t) = f(t, y(t)),$$

this gives the approximation

$$y_{n+1} = y_n + hf_n,$$

where $h = t_{n+1} - t_n$. The difference between $y(t+h)$ and y_{n+1} is $O(h^2)$. It takes $1/h$ steps of Euler's method with stepsize h to walk from t to $t+1$, so the error per unit time is $O(h^1)$. Therefore, we say that Euler is a **first-order method**.[9]

However we derive it, we define Euler's method in Algorithm 20.1.

[9]The order of a method is always one less than the exponent of h in the error formula for a single step.

Algorithm 20.1 Euler's Method

Given: y_0 and $t_0, t_1, \ldots, t_N$.
for $n = 0, \ldots, N-1$,
  $h_n = t_{n+1} - t_n$.
  $f_n = f(t_n, y_n)$.
  $y_{n+1} = y_n + h_n f_n$.
end

CHALLENGE 20.4.

Apply Euler's method to $y'(t) = 1$, $y(0) = 0$, using a stepsize of $h = .1$ to obtain approximate values for $y(0.1), y(0.2), \ldots, y(1.0)$.

Stability of Euler's method and the backward Euler method

In using a method like Euler's for solving ODEs, there are three sources of error:

- rounding error: especially if the steps get too small.
- local error: the error introduced assuming that y_n is the true value.
- global error: how far we have strayed from our original solution curve, assuming no rounding error.

Consider a single ODE. For Euler's method, Taylor series (Approach 3) tells us that if y_n is correct, then

$$y(t_{n+1}) - y_{n+1} = \frac{h_n^2}{2} y''(\xi),$$

where $h_n = t_{n+1} - t_n$. Therefore the local error is $\frac{h_n^2}{2} y''(\xi)$, which converges to zero as $h_n \to 0$.

But as we iterate with Euler's method, y_n has some error, so from the relations

$$y_{n+1} = y_n + h_n f(t_n, y_n)$$

and

$$y(t_{n+1}) = y(t_n) + h_n f(t_n, y(t_n)) + \frac{h_n^2}{2} y''(\xi_n),$$

we obtain the error formula

$$y(t_{n+1}) - y_{n+1} = y(t_n) - y_n + h_n(f(t_n, y(t_n)) - f(t_n, y_n)) + \frac{h_n^2}{2} y''(\xi_n).$$

In this expression, the plum-colored terms are global errors at times t_n and t_{n+1}, and the red term is the local error. See Figure 20.6 for an illustration of local and global errors in an ODE.

Using the mean value theorem, the blue term can be written as

$$h_n(f(t_n, y(t_n)) - f(t_n, y_n)) = h_n f_y(\eta)(y(t_n) - y_n),$$

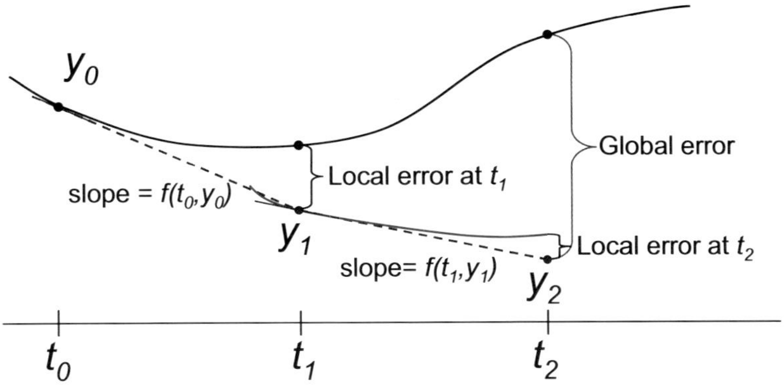

Figure 20.6. *Global errors accumulate. If we start on the solution curve at* y_0 *and take an Euler step, we obtain the value* y_1*, which is on a different curve in the solution family, and the local error is proportional to the stepsize squared. A second step takes us to* y_2*, which is on yet another solution curve. The error in moving from* y_1 *to* y_2 *is still proportional to the stepsize squared, but the global error can be much larger than the sum of the two local errors.*

where η is some point between t_n and t_{n+1}. Thus we have the expression

$$(\text{global error})_{n+1} = (1 + h_n f_y(\eta))\,(\text{global error})_n \; + (\text{local error})_n.$$

Therefore, the global errors are magnified if

$$|1 + h_n f_y(\eta)| > 1,$$

and we say that the Euler method is **unstable** in this case. The **stability interval for Euler's method**, the values of h_n for which the errors are not magnified at a given value of t, is defined by

$$-2 < h_n f_y(t) < 0$$

for a single equation. For a system of equations, the stability region contains the values of h_n for which the eigenvalues of $\boldsymbol{I} + h_n \boldsymbol{J}(t, y(t))$ are in the unit circle.

In contrast, in the method called the **backward Euler method** we define $\boldsymbol{y}_{n+1}$ as the solution to the equation

$$\boldsymbol{y}_{n+1} = \boldsymbol{y}_n + h\boldsymbol{f}(t_{n+1}, \boldsymbol{y}_{n+1}).$$

The geometry of the method is illustrated on the right in Figure 20.5, and we try the algorithm, summarized in Algorithm 20.2, in the next challenge.

Algorithm 20.2 The Backward Euler Method

Given: $\boldsymbol{y}_0$ and $t_0, t_1, \ldots t_N$.
for $n = 0, \ldots, N-1$
  Let $h_n = t_{n+1} - t_n$.
  Solve for $\boldsymbol{y}_{n+1}$ in the nonlinear equation

$$\boldsymbol{y}_{n+1} = \boldsymbol{y}_n + h_n f(\boldsymbol{t}_{n+1}, \boldsymbol{y}_{n+1}).$$

end

CHALLENGE 20.5.

Let $y'(t) = -y(t)$. Show that the backward Euler method computes

$$y_{n+1} = y_n - h_n y_{n+1}.$$

Solve this equation for y_{n+1} and compute several iterates for $y(0) = 1$ and $h_n = 0.1$.

To determine the error and stability for the backward Euler method, note that Taylor series tells us that

$$\boldsymbol{y}(t) = \boldsymbol{y}(t+h) - h\boldsymbol{y}'(t+h) + \frac{h^2}{2}\boldsymbol{y}''(\xi),$$

where $\xi \in [t, t+h]$. Thus Taylor series says that the local error is $\frac{h^2}{2}\boldsymbol{y}''(\xi)$. This is a first-order method, just like Euler's method.

To determine the stability of the backward Euler method we again consider an ODE with a single equation. We know that

$$y(t_{n+1}) = y(t_n) + h_n f(t_{n+1}, y(t_{n+1})) + \frac{h_n^2}{2} y''(\xi_n), \qquad n = 0, 1, 2, \ldots,$$

and

$$y_{n+1} = y_n + h_n f(t_{n+1}, y_{n+1}).$$

Therefore,

$$y(t_{n+1}) - y_{n+1} = y(t_n) - y_n + h_n(f(t_{n+1}, y(t_{n+1})) - f(t_{n+1}, y_{n+1})) + \frac{h_n^2}{2} y''(\xi_n).$$

Again, the plum-colored terms are global errors, and the red term is the local error. Using the mean value theorem, the blue term can be written as

$$h_n(f(t_{n+1}, y(t_{n+1})) - f(t_{n+1}, y_{n+1})) = h_n f_y(\eta)(y(t_{n+1}) - y_{n+1}).$$

Thus we have the expression

$$(1 - h_n f_y(\eta))\ (\text{global error})_{n+1} = (\text{global error})_n + (\text{local error})_n.$$

Therefore, the global errors are magnified if

$$|1 - h_n f_y(\eta)| < 1,$$

and we say that the backward Euler method is unstable in this case. The stability interval for the backward Euler method is

$$hf_y(t) < 0 \text{ or } hf_y(t) > 2$$

for a single equation. For a system of equations, the stability region is the region where all eigenvalues of $\boldsymbol{I} - h_n\boldsymbol{J}(t, \boldsymbol{y}(t))$ are outside the unit circle. For example, backward Euler is stable on the equation $y'(t) = -y(t)$ for all positive h.

Let's check stability for another method for solving ODEs.

CHALLENGE 20.6.
We want to solve the ODE

$$y'(t) + ay(t) = 0,$$

where $a > 0$ is a scalar, and $y(0)$ is given. We apply the **Crank–Nicholson method** to the ODE:

$$\frac{y^{n+1} - y^n}{h} + a\frac{y^{n+1} + y^n}{2} = 0,$$

where $y^n \approx y(nh)$ and $h > 0$ is the timestep. For what range of h values is the numerical method stable?

Figure 20.7 gives a geometric illustration of the importance of stability in solving ODEs.

20.2.2 Predictor-Corrector Methods

We could use our favorite nonlinear equation solver from Unit VI in Algorithm 20.2, but this is expensive. In rare occasions, as in Challenge 20.5, the nonlinear equation can be solved explicitly, but usually we use **functional iteration**, as shown in Algorithm 20.3. We could repeat the CE steps if the value of $\boldsymbol{y}_{n+1}$ has not converged (although reducing the stepsize is generally a safer cure). If repeated m times, we call this a **PE(CE)**^{m} scheme. Note that using this method usually fails to solve the nonlinear equation exactly, so this adds an additional source of error.

Let's see how this works on two examples.

Algorithm 20.3 Predictor-Corrector (PECE) Algorithm

Given: $\boldsymbol{y}_0$ and $t_0, t_1, \ldots t_N$.
for $n = 0, \ldots, N-1$
  P (predict): Guess $\boldsymbol{y}_{n+1}$ using a formula that only requires function values for times earlier than t_{n+1} (perhaps using Euler's method $\boldsymbol{y}_{n+1} = \boldsymbol{y}_n + h_n \boldsymbol{f}_n$).
  E (evaluate): Evaluate $\boldsymbol{f}_{n+1} = \boldsymbol{f}(t_{n+1}, \boldsymbol{y}_{n+1})$.
  C (correct): Guess $\boldsymbol{y}_{n+1}$ using a formula that requires function values for times including t_{n+1} (perhaps using the backward Euler method $\boldsymbol{y}_{n+1} = \boldsymbol{y}_n + h_n \boldsymbol{f}_{n+1}$).
  E (evaluate): Evaluate $\boldsymbol{f}_{n+1} = \boldsymbol{f}(t_{n+1}, \boldsymbol{y}_{n+1})$.
end

CHALLENGE 20.7.
Let

$$y'(t) = y^2(t) - 5t,$$
$$y(0) = 1.$$

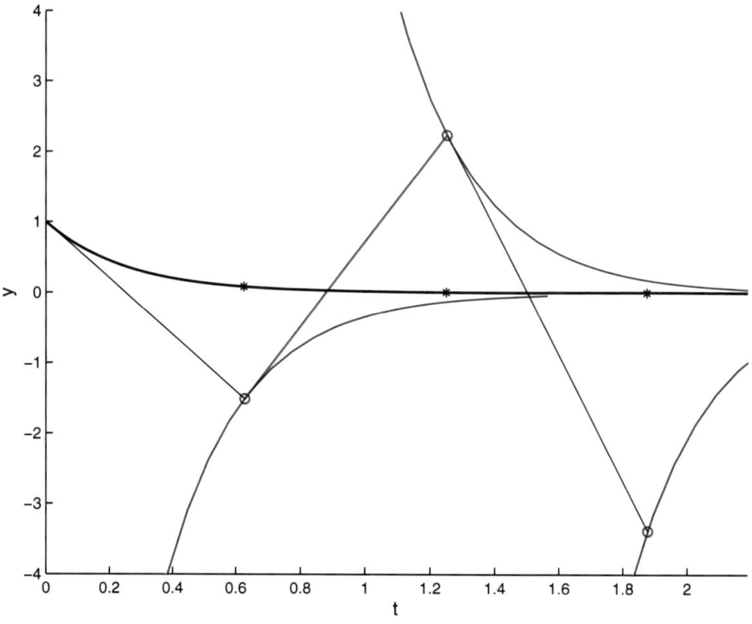

Figure 20.7. *A family of solutions to a stiff differential equation $y'(t) = -4y(t)$. The solution curve for $y(0) = 1$ is plotted in black, and other family members are plotted in red. We use Euler's method to approximate the true solution (marked with stars) by the values marked with circles. Notice that as we step with Euler's method, we move from one solution curve to another, and these curves can be qualitatively quite different. Therefore, our approximation, marked with blue circles, is quite different from the true solution. Lack of stability led to a very poor estimate.*

Apply a PECE scheme to this problem, using Euler and backward Euler with a stepsize $h = .1$, to obtain an approximation for $y(.1)$.

CHALLENGE 20.8.
Let

$$y'(t) = 10y^2(t) - 20,$$
$$y(0) = 1.$$

Apply a PECE scheme to this problem, using Euler and backward Euler with a stepsize $h = .1$, to obtain an approximation for $y(.1)$. What went wrong?

20.2.3 The Adams Family

Euler and backward Euler are the simplest examples of methods in the **Adams family**. Explicit methods like Euler's, which can be derived by polynomial interpolation at known function values, give **Adams–Bashforth** formulas, while those that use one unknown function value give **Adams–Moulton** formulas. To derive Adams–Bashforth formulas, notice that

$$y(t_{n+1}) = y(t_n) + \int_{t_n}^{t_{n+1}} f(t, y(t))\,\mathrm{d}t,$$

and we can approximate the integrand by polynomial interpolation at the points

$$(t_n, f_n), \dots, (t_{n-k+1}, f_{n-k+1})$$

for some integer k, as illustrated in Figure 20.8. For $k = 1$, this yields a fourth approach to deriving Euler's method: approximate the integral by the rectangle rule using the left endpoint so that

$$y_{n+1} = y_n + \int_{t_n}^{t_{n+1}} f(t, y(t))\,\mathrm{d}t \approx y_n + f(t_n, y_n)(t_{n+1} - t_n).$$

For Adams–Bashforth with $k = 2$, we interpolate the integrand by a polynomial of degree 1, with

$$p(t_n) = f_n,$$
$$p(t_{n-1}) = f_{n-1},$$

so letting $h_n = t_{n+1} - t_n$ and $h_{n-1} = t_n - t_{n-1}$, we obtain

$$\begin{aligned} y_{n+1} &= y_n + \int_{t_n}^{t_{n+1}} f_{n-1} + \frac{f_n - f_{n-1}}{t_n - t_{n-1}}(t - t_{n-1})\,\mathrm{d}t \\ &= y_n + h_n f_{n-1} + \frac{f_n - f_{n-1}}{h_{n-1}} \frac{(h_n + h_{n-1})^2 - h_{n-1}^2}{2}. \end{aligned}$$

For Adams–Moulton, we approximate the integrand by polynomial interpolation at the points

$$(t_{n+1}, f_{n+1}), \dots, (t_{n-k+2}, f_{n-k+2})$$

for some integer k. For $k = 1$, we derive the backward Euler method from the approximation:

$$y_{n+1} = y_n + \int_{t_n}^{t_{n+1}} f(t, y(t))\,\mathrm{d}t \approx y_n + f(t_{n+1}, y_{n+1})(t_{n+1} - t_n).$$

For Adams–Moulton with $k = 2$, we interpolate the integrand by a polynomial of degree 1, with

$$p(t_n) = f_n,$$
$$p(t_{n+1}) = f_{n+1},$$

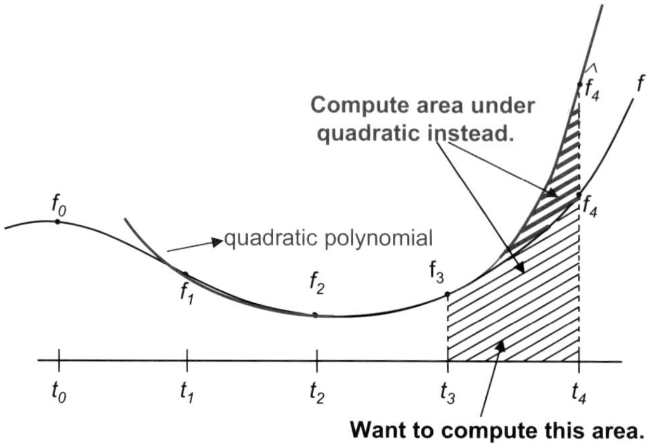

Figure 20.8. *Third-order Adams methods. Adams–Bashforth approximates the area under the curve from $t = t_3$ to $t = t_4$ by integrating the quadratic polynomial that interpolates the curve using the data f_1, f_2, and f_3. Adams–Moulton would interpolate f_2, f_3, and an estimate of f_4.*

so we obtain

$$\begin{aligned} y_{n+1} &= y_n + \int_{t_n}^{t_{n+1}} f_{n+1} + \frac{f_n - f_{n+1}}{t_n - t_{n+1}}(t - t_{n+1})\, \mathrm{d}t \\ &= y_n + \frac{h_n}{2}(f_{n+1} + f_n), \end{aligned}$$

which is a generalization of the **trapezoidal rule for integration**.

Adams methods use an Adams–Bashforth formula as a predictor and an Adams–Moulton formula as a corrector in Algorithm 20.3. Some sample Adams formulas are given in Table 20.1. Note that the two families share function evaluations, so for a **PECE** scheme, the cost per step is only 2 evaluations of f, independent of k.

To make a PECE scheme practical, we need to add **stepsize control** and **error estimation**.

20.2.4 Some Ingredients in Building a Practical ODE Solver

We have two tools to help us control the size of the local error:

- We can change the stepsize h.
- We can change the order k (where the local error is $O(h^{k+1})$).

Therefore, most ODE solvers ask the user for a tolerance parameter and try to keep the local error below this tolerance. Note, as we saw in Figure 20.6, that controlling the local error says nothing about control of global error, unless we know something about the stability of the ODE.

Table 20.1. *Some Adams formulas, assuming equal stepsizes h.*

Some Adams–Bashforth Formulas:	Order	Local Error
$y_{n+1} = y_n + hf_n$	$k=1$	$\frac{h^2}{2} y^{(2)}(\eta)$
$y_{n+1} = y_n + \frac{h}{2}(3f_n - f_{n-1})$	$k=2$	$\frac{5h^3}{12} y^{(3)}(\eta)$
$y_{n+1} = y_n + \frac{h}{12}(23f_n - 16f_{n-1} + 5f_{n-2})$	$k=3$	$\frac{3h^4}{8} y^{(4)}(\eta)$
Some Adams–Moulton Formulas:	Order	Local Error
$y_{n+1} = y_n + hf_{n+1}$	$k=1$	$-\frac{h^2}{2} y^{(2)}(\eta)$
$y_{n+1} = y_n + \frac{h}{2}(f_n + f_{n+1})$	$k=2$	$-\frac{h^3}{12} y^{(3)}(\eta)$
$y_{n+1} = y_n + \frac{h}{12}(5f_{n+1} + 8f_n - f_{n-1})$	$k=3$	$-\frac{h^4}{24} y^{(4)}(\eta)$

To control local error, we need to have a means of estimating it. For example, consider the use of these formulas:

$$\begin{array}{lll} \text{P} & y^{\text{P}}_{n+1} = y_n + \frac{h_n}{2}(3f_n - f_{n-1}) & \text{(Adams–Bashforth)}, \\ \text{E} & f_{n+1} = f(t_{n+1}, y^{\text{P}}_{n+1}), & \\ \text{C} & y^{\text{C}}_{n+1} = y_n + \frac{h_n}{2}(f_n + f_{n+1}) & \text{(Adams–Moulton)}, \\ \text{E} & f_{n+1} = f(t_{n+1}, y^{\text{C}}_{n+1}). & \end{array}$$

To estimate the local error in this method, we compute, using Table 20.1,

$$y^{\text{P}}_{n+1} - y(t_{n+1}) = \frac{5h_n^3}{12} y^{(3)}(\eta_1),$$

$$y^{\text{C}}_{n+1} - y(t_{n+1}) = -\frac{h_n^3}{12} y^{(3)}(\eta_2),$$

where $y(t_{n+1})$ is the solution to $y' = f(t,y)$ with $y(t_n) = y_n$ rather than with $y(0) = y_0$. Therefore we see that

$$y^{\text{P}}_{n+1} - y^{\text{C}}_{n+1} = \frac{h_n^3}{12}(5y^{(3)}(\eta_1) + y^{(3)}(\eta_2)) \approx \frac{6h_n^3}{12} y^{(3)}(\eta),$$

whose magnitude is about 6 times that of the local error in the corrector. (The points η_1, η_2, and η are all in the interval $[t_n, t_{n+1}]$.) This gives us a practical means for estimating local error, since $|y^{\text{P}}_{n+1} - y^{\text{C}}_{n+1}|$ is computable: we now know that if $|y^{\text{P}}_{n+1} - y^{\text{C}}_{n+1}|$ is small enough, then the local error should be small, but if it is big we should worry about the local error.

Practical methods for solving ODEs use such estimates for the local error to determine whether the current choice of stepsize h_n is adequate. If the local error estimate is larger than that requested by the user, the stepsize is usually halved, and the step is repeated. If the local error estimate is much lower than requested, then the stepsize is doubled for the next step. By using factors of two, old function values can be reused; see Figure 20.9.

Let's get some practice with the idea of error estimation and stepsize control.

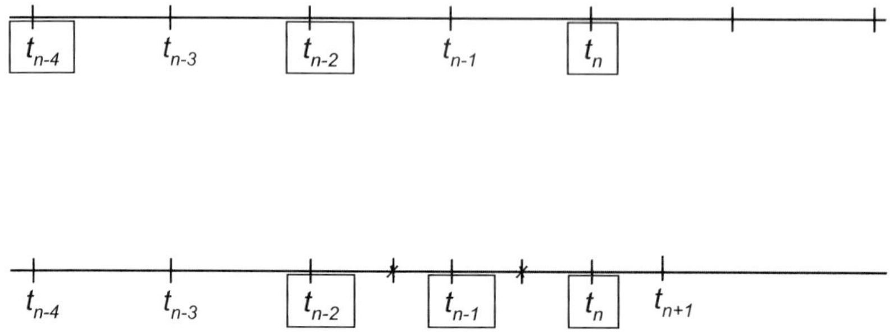

Figure 20.9. *(Top) Given data at t_{n-4}, t_{n-3}, t_{n-2}, t_{n-1}, and t_n, if we decide to increase the stepsize in our Adams formula by a factor of 2, we can reuse the function values at t_{n-4}, t_{n-2}, and t_n to obtain an approximation at t_{n+1}. (Bottom) If we reduce the stepsize by factor of two, then we interpolate to approximate the function values at the points marked with stars.*

CHALLENGE 20.9.

Suppose we have used a PECE method with predictor of order 4 (i.e., local error is proportional to $O(h^5)$, where h is the stepsize) and corrector of order 5. We want to keep the local error less than τ. Estimate the local error and explain how to alter the stepsize if necessary to achieve our local error criterion.

CHALLENGE 20.10.

Suppose we have used a PECE method with Adams–Bashforth and Adams–Moulton formulas of order 3 to form two estimates of $y(t_{n+1})$. How would you estimate the local error in the Adams–Moulton formula? How would you use that estimate to change h in order to keep the estimated local error less than a user-supplied local error tolerance τ without taking steps smaller than necessary?

We also change the order of the method to control error and get ourselves started. When we take our first step with an Adams method, we need to use Euler's method, because we have no history. When we take our second step, we can use the second-order Adams methods, because we have information about one old point. As we keep walking, we can use higher-order methods, by saving information about old function values. But there is a limit to the usefulness of history; generally, fifth-order methods are about as high as we go, and the order is kept smaller than this if the local error estimate indicates that the high-order derivatives are large.

CHALLENGE 20.11. (Extra)

- Write MATLAB statements to start an Adams algorithm. Begin using Euler's method, and gradually increase the order of the method to 3. Note that to control the error, the stepsizes are generally smaller when using lower-order methods.

POINTER 20.4. Controlling Error.

By changing h and k in our ODE solver, we control the local error. It is important to note that the user really cares about global error, not local error, but this depends on the stability of the differential equation. To estimate the global error, we might perform sensitivity analysis as discussed in the case study of Chapter 2, seeing how much the computed solution changes when the initial conditions are perturbed a bit. We might also experiment with perturbations to f, but random noise introduced in f causes large-magnitude derivatives and therefore small stepsizes in the ODE solver.

- Write MATLAB statements to estimate the error and decide whether to change h.
- Write MATLAB statements to decide whether to cut back to a lower-order method.

Note that PECE using the Adams formulas is relatively inexpensive, since, no matter what the order of the method, we only need two new function evaluations per step. Thus there is an advantage to using higher-order methods, since they allow a larger stepsize (and thus fewer steps) with little additional work. One pitfall to watch for is that since PECE provides an approximate solution to a nonlinear equation, we should suspect that we have not solved very accurately if $|y_{n+1}^P - y_{n+1}^C|$ is large; in this case, the stepsize should be reduced.

20.2.5 Solving Stiff Problems

The Adams family with PECE is good for nonstiff problems, but using this method to solve a stiff problem may result in artificial oscillation. An illustration is given in Figure 20.7. Instead, a **Gear family** of formulas is effective for stiff problems and can also be used for nonstiff problems, although it is usually uses somewhat smaller stepsizes than the Adams family. See Pointer 20.11.

20.2.6 An Alternative to Adams Formulas: Runge–Kutta

Runge–Kutta methods build on the idea we used in the Euler method of predicting a new value by walking along a tangent to the curve. They just do it in a more complicated way.

For example, suppose we let

$$\begin{aligned} k_1 &= h_n f(t_n, y_n), \\ k_2 &= h_n f(t_n + \alpha h, y_n + \beta k_1), \\ y_{n+1} &= y_n + a k_1 + b k_2, \end{aligned}$$

where we choose α, β, a, b to make the formula as accurate as possible.

We expand everything in Taylor series,

$$y(t_{n+1}) = y(t_n) + y'(t_n)h_n + y''(t_n)\frac{h_n^2}{2} + O(h_n^3), \tag{20.3}$$

and match as many terms as possible. Note that

$$y'(t_n) = f(t_n, y_n) \equiv f,$$
$$y''(t_n) = \frac{\partial f}{\partial t}(t_n, y_n) + \frac{\partial f}{\partial y}\frac{\partial y}{\partial t}(t_n, y_n) \equiv f_t + f_y f,$$

so (20.3) can be written

$$y(t_{n+1}) = y(t_n) + f h_n + (f_t + f_y f)\frac{h_n^2}{2} + O(h_n^3). \tag{20.4}$$

Now we expand k_1 and k_2:

$$\begin{aligned} k_1 &= h_n f(t_n, y_n), \\ k_2 &= h_n[f(t_n, y_n) + \alpha h_n f_t + \beta k_1 f_y + O(h_n^2)] \\ &= h_n f(t_n, y_n) + \alpha h_n^2 f_t + \beta k_1 h_n f_y + O(h_n^3), \end{aligned}$$

so our Runge–Kutta method satisfies

$$\begin{aligned} y_{n+1} &= y_n + ak_1 + bk_2 \\ &= y_n + ah_n f + b(h_n f + \alpha h_n^2 f_t + \beta h_n f h_n f_y) + O(h_n^3) \\ &= y_n + (a+b)h_n f + \alpha b h_n^2 f_t + \beta b h_n^2 f_y f + O(h_n^3). \end{aligned}$$

We want to match the coefficients in this expansion to the coefficients in the Taylor series expansion of y in (20.4), so we set

$$\begin{aligned} a + b &= 1, \\ \alpha b &= \frac{1}{2}, \\ \beta b &= \frac{1}{2}. \end{aligned}$$

There are many solutions (in fact, an infinite number). One of them is Heun's method:

$$a = \frac{1}{2}, \quad b = \frac{1}{2}, \quad \alpha = 1, \quad \beta = 1.$$

This choice gives a **second-order Runge–Kutta method**. Note that the work per step is 2 evaluations of $\boldsymbol{f}$.

The most useful Runge–Kutta method is one of order 4:

$$\begin{aligned} \boldsymbol{k}_1 &= h_n \boldsymbol{f}(t_n, \boldsymbol{y}_n), \\ \boldsymbol{k}_2 &= h_n \boldsymbol{f}\left(t_n + \frac{h_n}{2}, \boldsymbol{y}_n + \frac{\boldsymbol{k}_1}{2}\right), \\ \boldsymbol{k}_3 &= h_n \boldsymbol{f}\left(t_n + \frac{h_n}{2}, \boldsymbol{y}_n + \frac{\boldsymbol{k}_2}{2}\right), \\ \boldsymbol{k}_4 &= h_n \boldsymbol{f}(t_n + h_n, \boldsymbol{y}_n + \boldsymbol{k}_3), \\ \boldsymbol{y}_{n+1} &= \boldsymbol{y}_n + \frac{1}{6}(\boldsymbol{k}_1 + 2\boldsymbol{k}_2 + 2\boldsymbol{k}_3 + \boldsymbol{k}_4). \end{aligned}$$

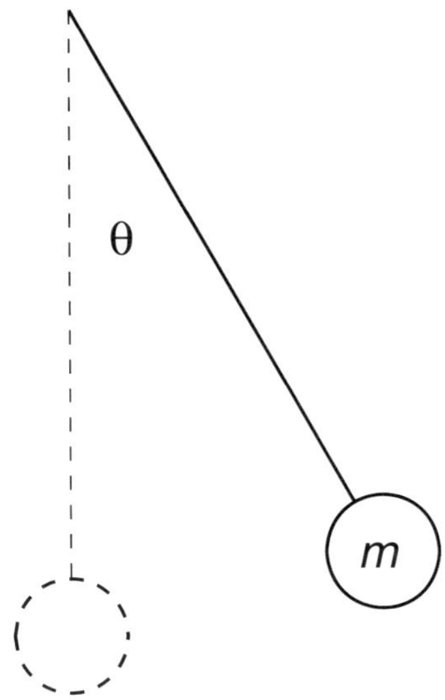

Figure 20.10. *Our simple harmonic oscillator, a pendulum. The pendulum is suspended at the origin and the string has length r. As the pendulum moves, its position is defined by $x(t) = p(t) = r\sin\theta(t)$, $y(t) = q(t) = -r\cos\theta(t)$ for some function $\theta(t)$.*

It requires 4 evaluations of f per step, and many pages for its derivation.

The fact that the Runge–Kutta methods use no old function values is both an advantage and a disadvantage. In contrast to PECE methods, it is easy to change the stepsize, since no old function values are needed. But PECE methods require only two function evaluations per iteration, regardless of order, so the Runge–Kutta methods use more work per iteration when the order is larger than 2.

20.3 Hamiltonian Systems

In some ODE systems, there is an associated **conservation law**, and if possible, we formulate the problem so that conservation is observed.

A **Hamiltonian system** is one for which there exists a scalar **Hamiltonian function** $H(\mathbf{y})$ so that

$$\mathbf{y}'(t) = \mathbf{D}\nabla_{\mathbf{y}} H(\mathbf{y}(t)), \tag{20.5}$$

where $\mathbf{D}$ is a block-diagonal matrix with blocks equal to

$$\begin{bmatrix} 0 & 1 \\ -1 & 0 \end{bmatrix}$$

and $\nabla_{\mathbf{y}} H(\mathbf{y}(t))$ denotes the gradient of H with respect to the $\mathbf{y}$ variables.

For example, consider the pendulum pictured in Figure 20.10. The functions $p(t)$ and $q(t)$ that define the position of the pendulum bob satisfy

$$q'(t) = \omega p(t), \tag{20.6}$$

$$p'(t) = -\omega q(t), \tag{20.7}$$

where $\omega > 0$ is a fixed parameter. The length r of the string does not change, so we might want to preserve this invariance, or conservation law, in our numerical method, requiring that the Hamiltonian of the system

$$H(t) = \frac{\omega}{2}(p^2(t) + q^2(t))$$

remain constant, as it does for the true solution. To verify (20.5) for this example, note that if $\mathbf{y}(t) = [q(t), p(t)]^T$, then

$$\nabla_{\mathbf{y}} H(\mathbf{y}(t)) = \begin{bmatrix} \omega q(t) \\ \omega p(t) \end{bmatrix},$$

so that

$$\mathbf{y}'(t) = \mathbf{D}\nabla_{\mathbf{y}} H(\mathbf{y}(t)) = \begin{bmatrix} 0 & 1 \\ -1 & 0 \end{bmatrix} \begin{bmatrix} \omega q(t) \\ \omega p(t) \end{bmatrix},$$

which is (20.6)-(20.7). Note that differentiating with respect to t gives

$$\begin{aligned} H'(t) &= \frac{\omega}{2}(2p(t)p'(t) + 2q(t)q'(t)) \\ &= \frac{\omega}{2}(2p(t)(-\omega q(t)) + 2q(t)(\omega p(t))) \\ &= 0, \end{aligned}$$

so $H(t)$ must be constant; in other words, the quantity $H(t)$ is conserved or invariant. We can verify this a different way by writing the general solution to the problem

$$\begin{bmatrix} q(t) \\ p(t) \end{bmatrix} = \begin{bmatrix} \cos\omega t & \sin\omega t \\ -\sin\omega t & \cos\omega t \end{bmatrix} \begin{bmatrix} q(0) \\ p(0) \end{bmatrix},$$

and computing $p(t)^2 + q(t)^2$. The eigenvalues of the matrix defining the solution are imaginary numbers, so a small perturbation of the matrix can cause the quantity $H(t)$ to either grow or shrink, and this does not produce a useful estimate of $p(t)$ and $q(t)$. Therefore, in solving systems involving Hamiltonians (conserved quantities), it is important to also build conservation into the numerical method whenever possible.

If, as in the pendulum example, an ODE of the form

$$\mathbf{y}'(t) = \mathbf{f}(t, \mathbf{y}(t)), \tag{20.8}$$

$$H(\mathbf{y}(t)) = 0, \tag{20.9}$$

is overdetermined, with more equations than unknown functions, then we can rewrite it as

$$\mathbf{y}'(t) = \mathbf{f}(t, \mathbf{y}(t)) - \mathbf{G}(\mathbf{y}(t))z(t), \tag{20.10}$$

$$H(\mathbf{y}(t)) = 0, \tag{20.11}$$

where $z(t)$ is another function (added so that the system is not overdetermined) and $\mathbf{G}(\mathbf{y}(t))$ is a function whose derivative matrix is bounded away from singularity for all t. If we solve system (20.10)–(20.11) exactly, then we get $z(t) = 0$ and we recover our original solution. But if we solve it numerically, the conservation law $H(\mathbf{y}(t)) = 0$ forces $z(t)$ to be nonzero to compensate for numerical errors.

For our pendulum, for example, we can choose $\mathbf{G}(\mathbf{y}) = \omega\mathbf{y}$ and rewrite our harmonic oscillator example as

POINTER 20.5. Warning about Hamiltonian Systems.

Sometimes, adding conservation to an ODE system makes the problem too expensive to solve; for example, if the solution is rapidly oscillating, the conservation law forces very small stepsizes.

$$q'(t) = \omega p(t) - \omega q(t)z(t), \tag{20.12}$$
$$p'(t) = -\omega q(t) - \omega p(t)z(t), \tag{20.13}$$
$$0 = \frac{\omega}{2}(p^2(t) + q^2(t) - r^2). \tag{20.14}$$

Let's consider another example of a Hamiltonian.

CHALLENGE 20.12.

Derive the Hamiltonian system for

$$H(y) = \frac{1}{2}y_1^2(t) + \frac{1}{2}y_2^2(t) + \frac{1}{2}y_3^2(t) + \frac{1}{2}y_4^2(t) + \frac{1}{2}y_1^2(t)y_2^2(t)y_3^2(t)y_4^2(t).$$

Adding an invariant, or conservation law, generally changes the ODE system to a system that includes nonlinear equations not involving derivatives as in (20.12)-(20.14) – a system of **differential-algebraic equations** (DAEs). Next we'll consider a little of the theory and computation of such DAEs.

20.4 Differential-Algebraic Equations

The general DAE has the form

$$\boldsymbol{F}(t, \widehat{\boldsymbol{y}}(t), \widehat{\boldsymbol{y}}'(t)) = \boldsymbol{0},$$

where $\boldsymbol{F}$ is a specified function. We'll consider a special case in which the problem can be written as

$$\boldsymbol{M}(t)\boldsymbol{y}'(t) = \boldsymbol{A}(t)\boldsymbol{y}(t) + \boldsymbol{f}(t). \tag{20.15}$$

The matrix in front of the derivatives is called the **mass matrix**. In the next challenge we convert a DAE to this **standard form**.

CHALLENGE 20.13.

Let the three variables $u(t)$, $v(t)$, and $w(t)$ be related by

$$u'(t) = 7u(t) - 6v(t) + 4t,$$
$$v'(t) = 4u(t) - 2v(t),$$
$$u(t) + v(t) + w(t) = 24.$$

Convert this to a system of the form $\boldsymbol{M}(t)\boldsymbol{y}'(t) = \boldsymbol{A}(t)\boldsymbol{y}(t) + \boldsymbol{f}(t)$.

POINTER 20.6. Existence and Uniqueness of Solutions to DAEs.

DAEs are a combination of ODEs and algebraic (linear or nonlinear) equations, so the subject of existence and uniqueness of solutions is somewhat complicated. We present a set of conditions for the DAE

$$\boldsymbol{M}(t)\boldsymbol{y}'(t) = \boldsymbol{A}(t)\boldsymbol{y}(t) + \boldsymbol{f}(t).$$

For any nonnegative integer ℓ, define the $(\ell+1)m \times (\ell+1)m$ matrix

$$\boldsymbol{P}_\ell(t) = \begin{bmatrix} \boldsymbol{M}(t) & & & & \\ \boldsymbol{M}'(t) - \boldsymbol{A}(t) & \boldsymbol{M}(t) & & & \\ \boldsymbol{M}''(t) - 2\boldsymbol{A}'(t) & 2\boldsymbol{M}'(t) - \boldsymbol{A}(t) & & & \\ \vdots & & \ddots & \ddots & \\ \boldsymbol{M}^{(\ell)}(t) - \ell\boldsymbol{A}^{(\ell-1)}(t) & \dots & \dots & \ell\boldsymbol{M}'(t) - \boldsymbol{A}(t) & \boldsymbol{M}(t) \end{bmatrix},$$

where the (i,j) block below the diagonal contains $\binom{i-1}{j-1}\boldsymbol{M}^{(i-j)} - \binom{i-1}{j}\boldsymbol{A}^{(i-j-1)}$. Define the $(\ell+1)m \times (\ell+1)m$ matrix

$$\boldsymbol{N}_\ell(t) = \begin{bmatrix} \boldsymbol{A}(t) & \boldsymbol{0} & \dots & \dots & \boldsymbol{0} \\ \boldsymbol{A}'(t) & \boldsymbol{0} & \dots & \dots & \boldsymbol{0} \\ \vdots & \vdots & & & \vdots \\ \boldsymbol{A}^{(\ell)}(t) & \boldsymbol{0} & \dots & \dots & \boldsymbol{0} \end{bmatrix}.$$

Suppose that for some value of ℓ, there are integers m_a and m_d, with $m_a + m_d = m$, such that for all values of t in the interval of interest:

- $\boldsymbol{P}_\ell(t)$ and $\boldsymbol{N}_\ell(t)$ are continuously differentiable.
- $\text{rank}(\boldsymbol{P}_\ell(t)) = (\ell+1)m - m_a$, and the m_a columns of some matrix $\boldsymbol{Z}(t)$ form a basis for the null space of $\boldsymbol{P}_\ell(t)^*$.
- Let $\widehat{\boldsymbol{N}}_\ell(t) = \boldsymbol{Z}(t)^*\boldsymbol{N}_\ell(t)[\boldsymbol{I}_n, \boldsymbol{0}, \dots, \boldsymbol{0}]^T$. Then $\text{rank}(\widehat{\boldsymbol{N}}_\ell(t)) = m_a$, and the m_d columns of some matrix $\boldsymbol{T}(t)$ form a basis for the null space of $\widehat{\boldsymbol{N}}_\ell(t)$.
- There exists $\boldsymbol{W}(t)$ of dimension $m \times m_d$ such that $\text{rank}(\boldsymbol{W}(t)^*\boldsymbol{M}(t)\boldsymbol{T}(t)) = m_d$.
- $\boldsymbol{Z}(t)$, $\boldsymbol{T}(t)$, and $\boldsymbol{W}(t)$ are continuously differentiable functions of t.

Then for every consistent initial condition the DAE has a unique solution. See Kunkel and Mehrmann [97, Theorem 3.52] for proof of this result as well as results for more general problems.

20.4.1 Some Basics

DAEs are classified by several parameters:

- m_a is the number of algebraic conditions in the DAE.
- m_d is the number of differential conditions in the DAE, and $m_a + m_d = m$.
- ℓ is the **strangeness** of the DAE.

Often a fourth parameter is considered: the **differential-index** of a DAE is the number of differentiations needed to convert the problem to an (explicit) system of ODEs. A system of ODEs has differential-index 0, and a system of algebraic equations $\boldsymbol{F}(\boldsymbol{y}) = \boldsymbol{0}$ has differential-index 1.

Existence and uniqueness of the solution to the DAE can be checked using the result in Pointer 20.6. For example, if $\boldsymbol{M}(t)$ in equation (20.15) has full rank for all t, then $\ell = 0$, $m_a = 0$, $\boldsymbol{Z} = [\ ]$, and $\boldsymbol{W} = \boldsymbol{T} = \boldsymbol{I}_m$, and we have a system of ODEs. Alternatively, if $\boldsymbol{M} = \boldsymbol{0}$ and $\boldsymbol{A}(t)$ is full rank, then $\ell = 0$, $m_a = m$, $\boldsymbol{Z} = \boldsymbol{I}_m$, and $\boldsymbol{W} = \boldsymbol{T} = [\ ]$, and we have a system of algebraic equations.

Let's consider a problem with strangeness $\ell = 1$.

CHALLENGE 20.14.

The **nonstationary Stokes equation** can be discretized in space to give the DAE

$$\begin{aligned} \boldsymbol{u}'(t) &= \boldsymbol{C}\boldsymbol{u}(t) + \boldsymbol{B}\boldsymbol{p}(t), \\ \boldsymbol{B}^T\boldsymbol{u}(t) &= \boldsymbol{0}, \end{aligned}$$

where $\boldsymbol{u}(t)$ is an $m_u \times 1$ vector of fluid velocities and $\boldsymbol{p}(t)$ is an $m_p \times 1$ vector of pressures, with $m_p < m_u$. Suppose $\boldsymbol{B}^T$ is a real full-rank matrix, and that the columns of the $m_u \times (m_u - m_p)$ matrix $\boldsymbol{X}$ are a basis for its null space. Verify the hypotheses of Pointer 20.6 for $\ell = 1$.

There is a major difference between DAEs and ODEs: For ODEs, it is easy to count how many initial conditions we need to uniquely determine the solution. For DAEs, it is not so simple; initial conditions may be inconsistent with the problem. For example, the equation

$$e^y = 5$$

is (trivially) a DAE and, of course, nonlinear equations like this require no initial conditions to specify the solution.

We now consider some numerical methods for solving DAEs.

20.4.2 Numerical Methods for DAEs

The main idea used to solve DAEs follows from what we know about data fitting. If we want to solve the DAE

$$\boldsymbol{F}(t, \boldsymbol{y}(t), \boldsymbol{y}'(t)) = \boldsymbol{0},$$

then we step from known values at $t = t_n, t_{n-1}, \dots, t_{n-k}$ to unknown values at $t = t_{n+1}$ using our favorite approximation scheme to replace $\boldsymbol{y}'(t_{n+1})$ by

$$\boldsymbol{y}'(t_{n+1}) \approx \sum_{j=0}^{k} \alpha_i \boldsymbol{y}(t_{n+1-j}).$$

POINTER 20.7. DAE **Software.**

The MATLAB ODE solvers, including `ode23s`, for example, handle some DAEs. There are several high-quality packages specifically designed for DAEs, including:

- SUNDIALS from Lawrence Livermore National Laboratory (`www.llnl.gov.CASC/sundials`),
- GELDA and GENDA by Peter Kunkel and Volker Mehrmann (`www.math.tu-berlin.de/numerik/mt.NumMat`).

Other software references can be found in [97].

This gives us a nonlinear equation to solve for our approximation $\mathbf{y}_{n+1} \approx \mathbf{y}(t_{n+1})$:

$$\mathbf{F}\left(t_{n+1}, \mathbf{y}_{n+1}, \sum_{j=0}^{k} \alpha_i \mathbf{y}(t_{n+1-j})\right) = \mathbf{0}.$$

In principle, we can solve this equation using our favorite method from Chapter 24 (Newton-like, homotopy, etc.). In practice, there are a few complications:

- Stability is an important consideration. Usually a stiff method is used to ensure numerical stability.
- The nonlinear equation may fail to have a solution.
- Even if a solution exists, the method for solving the nonlinear equation may fail to converge.
- Automatic control of order and stepsize is even more difficult than for ODEs.

The bottom line is that you should not try to write your own solver for DAEs. Use high-quality software, as indicated in Pointer 20.7. We will use a MATLAB DAE solver in Chapter 21.

20.5 Boundary Value Problems for ODEs

Up to now, the solution to our ODE has always been made unique by specifying values of the variables at some fixed time t_0. We consider in this section **boundary value problems**, for which values are specified at two different times.

For example, consider the problem of determining $u(t)$ for $t \in (0,1)$ given that

$$u''(t) = 6u'(t) - tu(t) + u^2(t), \tag{20.18}$$

$$u(0) = 5, \tag{20.19}$$

$$u(1) = 2. \tag{20.20}$$

POINTER 20.8. Existence and Uniqueness of Solutions to BVPs for ODEs.

Consider a problem that can be expressed as

$$\mathcal{A}u(t) = -(a(t)u'(t))' + b(t)u'(t) + c(t)u(t) = f(t) \text{ for } t \in (0,1), \tag{20.16}$$

$$u(0) = u_0, \qquad u(1) = u_1, \tag{20.17}$$

where $a(t)$, $b(t)$, $c(t)$, and $f(t)$ are given differentiable functions and u_0 and u_1 are given numbers. The interval is set to $t \in (0,1)$, but this can be changed by rescaling.

It is important to know that the problem is well posed, in the sense that a unique solution exists (with two continuous derivatives), and that small changes in the data lead to small changes in the solution. There are various conditions on the coefficients that guarantee this. One set of sufficient conditions is

- $a(t) \geq a_0$ for $t \in [0,1]$, where a_0 is a number greater than 0.
- $c(t) \geq 0$ for $t \in [0,1]$.
- $\int_0^1 |f(t)|^2 \, dt$ is finite.

For further information, consult a standard textbook such as [99, Chap. 2].

If we convert this to a system of first order equations, we let $y_{(1)}(t) = u(t)$, $y_{(2)}(t) = u'(t)$ and obtain

$$y'_{(1)}(t) = y_{(2)}(t) \tag{20.21}$$
$$y'_{(2)}(t) = 6y_{(2)}(t) - ty_{(1)}(t) + y^2_{(1)}(t), \tag{20.22}$$
$$y_{(1)}(0) = 5, \tag{20.23}$$
$$y_{(1)}(1) = 2. \tag{20.24}$$

So we have values of $y_{(1)}$ at 0 and 1. If we had values of $y_{(1)}$ and $y_{(2)}$ at 0, we could use our IVP methods. But now we have a **boundary value problem**. What can we do? There are three alternatives:

- Adapt our IVP methods to this problem in a method called **shooting**.
- Develop new methods using **finite differences**.
- Develop new methods using **finite elements**.

We consider the first two ideas in the following two sections, and finite element methods in the case study of Chapter 23.

Before studying these methods, though, we illustrate the use of the ideas in Pointers 20.8 and 20.9 to obtain information about a BVP without actually solving the problem.

POINTER 20.9. Bounds on Solutions to BVPs for ODEs.

Two facts about BVPs in the form (20.16)-(20.17) can help us compute bounds on the solutions without computing the solutions themselves.

First, the **maximum principle** tells us that if

- the solution u exists and has two continuous derivatives on $[0,1]$,
- $f(t) \leq 0$ for $t \in (0,1)$,

then we can bound the solution:

- If $c(t) = 0$, then
$$\max_{t\in[0,1]} u(t) = \max(u_0, u_1).$$
- If $c(t) \geq 0$ for $t \in (0,1)$, then
$$\max_{t\in[0,1]} u(t) \leq \max(u_0, u_1, 0).$$

Second, the **monotonicity theorem for ODEs** helps us compare solutions to different ODEs without computing them. In particular, if

- u satisfies $\mathcal{A}u(t) = f(t)$ for $t \in [0,1]$, with $u(0) = u_0$ and $u(1) = u_1$,
- v satisfies $\mathcal{A}v(t) = g(t)$ for $f \in [0,1]$, with $v(0) = v_0$ and $v(1) = v_1$,
- $f(t) \leq g(t)$ for $t \in [0,1]$,
- $u_0 \leq v_0$,
- $u_1 \leq v_1$,

then $u(t) \leq v(t)$ for $t \in [0,1]$.

For further information, consult a standard textbook such as [99, Chap. 2].

CHALLENGE 20.15.

Consider the differential equation

$$-u''(t) + 8.125\pi \cot((1+t)\pi/8)u'(t) + \pi^2 u(t) = -3\pi^2 \text{ on } \Omega = (0,1)$$

with boundary conditions $u(0) = -2.0761$, $u(1) = -2.2929$. Discuss the solution: Does it exist? Is it unique? What are upper and lower bounds on the solution? Justify each of your answers by citing a theorem and verifying its hypotheses. (Hint: One bound can be obtained by comparing the solution to $v(t) = -3$.)

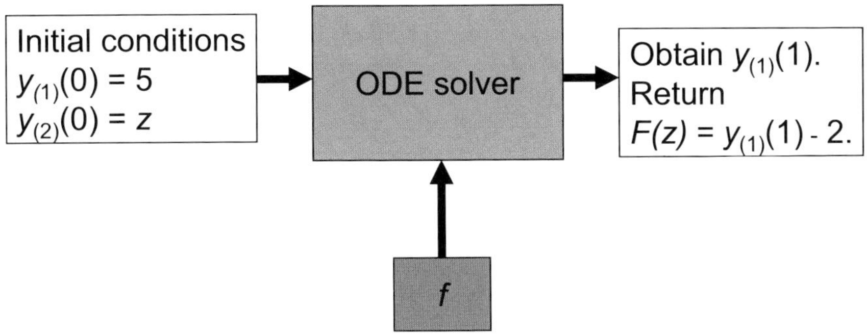

Figure 20.11. *The function evaluation $F(z)$ for the nonlinear equation solver in a shooting method to solve $\mathbf{y}'(t) = \mathbf{f}(t, \mathbf{y}(t))$, $y_{(1)}(0) = 5$, $y_{(1)}(1) = 2$. We return the value $F(z) = y_{(1)}(1) - 2$. When the computed value of $y_{(1)}(1) \approx 2$, then we have the correct initial condition z.*

20.5.1 Shooting Methods

When in doubt, guess. The idea behind shooting methods is to guess at the missing initial values, solve the IVP using our favorite method, and then use the results to improve our guess.

In fact, we recognize this as a nonlinear system of equations: to solve our example problem (20.18)–(20.20), we want to solve the nonlinear equation

$$F(z) = 0,$$

where z is the value we give to $y_{(2)}(0)$ and $F(z)$ is the difference between the value that our (IVP) ODE solver returns for $y_{(1)}(1)$ and the desired value, 2. So a **shooting method** involves using a nonlinear equation solver, as illustrated in Figure 20.11. In doing this we evaluate $F(z)$ by applying our favorite IVP-ODE solver to (20.21)–(20.23). Once we find the initial value z, then the IVP-ODE solver can estimate values $\boldsymbol{u}(t)$ for any t.

Some warnings:

- If the IVP is difficult to solve (for example, stiff), then it is difficult to get an accurate estimate of z.

- Our function evaluation for the nonlinear equation $F(z) = 0$ is **noisy**: it includes all of the rounding error and the global discretization error introduced by the IVP-ODE solver. The resulting wiggles in the values of F can cause the nonlinear equation solver to have trouble finding an accurate solution, and can also introduce multiple solutions where there is really only one; see Figure 20.12.

- If the interval of integration is long, these difficulties can be overwhelming and we need to go to more complicated methods; for example, **multiple shooting**.

Let's see how shooting works.

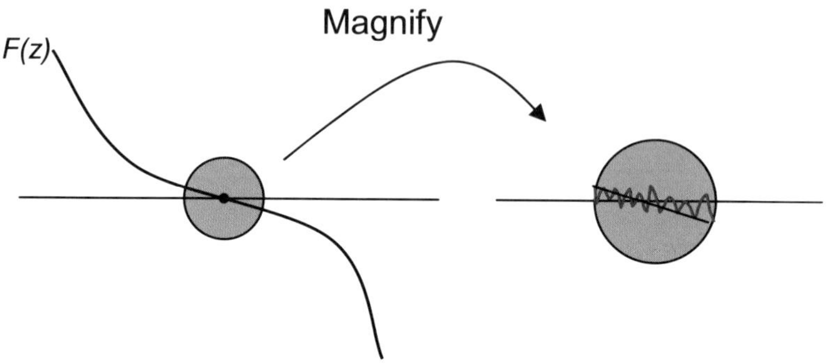

Figure 20.12. *In a shooting method, the evaluation of the function is noisy. Instead of computing points on the (true) black curve, we compute points on the red curve. Thus the computed function may have several zeros near the true solution.*

CHALLENGE 20.16.
Let

$$\begin{aligned} a''(t) &= a^2(t) - 5a'(t), \\ a(0) &= 5, \\ a(1) &= 2. \end{aligned}$$

Convert this to a system of the form $\mathbf{y}' = \mathbf{f}(t, \mathbf{y}(t))$ and write a MATLAB program that uses a shooting method to solve it.

CHALLENGE 20.17.
Let

$$u''(t) = -\left(\frac{\pi}{2}\right)^2 (u(t) - t^2) + 2,$$

with $u(0) = u(1) = 1$. Write a MATLAB program to solve this problem using the shooting method.

CHALLENGE 20.18. (Extra)
Write a program to solve the BVP (20.18)-(20.20) using `ode45` and `fzero`.

20.5.2 Finite Difference Methods

Finite difference methods provide an alternative to shooting methods for BVPs. We derive two finite difference formulas in the following exercise.

CHALLENGE 20.19.

Suppose u has 4 continuous derivatives. Prove that

$$u'(t) = \frac{u(t+h)-u(t-h)}{2h} + O(h^2),$$
$$u''(t) = \frac{u(t-h)-2u(t)+u(t+h)}{h^2} + O(h^2)$$

for small values of h.

Now return to our example problem in its original form (20.18)-(20.20). Given a large number n (for example, $n = 100$), let $h = 1/n$ and define

$$u_j \approx u(jh), \qquad j = 0,\dots,n.$$

Then using the formulas in Challenge 20.19 we can approximate our original equation

$$u''(t) = 6u'(t) - tu(t) + u^2(t)$$

at $t = t_j$ $(0 < j < n)$ by

$$\frac{u_{j-1} - 2u_j + u_{j+1}}{h^2} = 6\frac{u_{j+1} - u_{j-1}}{2h} - t_j u_j + u_j^2.$$

Since we already know that

$$u_0 \approx u(0) = 5,$$
$$u_n \approx u(1) = 2,$$

we have a system of $n-1$ nonlinear equations in $n-1$ unknowns with a tridiagonal coefficient matrix. Defining

$$\begin{aligned}
a_{jj} &= -\tfrac{2}{h^2} + jh, & j &= 1,\dots,n-1,\\
a_{j,j+1} &= \tfrac{1}{h^2} - \tfrac{3}{h}, & j &= 1,\dots,n-2,\\
a_{j,j-1} &= \tfrac{1}{h^2} + \tfrac{3}{h}, & j &= 2,\dots,n-1,\\
a_{jk} &= 0, & k &\neq j, j-1, j+1,
\end{aligned}$$

we have

$$\begin{bmatrix}
a_{11} & a_{12} & & & & \\
a_{21} & a_{22} & a_{23} & & & \\
 & \cdot & \cdot & \cdot & & \\
 & & \cdot & \cdot & \cdot & \\
 & & & a_{n-2,n-3} & a_{n-2,n-2} & a_{n-2,n-1} \\
 & & & & a_{n-1,n-2} & a_{n-1,n-1}
\end{bmatrix}
\begin{bmatrix}
u_1 \\ u_2 \\ \cdot \\ \cdot \\ u_{n-2} \\ u_{n-1}
\end{bmatrix}$$

$$= \begin{bmatrix} -\frac{15}{h} - \frac{5}{h^2} \\ 0 \\ \cdot \\ \cdot \\ 0 \\ \frac{6}{h} - \frac{2}{h^2} \end{bmatrix} + \begin{bmatrix} u_1^2 \\ u_2^2 \\ \cdot \\ \cdot \\ u_{n-2}^2 \\ u_{n-1}^2 \end{bmatrix},$$

or

$$\boldsymbol{Au} = \boldsymbol{b} + \begin{bmatrix} u_1^2 \\ u_2^2 \\ \cdot \\ \cdot \\ u_{n-2}^2 \\ u_{n-1}^2 \end{bmatrix}.$$

Now we can use our favorite method for nonlinear equations (Unit VI). If we don't have nonlinear terms, all we need to do is to solve a linear system (Unit II or VII).

Next we get some practice in forming the system of linear or nonlinear equations for our finite difference method.

CHALLENGE 20.20.

Let

$$u''(t) = u'(t) + 6u(t)$$

with $u(0) = 2$ and $u(1) = 3$. Let $h = 1/5$, and write a set of finite difference equations that approximate the solution to this problem at $t = jh$, $j = 0, \ldots, 5$.

CHALLENGE 20.21.

Suppose we solve the problem

$$u''(t) = u'(t) - tu(t) + e^{u(t)},$$
$$u(0) = 1,$$
$$u(1) = 0,$$

using the finite difference method, approximating $u_i \approx u(ih)$, where $h = .01$, $i = 0, \ldots, 100$. We can use a nonlinear equation solver on the system $\boldsymbol{F}(\boldsymbol{u}) = \boldsymbol{0}$, where there are 99 unknowns and 99 equations. Write the equations for $\boldsymbol{F}(\boldsymbol{u})$.

20.6 Summary

Some tips for solving the problems considered in this chapter are given in Pointer 20.10. In the next chapter we see how differential equations arise in models for problems such as the spread of an epidemic.

POINTER 20.10. Some Tips for Solving Differential Equations.

- IVPs for ODEs that arise in practice can be very difficult to solve.
  - When in doubt, use a stiff method.
  - If there is a **conservation law** or Hamiltonian, make sure to incorporate it into the formulation. (Otherwise, your user may be very unhappy with the numerical results.) But be aware that if you don't do this in a smart way, it may cause the ODE solver to take very small steps.
- We have just touched on the existence, uniqueness, and stability theory for ODEs and DAEs. If you need to solve an important problem, be ready to study these issues further before you go to the computer.
- Numerical solution of DAEs is still an evolving science, so watch the literature if you are working in this field.
- For BVPs, finite difference methods and finite element methods (Chapter 23) are the most commonly used methods.

POINTER 20.11. Further Reading.

More information on numerical solution of ODEs can be found, for example, in Chapter 9 of Van Loan's book [148]. More Adams formulas are listed in Van Loan [148, p. 354].

The Gear family of formulas is discussed, for example, in [54] and used in the MATLAB stiff ODE solvers whose names end in 's'.

DAEs are discussed in the books by Brenan, Campbell, and Petzold [17] and by Kunkel and Mehrmann [97].

Chapter 21 / Case Study

More Models of Infection: It's Epidemic

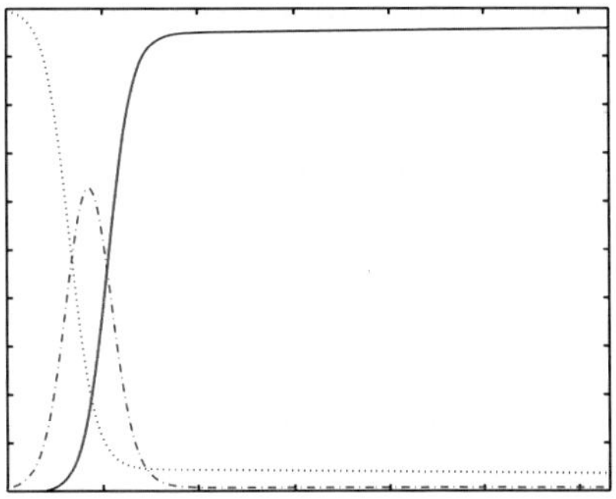

In the case study of Chapter 19, we studied a model of the spread of an infection through a hospital ward. The ward was small enough that we could track each patient as an individual.

When the size of the population becomes large, it becomes impractical to use that kind of model, so in this case study we turn our attention to models which study the population as a whole.

As before, we divide the population into three groups: At day t, $I(t)$ is the proportion of the population that is infected and $S(t)$ is the proportion of the population that has never been infected. These quantities satisfy $0 \le I(t) \le 1$ and $0 \le S(t) \le 1$ for $t \ge 0$. The third part, $R(t)$, is the proportion of the population that was once infected but has now recovered, and it can be derived from these two: $R(t) = 1 - I(t) - S(t)$.

Models without Spatial Variation

In the models we studied before, the probability of an individual becoming infected depended on the status of the individual's neighbors. In the models in this section, we consider all individuals to be neighbors. This is equivalent to assuming a **well-mixed** model, in which all members of the population have contact with all others.

How might we model the three groups in the population? If the infection (or at least the contagious phase of the infection) lasts k days, then we might assume as an approximation that the rate of recovery is equal to the number infected divided by k. Thus, on average, $1/k$ of the infected individuals recover each day.

Let τ be the proportion of encounters between an infected individual and a susceptible individual that transmit the infection. Then the rate of new infections should increase as any of the parameters I, S, or τ increases, so we model this rate as $\tau I(t)S(t)$.

Next, we take the limit as the "timestep" Δt goes to zero, obtaining a system of ODEs. This gives us a simple but interesting Model 1:

$$\frac{dI(t)}{dt} = \tau I(t)S(t) - I(t)/k,$$
$$\frac{dS(t)}{dt} = -\tau I(t)S(t),$$

POINTER 21.1. Software.

The MATLAB function `ode23s` provides a good solver for the ODEs of Challenge 21.1. Most ODE software provides a mechanism for stopping the integration when some quantity goes to zero; in `ode23s` this is done by using the `Events` property in an option vector.

For Challenge 21.2, some ODE software, including `ode23s`, can be used to solve some DAEs; in MATLAB, this is done using the `Mass` property in the option vector.

In MATLAB, some DDEs can be solved using `dde23`.

$$\frac{dR(t)}{dt} = I(t)/k.$$

We start the model by assuming some proportion of infected individuals; for example, $I(0) = 0.005$, $S(0) = 1 - I(0)$, $R(0) = 0$.

CHALLENGE 21.1.

Solve Model 1 using `ode23s` for $k = 4$ and $\tau = .8$ for $t > 0$ until either $I(t)$ or $S(t)$ drops below 10^{-5}. Plot $I(t)$, $S(t)$, and $R(t)$ on a single graph. Report the proportion of the population that became infected and the maximum difference between $I(t) + S(t) + R(t)$ and 1.

Instead of using the equation $dR/dt = I/k$, we could have used the conservation principle

$$I(t) + S(t) + R(t) = 1$$

for all time. Substituting this for the dR/dt equation gives us an equivalent system of differential-algebraic equations (DAEs), studied in Section 20.4, and we call this Model 2.

CHALLENGE 21.2.

Redo Challenge 21.1 using Model 2 instead of Model 1. One way to do this is to differentiate the conservation principle and express the three equations of the model as $\boldsymbol{M}\boldsymbol{u}' = \boldsymbol{f}(t, \boldsymbol{u})$, where $\boldsymbol{M}$ is a 3×3 matrix and $\boldsymbol{u}$ is a function of t with three components. Another way is to use a DAE formulation.

There are many limitations in the model, but one of them is that the recovery rate is proportional to the current number of infections. This means that we are not very faithful to the hypothesis that each individual is infectious for k days. One way to model this more closely is to use a **delay differential equation** (DDE). We modify Model 1 by specifying that the rate of recovery at time t is equal to the rate of new infections at time $t - k$. This gives us a new model, Model 3:

$$\frac{dI(t)}{dt} = \tau I(t)S(t) - \tau I(t-k)S(t-k),$$

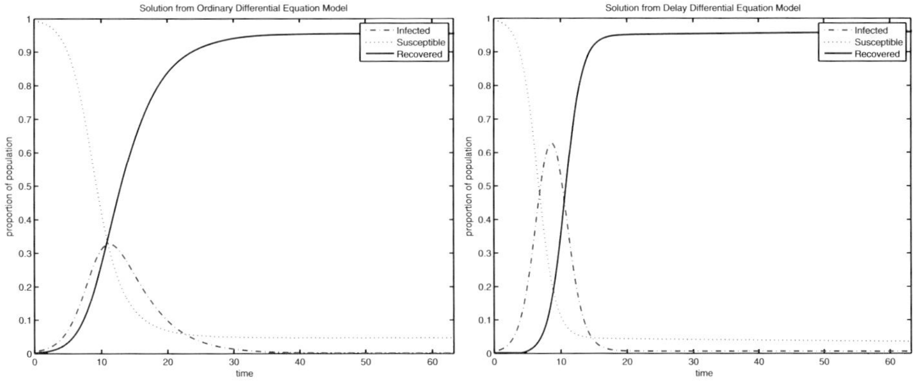

Figure 21.1. *Results of our models. (Left) Proportion of individuals infected by the epidemic from the* ODE *Model 1 or the* DAE *Model 2. (Right) Proportion of individuals infected by the epidemic from the* DDE *Model 3.*

$$\frac{dS(t)}{dt} = -\tau I(t)S(t),$$
$$\frac{dR(t)}{dt} = \tau I(t-k)S(t-k).$$

One disadvantage of this model is that we need to specify initial conditions not just at $t = 0$ but for $-k \le t \le 0$, so it requires a lot more information. A second disadvantage is that the functions I, S, and R are likely to have discontinuous derivatives (for example, at $t = 0$ and $t = k$, as we switch from dependence on the initial conditions to dependence only on the integration history). This causes solvers to do extra work at these points of discontinuity.

CHALLENGE 21.3.

Redo Challenge 21.1 with MATLAB's `dde23` using Model 3 instead of Model 1. For $t < 0$, use the initial conditions

$$I(t) = 0, \quad S(t) = 1, \quad R(t) = 0,$$

and let $I(0) = 0.005$, $S(0) = 1 - I(0)$, $R(0) = 0$. Note that these conditions match our previous ones, but have a jump at $t = 0$. Compare the results of the three models, as illustrated in Figure 21.1.

Models that Include Spatial Variation

Epidemics vary in space as well as time. They usually start in a single location and then spread, based on interaction of infected individuals with neighbors, as in the models of Chapter 19. The models of the previous section lose this characteristic. To recover it, we now let S, I, and R depend on a spatial coordinate (x, y) as well as t and see what such a model predicts.

Since people move in space, we introduce a **diffusion term** that allows infected individuals to affect susceptible individuals that are close to them in space. Diffusion adds a term $\delta((\partial^2 I)/(\partial x^2)+(\partial^2 I)/(\partial y^2))S$ to dI/dt, and subtracts the same term from dS/dt. This produces differential equations analogous to Model 1:

$$\begin{aligned}
\frac{\partial I(t,x,y)}{\partial t} &= \tau I(t,x,y)S(t,x,y) - I(t,x,y)/k \\
&\quad +\delta\left(\frac{\partial^2 I(t,x,y)}{\partial x^2}+\frac{\partial^2 I(t,x,y)}{\partial y^2}\right)S(t,x,y), \\
\frac{\partial S(t,x,y)}{\partial t} &= -\tau I(t,x,y)S(t,x,y) - \delta\left(\frac{\partial^2 I(t,x,y)}{\partial x^2}+\frac{\partial^2 I(t,x,y)}{\partial y^2}\right)S(t,x,y), \\
\frac{\partial R(t,x,y)}{\partial t} &= I(t,x,y)/k.
\end{aligned}$$

We assume that the initial values $I(0,x,y)$ and $S(0,x,y)$ are given, that we study the problem for $0 \le x \le 1$, $0 \le y \le 1$, and $t \ge 0$, and that there is no diffusion across the boundaries $x = 0$, $x = 1$, $y = 0$, and $y = 1$.

To solve this problem, Model 4, we **discretize** and approximate the solution at the points of a grid of size $n \times n$. Let $h = 1/(n-1)$ and let $x_i = ih$, $i = 0,\dots,n-1$, and $y_j = jh$, $j = 0,\dots,n-1$. Our variables are our approximations $I(t)_{ij} \approx I(t,x_i,y_j)$ and similarly for $S(t)_{ij}$ and $R(t)_{ij}$.

CHALLENGE 21.4.

(a) Use Taylor series expansions to show that we can approximate

$$\frac{\partial^2 I(t,x_i,y_j)}{\partial x^2} = \frac{I(t,x_{i-1},y_j) - 2I(t,x_i,y_j) + I(t,x_{i+1},y_j)}{h^2} + O(h^2).$$

A similar expression can be derived for $\partial^2 I(t,x_i,y_j)/\partial y^2$.

(b) Form a vector $\widehat{\boldsymbol{I}}(t)$ from the approximate values of $I(t)$ by ordering the unknowns as $I_{00}, I_{01}, \dots, I_{0,n-1}, I_{10}, I_{11}, \dots, I_{1,n-1}, \dots, I_{n-1,0}, I_{n-1,1}, \dots, I_{n-1,n-1}$, where $I_{ij}(t) = I(t,x_i,y_j)$. In the same way, form the vectors $\widehat{\boldsymbol{S}}(t)$ and $\widehat{\boldsymbol{R}}(t)$ and derive the matrix $\boldsymbol{A}$ so that our discretized equations become Model 4:

$$\begin{aligned}
\frac{\partial \widehat{\boldsymbol{I}}(t)}{\partial t} &= \tau\widehat{\boldsymbol{I}}(t).*\widehat{\boldsymbol{S}}(t) - \widehat{\boldsymbol{I}}(t)/k + \delta(\boldsymbol{A}\widehat{\boldsymbol{I}}(t)).*\widehat{\boldsymbol{S}}(t), \\
\frac{\partial \widehat{\boldsymbol{S}}(t)}{\partial t} &= -\tau\widehat{\boldsymbol{I}}(t).*\widehat{\boldsymbol{S}}(t) - \delta(\boldsymbol{A}\widehat{\boldsymbol{I}}(t)).*\widehat{\boldsymbol{S}}(t), \\
\frac{\partial \widehat{\boldsymbol{R}}(t)}{\partial t} &= \widehat{\boldsymbol{I}}(t)/k,
\end{aligned}$$

where the notation $\widehat{\boldsymbol{I}}(t).*\widehat{\boldsymbol{S}}(t)$ means the vector formed by the product of each component of $\widehat{\boldsymbol{I}}(t)$ with the corresponding component of $\widehat{\boldsymbol{S}}(t)$. To form the approximation near the boundary, assume that the (Neumann) boundary conditions imply that $I(t,-h,y) = I(t,h,y)$,

$I(t,1+h,y) = I(t,1-h,y)$ for $0 \le y \le 1$, and similarly for S and R. Make the same type of assumption at the two other boundaries.

There are two ways to use this model. First, suppose we fix the timestep Δt and use Euler's method to approximate the solution; this means we approximate the solution at $t + \Delta t$ by the solution at t, plus Δt times the derivative at t. This gives us an iteration

$$\begin{aligned}
\widehat{\boldsymbol{I}}(t+\Delta t) &= \widehat{\boldsymbol{I}}(t) + \Delta t(\tau \widehat{\boldsymbol{I}}(t).*\widehat{\boldsymbol{S}}(t) - \widehat{\boldsymbol{I}}(t)/k + \delta(\boldsymbol{A}\widehat{\boldsymbol{I}}(t)).*\widehat{\boldsymbol{S}}(t)),\\
\widehat{\boldsymbol{S}}(t+\Delta t) &= \widehat{\boldsymbol{S}}(t) + \Delta t(-\tau \widehat{\boldsymbol{I}}(t).*\widehat{\boldsymbol{S}}(t) - \delta(\boldsymbol{A}\widehat{\boldsymbol{I}}(t)).*\widehat{\boldsymbol{S}}(t)),\\
\widehat{\boldsymbol{R}}(t+\Delta t) &= \widehat{\boldsymbol{R}}(t) + \Delta t(\widehat{\boldsymbol{I}}(t)/k).
\end{aligned}$$

This model is very much in the spirit of the models we considered in the case study of Chapter 19, except that it is deterministic rather than stochastic.

Alternatively, we could apply a more accurate ODE solver to this model, and we investigate this in the next challenge.

CHALLENGE 21.5.

(a) Set $n = 11$ (so that $h = 0.1$), $k = 4$, $\tau = 0.8$ and $\delta = 0.2$ and use an ODE solver to solve Model 4. For initial conditions, set $S(0,x,y) = 1$ and $I(0,x,y) = R(0,x,y) = 0$ at each point (x,y), except that $S(0,0.5,0.5) = I(0,0.5,0.5) = .5$. (For simplicity, you need only use I and S in the model, and you may derive $R(t)$ from these quantities.) Stop the simulation when either the average value of $\widehat{\boldsymbol{I}}(t)$ or $\widehat{\boldsymbol{S}}(t)$ drops below 10^{-5}. Form a plot similar to that of Challenge 21.1 by plotting the average value of $I(t)$, $S(t)$, and $R(t)$ vs time. Compare the results.

(b) Let's vaccinate the susceptible population at a rate

$$\frac{\nu S(t,x,y) I(t,x,y)}{I(t,x,y) + S(t,x,y)}.$$

This rate is the derivative of the vaccinated population $V(t,x,y)$ with respect to time, and this term is subtracted from $\partial S(t,x,y)/\partial t$. So now we model four segments of the population: susceptible $S(t)$, infected $I(t)$, recovered $R(t)$, and vaccinated $V(t)$. Your program can track three of these and derive the fourth from the conservation principle $S(t) + I(t) + R(t) + V(t) = 1$. Run this model with $\nu = 0.7$ and compare the results with those of Model 4.

POINTER 21.2. Further Reading.

Model 1 is the SIR model of Kermack and McKendrick, introduced in 1927. It is discussed, for example, by Britton [21].

DDEs such as those in Challenge 21.3 arise in many applications, including circuit analysis. To learn more about these problems, consult a textbook such as that by Bellman and Cooke [12] or by Hale and Lunel [67].

Stochastic differential equations are an active area of research. Higham [77] gives a good introduction to computational aspects and supplies references for further investigation.

The differential equations leading to Model 4 are presented, for example, by Callahan [24], following a model with one space dimension given in [87].

If you want to experiment further with Model 4, incorporate the delay recovery term in place of $-\widehat{\boldsymbol{I}}(t)/k$.

CHALLENGE 21.6. (Extra)

Include a delay in Model 4. Solve the resulting DDE model numerically and compare with the previous results.

In the models we used in the case study of Chapter 19, we incorporated some randomness to account for factors that were not explicitly modeled. We could also put randomness into our differential equation models, resulting in **stochastic differential equations**. See Pointer 21.2 for references on this subject.

Chapter 22 / Case Study

Robot Control: Swinging Like a Pendulum

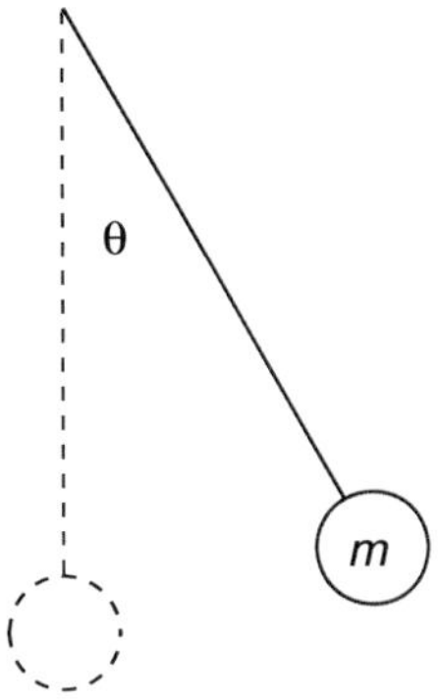

(coauthored by Yalin E. Sagduyu)

Suppose we have a robot arm with a single joint, simply modeled as a damped driven pendulum. It is amazing how such a trivial system illustrates so many difficult concepts! In this project, we study the stability and behavior of this robot arm and develop a strategy to move the arm from one position to another using minimal energy.

The Model

We assume that the pendulum of length ℓ has a bob of mass m. Figure 22.1 shows the pendulum's position at some time t, with the variable $\theta(t)$ denoting the angle that the pendulum makes with the vertical axis at that time. The angle is measured in radians. The acceleration of the pendulum is proportional to the angular displacement from vertical, and we model the drag due to friction with the air as being proportional to velocity. This yields a second-order ordinary differential equation (ODE) for $t \geq 0$:

$$m\ell \frac{\mathrm{d}^2\theta(t)}{\mathrm{d}t^2} + c\frac{\mathrm{d}\theta(t)}{\mathrm{d}t} + mg\sin(\theta(t)) = u(t), \tag{22.1}$$

where g is the gravitational acceleration on an object at the surface of the earth, and c is the damping (frictional) constant. The term $u(t)$ defines the external force applied to the pendulum. In this project, we consider what happens in three cases: no external force, constant external force driving the pendulum to a final state, and then a force designed to minimize the energy needed to drive the robot arm from an initial position to an angle θ_f.

Stability and Controllability of the Robot Arm

The solution to equation (22.1) depends on relations among m, ℓ, c, g, and $u(t)$ and ranges from fixed amplitude oscillations for the **undamped** case ($c = 0$) to decays (oscillatory or

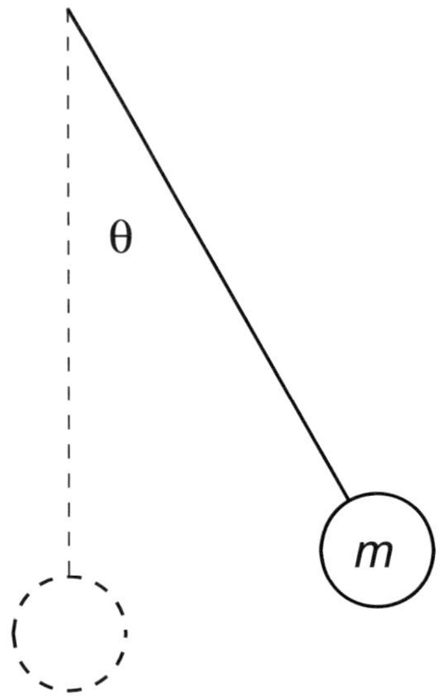

Figure 22.1. *We move this robot arm (pendulum), shown here at a position $\theta(t)$.*

strict) for the **damped** case ($c > 0$). Unfortunately, there is no simple analytical solution to the pendulum equation in terms of elementary functions unless we linearize the term $\sin(\theta(t))$ in (22.1) as $\theta(t)$, an approximation that is only valid for small values of $\theta(t)$. Despite the limitations of the linear approximation, the linearization enables us to find analytical solutions and also to apply the results of **linear control theory** to the specific problem of robot-arm control.

CHALLENGE 22.1.

Consider the **undriven damped pendulum** modeled by (22.1) with $u(t) = 0$ and $c > 0$. Linearize the second-order nonlinear differential equation using the approximation $\sin(\theta(t)) \approx \theta(t)$. Use the method in Section 20.1.1 to transform this equation into a first order system of ordinary differential equations of the form $\mathbf{y}' = \mathbf{A}\mathbf{y}$, where $\mathbf{A}$ is a 2×2 matrix, and the two components of the vector $\mathbf{y}(t)$ represent $y_{(1)}(t) = \theta(t)$, and $y_{(2)}(t) = \mathrm{d}\theta(t)/\mathrm{d}t$. Determine the eigenvalues of $\mathbf{A}$. Show that the damped system is **stable**, whereas the undamped system is not. (Recall from Section 20.1.2 that stability means that the real part of each eigenvalue of $\mathbf{A}$ is negative.) Use the eigenvalue information to show how the solutions behave in the damped and undamped systems.

Stability of the original differential equation (22.1) is more difficult to analyze than the stability of the linearized approximation to it, as we saw in Section 20.1.2. **Lyapunov stability** occurs when the total energy of an unforced (undriven), dissipative mechanical system decreases as the state of the system evolves in time. Therefore, the **state vector** $\mathbf{y}^T = [\theta(t), \mathrm{d}\theta(t)/\mathrm{d}t]$ approaches a constant value (**steady state**) corresponding to zero energy as time increases (i.e., $\mathbf{y}'(t) \to \mathbf{0}$ as $t \to \infty$). According to Lyapunov's formulation, the equilibrium point $\boldsymbol{y} = \boldsymbol{0}$ of a system described by the equation $\boldsymbol{y}' = \boldsymbol{f}(t,\boldsymbol{y})$ is globally asymptotically stable if $\lim_{t\to\infty} \boldsymbol{y}(t) = \boldsymbol{0}$ for any choice of $\boldsymbol{y}(0)$.

Let $\boldsymbol{y}' = \boldsymbol{f}(t,\boldsymbol{y})$ and let $\bar{\boldsymbol{y}}$ be a steady state solution of this differential equation. Terminology varies in the literature, but we use these definitions:

- A **positive definite Lyapunov function** v at $\bar{\boldsymbol{y}}(t)$ is a continuously differentiable function into the set of nonnegative numbers. It satisfies $v(\bar{\boldsymbol{y}}) = 0$, $v(\boldsymbol{y}(t)) > 0$ and

$$\frac{\mathrm{d}}{\mathrm{d}t} v(\boldsymbol{y}(t)) \leq 0$$

  for all $t > 0$ and all $\boldsymbol{y}$ in a neighborhood of $\bar{\boldsymbol{y}}$.

- An **invariant set** is a set for which the solution to the differential equation remains in the set when the initial state is in the set.

Suppose v is a positive definite Lyapunov function for a steady state solution $\bar{\boldsymbol{y}}$ of $\boldsymbol{y}' = \boldsymbol{f}(t,\boldsymbol{y})$. Then $\bar{\boldsymbol{y}}$ is **stable**. If in addition $\{\boldsymbol{y} : \mathrm{d}v(\boldsymbol{y}(t))/\mathrm{d}t = 0\}$ contains no invariant sets other than $\bar{\boldsymbol{y}}$, then $\bar{\boldsymbol{y}}$ is **asymptotically stable** [13]. This result guides our analysis.

Finding a Lyapunov function for a given problem can be difficult, but success yields important information. For unstable systems, small perturbations in the application of the external force can cause large changes in the behavior of the solution to the equation and therefore to the pendulum behavior, so the robot arm might behave erratically. Therefore, in practice we need to ensure that the system is stable.

CHALLENGE 22.2.

Consider the function

$$v(\theta, \mathrm{d}\theta/\mathrm{d}t) = \frac{(1-\cos\theta)g}{\ell} + \frac{1}{2}\left(\frac{\mathrm{d}\theta}{\mathrm{d}t}\right)^2$$

for the pendulum described by (22.1). Show that it is a valid positive definite Lyapunov function for the undriven model. Investigate the stability of the solution $\theta(t) = 0, \mathrm{d}\theta(t)/\mathrm{d}t = 0$ for undamped systems and damped systems.

Consider the first-order system described by $\boldsymbol{y}' = \boldsymbol{A}\boldsymbol{y} + \boldsymbol{B}\boldsymbol{u}$, where $\boldsymbol{A}$ is an $n \times n$ matrix and $\boldsymbol{B}$ is an $n \times m$ matrix. The matrices $\boldsymbol{A}$ and $\boldsymbol{B}$ may depend on time t, but in our example they do not. The system is **controllable** on $t \in [0, t_f]$ if given any initial state $\boldsymbol{y}(0)$ there exists a continuous function $\boldsymbol{u}(t)$ such that the solution of $\boldsymbol{y}' = \boldsymbol{A}\boldsymbol{y} + \boldsymbol{B}\boldsymbol{u}$ satisfies $\boldsymbol{y}(t_f) = \boldsymbol{0}$. For controllability on any time interval, it is necessary and sufficient that the $n \times nm$ controllability matrix $[\boldsymbol{B}, \boldsymbol{A}\boldsymbol{B}, \ldots, \boldsymbol{A}^{n-1}\boldsymbol{B}]$ have rank n on that interval, and the rank can be computed using the methods in Chapter 5.

Controllability of the robot arm means that we can specify a force that drives it to any desired position. We investigate the controllability of the linearized pendulum model.

CHALLENGE 22.3.

Consider the linearized version of the driven (forced), damped pendulum system with constant force term u. Transform the corresponding differential equation to a first order ODE system of the form $\boldsymbol{y}' = \boldsymbol{A}\boldsymbol{y} + \boldsymbol{B}u$. Specify the matrices $\boldsymbol{A}$ and $\boldsymbol{B}$ and show that the system is controllable for both the damped and undamped cases.

Numerical Solution of the Initial Value Problem

We now develop some intuition for the behavior of the original model and the linearized model by comparing them under various experimental conditions.

For the numerical investigations in Challenges 22.4–22.6, assume that $m = 1$ kg, $\ell = 1$ m, and $g = 9.81\text{m/sec}^2$, with $c = 0$ for the undamped case and $c = 0.5$ kg-m/sec for the damped case.

First we investigate the effects of damping and of applied forces.

CHALLENGE 22.4.

For the initial conditions $\theta(0) = \pi/4$ and $d\theta(0)/dt = 0$, use an ODE solver to find the numerical solutions on the interval $t = [0, 30]$ for the nonlinear model (22.1) for

1. an undamped ($c = 0$), undriven ($u = 0$) pendulum,
2. a damped ($c > 0$), undriven ($u = 0$) pendulum, and
3. a damped ($c > 0$), driven pendulum with the applied forces $u = mg\sin(\theta_f)$, where $\theta_f = \pi/8, \pi/4, \pi/3$.

Repeat the same experiments for the linearized model of the pendulum and discuss the difference in behavior of the solutions. It helps to plot $\theta(t)$ for the corresponding linear and nonlinear models in the same figure, as in Figure 22.2.

Missing Data: Solution of the Boundary Value Problem

In Challenge 22.4, we solved the **initial value problem**, in which values of θ and $d\theta/dt$ were given at time $t = 0$. In many cases, we do not have the initial value for $d\theta/dt$; this value might not be **observable**. The missing initial condition prevents us from applying standard methods to solve initial value problems. Instead, we might have the value $\theta(t_B) = \theta_B$ at some other time t_B. Next we investigate two solution methods for this **boundary value problem**.

Recall from Section 20.5.1 that the idea behind the **shooting method** is to guess at the missing initial value $z = d\theta(0)/dt$, integrate equation (1) using our favorite method, and then use the results at the final time t_B to improve the guess. To do this systematically, we use a nonlinear equation solver to solve the equation $\rho(z) \equiv \theta_z(t_B) - \theta_B = 0$, where $\theta_z(t_B)$

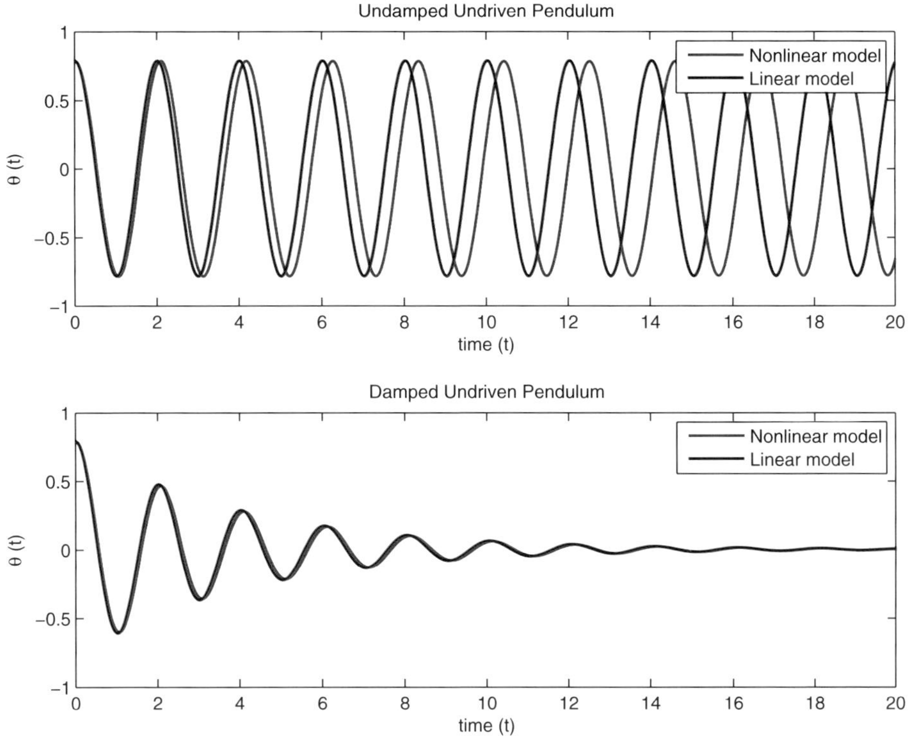

Figure 22.2. *The linear and nonlinear undriven models from Challenge 22.4.*

is the value reported by an initial value problem ODE solver for $\theta(t_B)$, given the initial condition $z = d\theta(0)/dt$.

The **finite difference method** of Section 20.5.2 is an alternate method to solve a boundary value problem. Choose a small time increment $h > 0$ and replace the first derivative in the linearized model of problem (1) by

$$\frac{d\theta(t)}{dt} \approx \frac{\theta(t+h) - \theta(t-h)}{2h}$$

and second derivative by

$$\frac{d^2\theta(t)}{dt^2} \approx \frac{\theta(t+h) - 2\theta(t) + \theta(t-h)}{h^2}.$$

Let $n = t_B/h$, and write the equation for each value $\theta_j \approx \theta(jh)$, $j = 1, \ldots, n-1$. The **boundary conditions** can be stated as $\theta_0 = \theta(0)$, $\theta_n = \theta_B$. This method transforms the linearized version of the second-order differential equation (22.1) to a system of $n-1$ linear equations with $n-1$ unknowns. Assuming the solution to this linear system exists, we then use our favorite linear system solver to solve these equations.

CHALLENGE 22.5.

Consider the linearized model with constant applied force $u(t) = mg\sin(\pi/8)$ and damping constant $c = 0.5$. Suppose that we are given the boundary conditions $\theta(0) = \pi/32$ and $\theta(10) = \theta_B$, where θ_B is the value of the solution when $\mathrm{d}\theta(0)/\mathrm{d}t = 0$.

Apply the shooting method to find the solutions to the damped, driven linearized pendulum equation on the time interval $t = [0, 10]$. Try different initial guesses for $\mathrm{d}\theta(0)/\mathrm{d}t$ and compare the results.

Now use the finite difference method to solve this boundary value problem with $h = 0.01$. Use your favorite linear system solver to solve the resulting linear system of equations.

Compare the results of the shooting and finite difference methods with the solution to the original initial value problem.

Controlling the Robot Arm

Finally, we investigate how to design a forcing function that drives the robot arm from an initial position to some other desired position with the least expenditure of energy. We measure energy as the integral of the absolute force applied between time 0 and the convergence time t_c at which the arm reaches its destination:

$$e_f = \int_0^{t_c} |u(t)|\mathrm{d}t\,.$$

CHALLENGE 22.6.

Consider the damped, driven pendulum with applied force

$$u(t) = mg\sin(\theta_f) + m\ell b\,\mathrm{d}\theta(t)/\mathrm{d}t,$$

where $\theta_f = \pi/3$. This force is a particular **closed-loop control** with control parameter b, and it drives the pendulum position to θ_f. The initial conditions are given as $\theta(0) = \pi/4$ and $\mathrm{d}\theta(0)/\mathrm{d}t = 0$. Assume $c = 0.5$ as the damping constant, $t_c = 5$ sec as the time limit for achieving the position θ_f, and $h = 0.01$ as the time step length for numerical solutions.

We call a parameter b successful if the pendulum position satisfies $|\theta(t) - \theta_f| < 10^{-3}$ for $5 \le t \le 10$. Approximate the total energy by

$$\hat{e}_f \approx \sum_{k=1}^{5/h} |u(kh)|h\,.$$

Write a function that evaluates $\hat{e}_f$. The input to the function should be the control parameters b and the output should be the approximate total consumed energy $\hat{e}_f$.

For stability of the closed-loop control system, we impose the constraint $b < c/(m\ell)$, which make the real parts of the eigenvalues (of the linearized version) of the system strictly negative.

POINTER 22.1. Further Reading.

For Challenge 22.1, consult Section 20.1.1 for converting second derivative equations to a system of equations involving only first derivatives. An elementary linear algebra textbook [102] discusses computation of the eigenvalues and eigenvectors of a 2×2 matrix, and Section 20.1.2 gives an example of how to solve linear ODE systems once the eigensystem is known.

An ODE textbook [101] can serve as a reference on Lyapunov stability, used in Challenge 22.2. Control theory textbooks [116, 128] discuss stability plus the concept of controllability used in Challenge 22.3.

Challenge 22.6 relies on an ODE solver and a function for minimization of a function of a single variable under bound constraints (e.g., MATLAB's `fminbnd`).

Now use your favorite constrained minimization solver to select the control parameter b to minimize the energy function $\hat{e}_f(b)$. Display the optimal parameter and graph the resulting $\theta(t)$.

Chapter 23 / Case Study

Finite Differences and Finite Elements: Getting to Know You

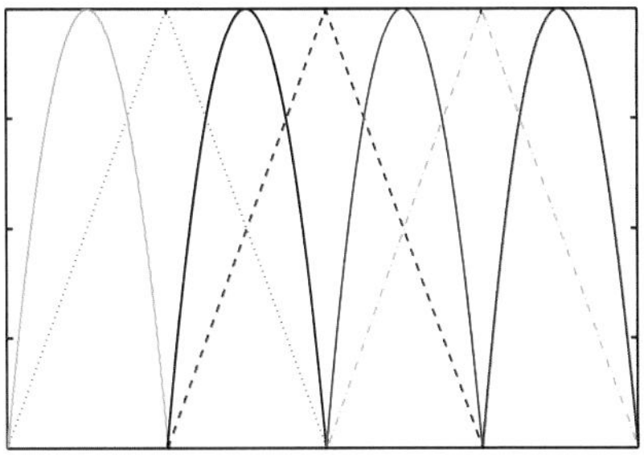

Numerical solution of differential equations relies on two main methods: finite differences and finite elements. In this case study, we explore the nuts and bolts of the two methods for a simple **two-point boundary value problem**:

$$-(a(t)u'(t))' + c(t)u(t) = f(t) \text{ for } t \in (0,1),$$

with the functions a, c, and f given and $u(0) = u(1) = 0$. We assume that $a(t) \geq a_0$, where a_0 is a positive number, and $c(t) \geq 0$ for $t \in [0,1]$. Check Pointer 20.8 for conditions that guarantee the existence of a unique solution for our problem.

The Finite Difference Method

Let's rewrite our equation as

$$-a(t)u''(t) - a'(t)u'(t) + c(t)u(t) = f(t) \tag{23.1}$$

and approximate each derivative of u by a finite difference:

$$u'(t) = \frac{u(t) - u(t-h)}{h} + O(h),$$
$$u''(t) = \frac{u(t-h) - 2u(t) + u(t+h)}{h^2} + O(h^2).$$

(We'll compute $a'(t)$ analytically, so we won't need an approximation to it.)

The finite difference approach is to choose **grid points** $t_j = jh$, where $h = 1/(M-1)$ for some large integer M, and solve for $u_j \approx u(t_j)$ for $j = 1, \ldots, M-2$. We write one equation for each unknown, by substituting our finite difference approximations for u'' and u' into (23.1), and then evaluating the equation at $t = t_j$.

CHALLENGE 23.1.

Let $M = 6$, $a(t) = 1$, and $c(t) = 0$ and write the four finite difference equations for u at $t = .2, .4, .6$, and $.8$.

POINTER 23.1. Some Notes on the Challenges.

To debug your programs, it is helpful to experiment with the simplest test problem and a small number of grid points. Look ahead to Challenge 23.6 for sample problems.

Challenge 23.2 uses the MATLAB function `spdiags` to construct a sparse matrix. If you have never used sparse matrices in MATLAB, print the matrix `A` to see that the data structure for it contains the row index, column index, and value for each nonzero element. If you have never used `spdiags`, type `help spdiags` to see the documentation, and then try it on your own data to see exactly how the matrix elements are defined.

Use MATLAB's `quad` to compute the integrals for the entries in the matrix and right-hand side for the finite element formulations.

Before tackling the programming for Challenge 23.5 and 23.6, take some time to understand exactly where the nonzeros are in the matrix, and exactly what intervals of integration should be used. The programs are short, but it is easy to make mistakes if you don't understand what they compute.

In Challenge 23.7, we measure **work** by counting the number of multiplications. One alternative is to count the number of floating-point computations, but this usually gives a count of about twice the number of multiplications, since typically multiplications and additions are paired in computations. Computing time is another very useful measure of work, but it can be contaminated by the effects of other users or other processes on the computer.

In determining and understanding the convergence rate in Challenge 23.7, plotting the solutions or the error norms might be helpful.

Notice that the matrix constructed in Challenge 23.1 is tridiagonal. (See Pointer 5.2.) The full matrix requires $(M-2)^2$ storage locations, but, if we are careful, we can instead store all of the data in $O(M)$ locations by agreeing to store only the nonzero elements, along with their row and column indices. This is a standard technique for storing **sparse matrices**, those whose elements are mostly zero; see Chapter 27.

Let's see how this finite difference method is implemented.

CHALLENGE 23.2.

The MATLAB function `finitediff1.m`, found on the website, implements the finite difference method for our equation. The inputs are the parameter M and the functions `a`, `c`, and `f` that define the equation. Each of these functions takes a vector of points as input and returns a vector of function values. (The function `a` also returns a second vector of values of a'.) The outputs of `finitediff1.m` are a vector `ucomp` of computed estimates of u at the grid points `tmesh`, along with the matrix `A` and the right-hand side `g` from which `ucomp` was computed, so that `A ucomp = g`.

Add documentation to the function `finitediff1.m` so that a user could easily use it, understand the method, and modify the function if necessary. (See Section 4.1 for advice on documentation.)

There is a mismatch in `finitediff1.m` between our approximation to u'', which is second order in h, and our approximation to u', which is only first order. We can compute a better solution, for the same cost, by using a second-order (central difference) approximation to u', so next we make this change to our function.

CHALLENGE 23.3.

Define a central difference approximation to the first derivative by

$$u'(t) \approx \frac{u(t+h) - u(t-h)}{2h}.$$

Modify the function of Challenge 23.2 to produce a function `finitediff2.m` that uses this second-order accurate central difference approximation (studied in Challenge 20.19) in place of the first-order approximation.

The Finite Element Method

We'll use a **Galerkin method** with finite elements to solve our problem. In particular, we notice that

$$-(a(t)u'(t))' + c(t)u(t) = f(t) \text{ for } t \in (0,1)$$

implies that

$$\int_0^1 (-(a(t)u'(t))' + c(t)u(t))v(t)\, dt = \int_0^1 f(t)v(t)\, dt$$

for all functions v. Now we use integration by parts on the first term, recalling that our boundary values are zero, to obtain the equation

$$\int_0^1 a(t)u'(t)v'(t) + c(t)u(t)v(t)\, dt = \int_0^1 f(t)v(t)\, dt,$$

for all functions v. If a, c, and f are smooth functions (i.e., their first few derivatives exist), then the solution to our differential equation satisfies the boundary conditions and has a first derivative, with the integral of $(u'(t))^2$ on $[0,1]$ finite. We call the space of all such functions H_0^1, and that is also the space we draw v from.

How does this help us solve the differential equation? We choose a subspace S_h of H_0^1 that contains functions that are good approximations to every function in H_0^1, and we look for a function $u_h \in S_h$ so that

$$\int_0^1 a(t)u_h'(t)v_h'(t) + c(t)u_h(t)v_h(t)\, dt = \int_0^1 f(t)v_h(t)\, dt$$

for all functions $v_h \in S_h$. This gives us an approximate solution to our problem.

A common choice for S_h is the set of functions that are continuous and linear on each interval $[t_{j-1}, t_j] = [(j-1)h, jh]$, $j = 1, \ldots, M-1$ (piecewise linear elements), where

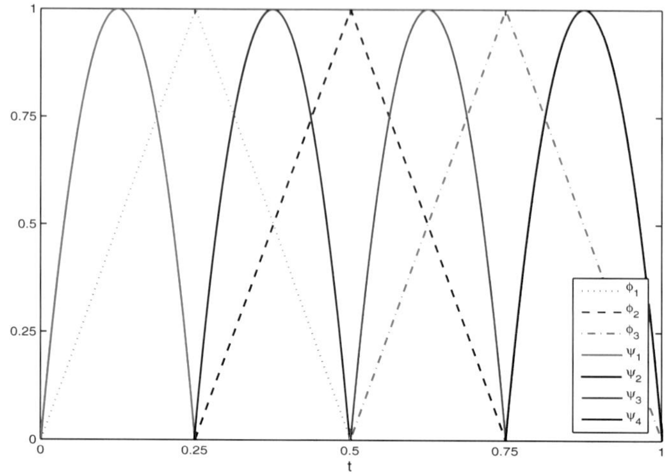

Figure 23.1. *The nonzero pieces of the three linear (ϕ) and four quadratic (ψ) basis functions for three interior grid points and four subintervals ($M = 5$).*

$h = 1/(M-1)$. We can construct our solution using any basis for S_h, but one basis is particularly convenient: the set of **hat functions** ϕ_j, $j = 1, \ldots, M-2$, where

$$\phi_j(t) = \begin{cases} \frac{t-t_{j-1}}{t_j-t_{j-1}}, & t \in [t_{j-1}, t_j], \\ \frac{t-t_{j+1}}{t_j-t_{j+1}}, & t \in [t_j, t_{j+1}], \\ 0 & \text{otherwise}. \end{cases}$$

These are designed to satisfy $\phi_j(t_j) = 1$ and $\phi_j(t_k) = 0$ if $j \neq k$; see Figure 23.1 for an illustration. Note that ϕ_j is nonzero only on the interval (t_{j-1}, t_{j+1}) (i.e., it has **small support**), but it is defined everywhere.

Then we can express our approximate solution u_h as

$$u_h(t) = \sum_{j=1}^{M-2} u_j \phi_j(t)$$

for some coefficients u_j, which happen to be approximate values for $u(t_j)$.

Define

$$\begin{aligned} \mathrm{a}(u,v) &= \int_0^1 (a(t)u'(t)v'(t) + c(t)u(t)v(t))\,\mathrm{d}t, \\ (f,v) &= \int_0^1 f(t)v(t)\,\mathrm{d}t. \end{aligned}$$

Then our solution u satisfies

$$\mathrm{a}(u,v) = (f,v)$$

for all $v \in H_0^1$, and we demand that our approximate solution $u_h \in S_h$ satisfy

$$a(u_h, v_h) = (f, v_h)$$

for all $v_h \in S_h$. In Challenge 23.4, we reduce this to a linear system of equations that can be solved for the coefficients u_j, and we implement our ideas in Challenge 23.5.

CHALLENGE 23.4.

(a) Since the functions ϕ_j form a basis for S_h, any function $v_h \in S_h$ can be written as

$$v_h(t) = \sum_{j=1}^{M-2} v_j \phi_j(t)$$

for some coefficients v_j. Show that if

$$a(u_h, \phi_j) = (f, \phi_j) \tag{23.2}$$

for $j = 1, \ldots, M-2$, then

$$a(u_h, v_h) = (f, v_h)$$

for all $v_h \in S_h$.

(b) Putting the unknowns u_j in a vector $\boldsymbol{u}$ we can write the system of equations resulting from (23.2) as $\boldsymbol{A}\boldsymbol{u} = \boldsymbol{g}$, where the (j,k) entry in $\boldsymbol{A}$ is $a(\phi_j, \phi_k)$ and the jth entry in $\boldsymbol{g}$ is (f, ϕ_j). Write this system of equations for $M = 6$, $a(t) = 1$, $c(t) = 0$, and compare with your solution to Challenge 23.1.

CHALLENGE 23.5.

Write a function `fe_linear.m` that has the same inputs and outputs as `finitediff1.m` but computes the finite element approximation to the solution using piecewise linear elements. Remember to store $\boldsymbol{A}$ as a sparse matrix.

It can be shown that the computed solution is within $O(h^2)$ of the exact solution, if the data is smooth enough. Better accuracy can be achieved if we use **higher-order elements**; for example, piecewise quadratic elements would produce a result within $O(h^3)$ for smooth data. A convenient basis for this set of elements is the piecewise linear basis plus $M-1$ quadratic functions ψ_j that are zero outside $[t_{j-1}, t_j]$ and satisfy

$$\begin{aligned} \psi_j(t_j) &= 0, \\ \psi_j(t_{j-1}) &= 0, \\ \psi_j(t_{j-1} + h/2) &= 1 \end{aligned}$$

for $j = 1, \ldots, M-1$. See Figure 23.1 for an illustration.

CHALLENGE 23.6.

Write a function `fe_quadratic.m` that has the same inputs and outputs as `finitediff1.m` but computes the finite element approximation to the solution using piecewise quadratic elements. In order to keep the number of unknowns comparable to the number in the previous functions, let the number of intervals be $m = \lfloor M/2 \rfloor$. When $M = 10$, for example, we have 5 quadratic basis functions (one for each subinterval) and 4 linear ones (one for each interior grid point). If you order the basis elements as $\psi_1, \phi_1, \ldots, \psi_{m-1}, \phi_{m-1}, \psi_m$ then the matrix $\boldsymbol{A}$ has 5 nonzero bands including the main diagonal. Compute one additional output `uval` which is the finite element approximation to the solution at the $m-1$ interior grid points and the m midpoints of each interval, where the $2m-1$ equally spaced points are ordered smallest to largest. (In our previous methods, this was equal to `ucomp`, but now the values at the midpoints of the intervals are a linear combination of the linear and quadratic elements.)

Now we have four solution algorithms, so we define a set of functions for experimentation:

$$
\begin{aligned}
u_1(t) &= t(1-t)e^t, \\
u_2(t) &= \begin{cases} u_1(t) & \text{if } t \le 2/3, \\ t(1-t)e^{2/3} & \text{if } t > 2/3, \end{cases} \\
u_3(t) &= \begin{cases} u_1(t) & \text{if } t \le 2/3, \\ t(1-t) & \text{if } t > 2/3, \end{cases} \\
a_1(t) &= 1, \\
a_2(t) &= 1 + t^2, \\
a_3(t) &= \begin{cases} a_2(t) & \text{if } t \le 1/3, \\ t + 7/9 & \text{if } t > 1/3, \end{cases} \\
c_1(t) &= 0, \\
c_2(t) &= 2, \\
c_3(t) &= 2t.
\end{aligned}
$$

For each particular choice of u, a, and c, we define f using (23.1).

CHALLENGE 23.7.

Use your four algorithms to solve 7 problems:

- a_1 with c_j $(j = 1, 2, 3)$ and true solution u_1.
- a_j $(j = 2, 3)$ with c_1 and true solution u_1.
- a_1 and c_1 with true solution u_j $(j = 2, 3)$.

Compute three approximations for each algorithm and each problem, with the number of unknowns in the problem chosen to be 9, 99, and 999. For each approximation, print

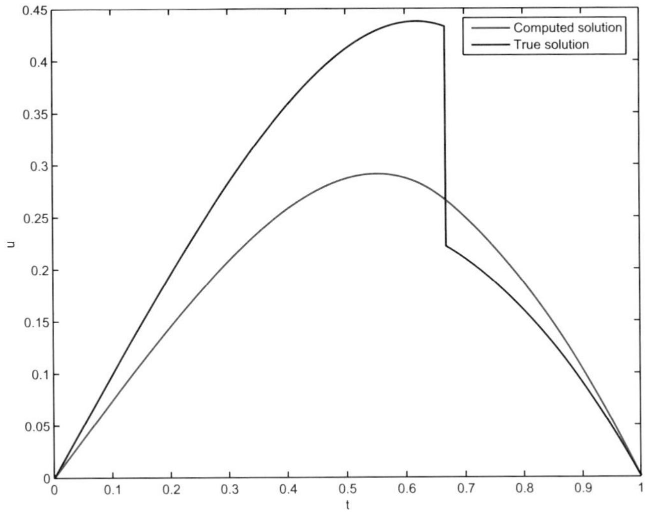

Figure 23.2. *The "solution" to the seventh test problem. We compute an accurate answer to a different problem.*

$\|\boldsymbol{u}_{computed} - \boldsymbol{u}_{true}\|_\infty$, where $\boldsymbol{u}_{true}$ is the vector of true values at the points jz, where $z = 1/10, 1/100$, or $1/1000$, respectively.

Discuss the results:

- How easy is it to program each of the four methods? Estimate how much work MATLAB does to form and solve the linear systems. (The work to solve the tridiagonal systems should be about $5M$ multiplications, and the work to solve the 5-diagonal systems should be about $11M$ multiplications, so you just need to estimate the work in forming each system.)

- For each problem, note the observed convergence rate r: if the error drops by a factor of 10^r when M is increased by a factor of 10, then the observed convergence rate is r.

- Explain any deviations from the theoretical convergence rate: $r = 1$ and $r = 2$ for the two finite difference implementations, and $r = 2$ and $r = 3$ for the finite element implementations when measuring $(u - u_h, u - u_h)^{1/2}$. The solution to the last problem is shown in Figure 23.2.

In doing this work, we begin to understand the complexities of implementation of finite difference and finite element methods. We have left out many features that a practical implementation should contain. In particular, the algorithm should be adaptive, estimating the error on each grid interval and subdividing the intervals (or raising the order of polynomials) where the error is too high. This method can handle partial differential equations, too. Luckily, there are good implementations of these methods for two- and three-dimensional domains, so we don't need to write our own.

POINTER 23.2. Further Reading.

A good introduction to the theory of finite difference and finite element methods is given by Gockenbach [59]; for a more advanced treatment, see, for example, Larsson and Thomée [99, Chap. 2, Section 4.1, Section 5.1].

Unit VI

Nonlinear Equations and Continuation Methods

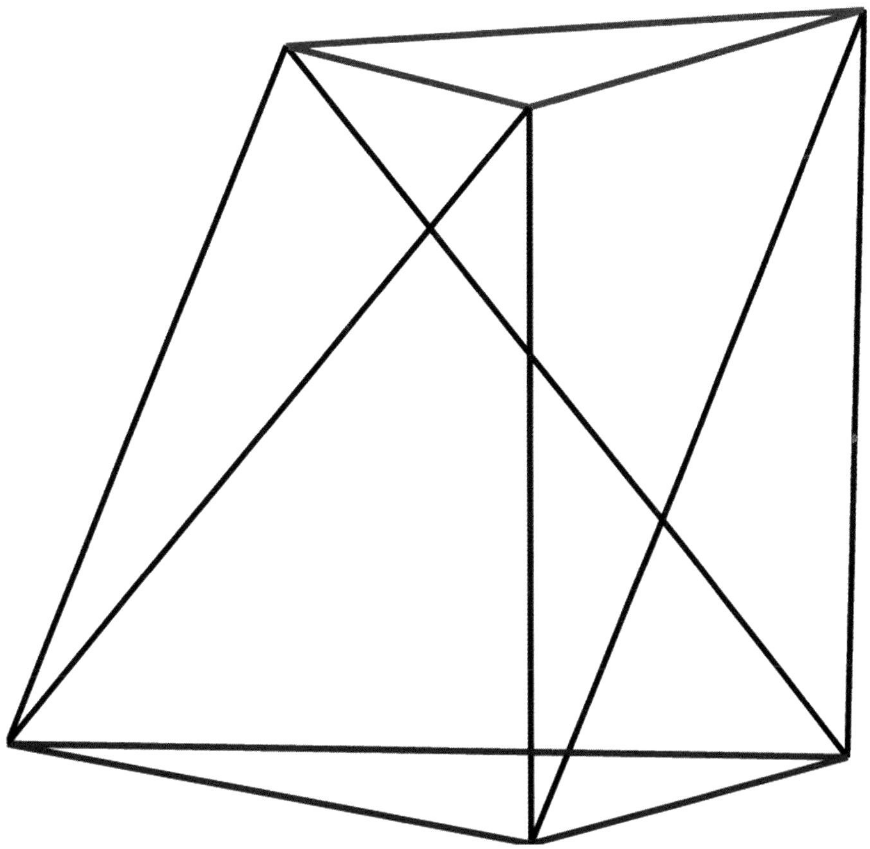

The basic techniques for solving systems of nonlinear equations are drawn from those for solving linear equations, optimization problems, and differential equations. In this unit, we survey these techniques. First we survey Newton-like methods, related to those discussed in Unit III for optimization problems. Then we consider a class of methods called continuation methods that are useful for particularly difficult problems.

Continuation methods are applied to a truss problem in Chapter 25. Then in Chapter 26 we solve a system of nonlinear equations to determine some parameters related to the life cycle of a flour beetle.

BASICS: To understand this unit, the following background is helpful:

- Properties of polynomial functions of a single variable x, such as how many roots they have and the use of polynomials in modeling other functions.
- Numerical solution of a single nonlinear equation.

This background material can be found in a standard introductory textbook (e.g., [32, 148]).

MASTERY: After you have worked through this unit, you should be able to do the following:

- Describe the methods used in `fzero`.
- Formulate a nonlinear equation as a least squares problem.
- Recognize polynomial equations and apply appropriate algorithms.
- Apply Newton's method and Newton-like methods, with linesearch or trust region, to solve nonlinear equations.
- Apply fixed-point iterations to solve nonlinear equations, and realize that they are very slow.
- Define homotopy.
- Write a program to find the solution to a nonlinear equation using a continuation method. Solve the intermediate problems using a nonlinear equation solver, or follow the path using an ODE solver.
- Explain why continuation methods sometimes solve problems that standard nonlinear equation solvers can't.
- Identify cases in which continuation methods run into trouble (bifurcation, etc.).
- Use inverse interpolation to solve a single nonlinear equation.
- Use the theorems that guarantee a globally convergent continuation method.
- Determine the stability of a dynamical system (system of ODEs).
- Solve nonlinear least squares problems.

Chapter 24

Nonlinear Systems

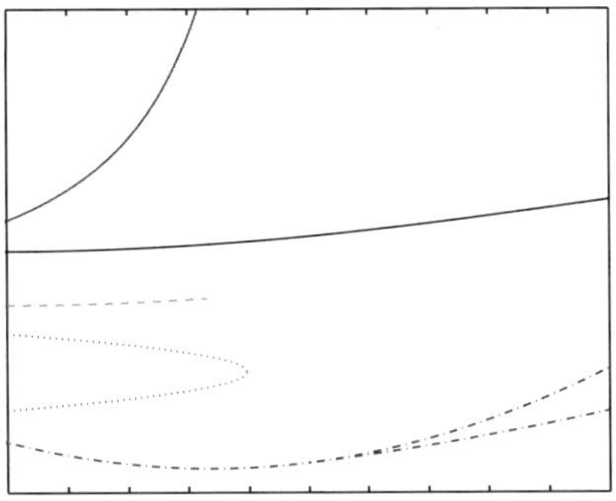

24.1 The Problem

In this chapter we focus on solving nonlinear systems of equations.

Nonlinear system of equations: Given a function $\boldsymbol{F} : \mathcal{R}^n \to \mathcal{R}^n$, find a point $\boldsymbol{x} \in \mathcal{R}^n$ such that

$$\boldsymbol{F}(\boldsymbol{x}) = \boldsymbol{0}.$$

We call such a point a solution of the equation or a **zero** of $\boldsymbol{F}$.

We'll assume that $n > 1$; see Pointer 24.1 for methods for solving nonlinear equations with a single variable. An important special case is **polynomial root finding**, in which $\boldsymbol{F}$ is a **polynomial** in the variables $\boldsymbol{x}$. An example is the system

$$x^2 y^3 + xy = 2,$$
$$2xy^2 + x^2 y + xy = 0.$$

In this case, $\boldsymbol{x} = [x, y]^T$. If a system is polynomial, then this feature should be exploited in the solution algorithm, and there is special purpose software that finds all of the solutions, including the complex ones, reliably. See Pointers 24.1 and 24.4.

Solving nonlinear equations is a close kin to solving optimization problems. The main difference is that there is no natural **merit function** $f(\boldsymbol{x})$ to use to measure our progress.

From our study of optimization problems, we already have many techniques that help us in solving nonlinear equations. The main algorithms for both problems are **Newton's method** and **Newton-like methods**, but there are some differences in how they are used. We know that every solution to a differentiable optimization problem $\min_{\boldsymbol{x}} f(\boldsymbol{x})$ is the solution to the nonlinear system of equations formed by setting the gradient $\boldsymbol{g}$ of f equal to zero, and this is a system of nonlinear equations. General systems of nonlinear equations lack two important features of the system $\boldsymbol{g}(\boldsymbol{x}) = \boldsymbol{0}$:

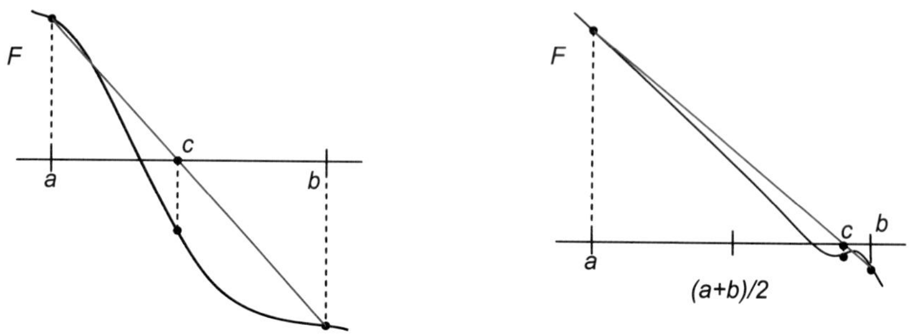

Figure 24.1. *To solve the nonlinear equation $F(z) = 0$, the* `zeroin` *algorithm uses the secant method when sufficient progress is made, and bisection otherwise. We begin with $[a,b]$ as an interval bracketing the root, so that $F(a)F(b) < 0$. On the left, the secant estimate c is the point where the line connecting the points $(a, F(a))$ and $(b, F(b))$ intersects the horizontal axis. We have a new bracketing interval $[a,c]$, since $F(a)F(c) < 0$, and it is much shorter than the old interval. On the right, the secant method yields a bracketing interval $[a,c]$, which is close in length to the initial interval, so* `zeroin` *would use bisection instead, yielding an interval $[(a+b)/2, b]$ of half the length of the original interval.*

POINTER 24.1. Software for One-Dimensional Problems.

The one-dimensional case ($n = 1$) is covered in elementary textbooks. The best software for the general problem is some variant on Richard Brent's `zeroin` [19], which determines an interval that contains the solution and then reduces its length through a combination of bisection and the secant method, as illustrated in Figure 24.1. The MATLAB algorithm related to `zeroin` is called `fzero`. Sometimes a simple approach such as inverse interpolation, considered in Challenge 24.5, is a useful method. If $F(x)$ is a polynomial, then very effective special purpose algorithms are available to find all of the roots; see, for example, [153, 86].

- Instead of the Hessian matrix of f, we have the **Jacobian matrix $\boldsymbol{J}$** of first derivatives of $\boldsymbol{F}$ with elements

$$J_{ik}(\boldsymbol{x}) = \frac{\partial F_i(\boldsymbol{x})}{\partial x_k}.$$

  Unlike the Hessian matrix, the Jacobian matrix is generally not symmetric.

- Linesearches are more difficult to guide, since we have no function $f(\boldsymbol{x})$ by which to measure progress. Some methods use $\|\boldsymbol{F}(\boldsymbol{x})\|$ as a merit function, but there are difficulties with this approach.

Our plan is to recall our discussion of nonlinear least squares problems and then to study Newton-like methods (for easy systems of nonlinear equations) and continuation methods (for hard problems). We'll compare these methods in the case study of Chapter 25.

POINTER 24.2. Existence, Uniqueness, and Sensitivity of Solutions to Nonlinear Equations.

There is no obvious property of $\boldsymbol{F}(\boldsymbol{x}) = \boldsymbol{0}$ that tells us whether or not a real solution exists. For example, consider the equation $F_1(\boldsymbol{x}) = x_1^2 + x_2^2 - r^2$, defining a circle centered at the origin, and the equation $F_2(\boldsymbol{x}) = x_1^2 + 2x_2^2 - c^2$, defining an ellipse with the same center. Depending on the choice of r and c, we can have no (real) point that satisfies both equations (e.g., $r = 10, c = 1$), 2 points (e.g., $r = c = 1$), or 4 points (e.g., $r = 1, c = 2$). There are always 4 complex solution points, however, since this is a polynomial system of equations, and a theorem of Bezoit tells us that if there are not an infinite number of solutions, then the number of real and complex solutions (counting multiplicities) is determined by multiplying the degrees of each equation. Since the highest-order term in each is 2, the degree of our system is 4.

One theory of existence of real solutions is based on **contraction maps**. If the function $\widehat{\boldsymbol{F}}(\boldsymbol{x})$ has a Lipschitz constant less than 1 for all $\boldsymbol{x}$ in a closed set, then there is a unique point in that set satisfying $\boldsymbol{x} = \widehat{\boldsymbol{F}}(\boldsymbol{x})$. Therefore if $\widehat{\boldsymbol{F}}(\boldsymbol{x}) = \boldsymbol{F}(\boldsymbol{x}) + \boldsymbol{x}$ is a contraction map on a closed set, then $\boldsymbol{F}(\boldsymbol{x}) = \boldsymbol{0}$ has a unique solution in that set.

Sensitivity of solutions can be understood through Taylor series expansion. Suppose $\boldsymbol{x}$ solves $\boldsymbol{F}(\boldsymbol{x}) = \boldsymbol{0}$. Then

$$\boldsymbol{F}(\boldsymbol{y}) = \boldsymbol{F}(\boldsymbol{x}) + \boldsymbol{J}(\boldsymbol{\eta})(\boldsymbol{y} - \boldsymbol{x}) = \boldsymbol{J}(\boldsymbol{\eta})(\boldsymbol{y} - \boldsymbol{x}),$$

where $\boldsymbol{\eta}$ is a point between $\boldsymbol{x}$ and $\boldsymbol{y}$, so

$$\|(\boldsymbol{y} - \boldsymbol{x})\| = \|(\boldsymbol{J}(\boldsymbol{\eta}))^{-1}\boldsymbol{F}(\boldsymbol{y})\|.$$

If the norm of the residual $\boldsymbol{F}(\boldsymbol{y})$ is small, then $\|(\boldsymbol{y} - \boldsymbol{x})\|$ will be small when $\boldsymbol{J}(\boldsymbol{\eta})$ is well conditioned, so sensitivity of the zero to perturbations in $\boldsymbol{F}$ depends on the conditioning of the Jacobian in the neighborhood of the solution.

24.2 Nonlinear Least Squares Problems

We can solve $\boldsymbol{F}(\boldsymbol{x}) = \boldsymbol{0}$ by solving the **nonlinear least squares problem**

$$\min_{\boldsymbol{x}} \|\boldsymbol{F}(\boldsymbol{x})\|_2^2$$

using any of our methods from Chapter 9. We seek a minimizer that gives a function value equal to zero. (There may be local minimizers that give a function value greater than zero.)

The advantages of this approach are that it uses all of our old machinery, and it can also be used for **overdetermined** systems in which the number of equations is greater than the number of variables. Often our system of equations arises from trying to fit a model to a set of data that have some errors; for example, in Section 5.3.4 we tried to fit a straight line $x_1 + tx_2$ to a set of 10 measurements of pollution data. This gave us a **linear least squares problem**. If the variables in the model appear nonlinearly, for example as $x_1 t^{x_2} + x_3$, then our problem becomes nonlinear least squares, and we again try to make the model match the data as closely as possible. We studied such problems in the case studies of Chapter 12 and 13. Alternate formulations of the data fitting problem replace the 2-norm with a 1-norm or ∞-norm.

The main disadvantage of the least squares approach is that derivatives are rather expensive: if $f(\boldsymbol{x}) = \|\boldsymbol{F}(\boldsymbol{x})\|^2$, then

$$\begin{aligned} \boldsymbol{g}(\boldsymbol{x}) &= 2\boldsymbol{J}(\boldsymbol{x})^T \boldsymbol{F}(x), \\ \boldsymbol{H}(\boldsymbol{x}) &= 2\boldsymbol{J}(\boldsymbol{x})^T \boldsymbol{J}(\boldsymbol{x}) + \boldsymbol{Z}(\boldsymbol{x}), \end{aligned}$$

where $\boldsymbol{Z}(\boldsymbol{x})$ involves second derivatives of $\boldsymbol{F}$. Thus evaluation of the gradient of f generally requires $O(n^2)$ operations while the Hessian requires $O(n^3)$.

24.3 Newton-like Methods

Newton-like methods are the most popular algorithms for solving nonlinear equations.

24.3.1 Newton's Method for Nonlinear Equations

Recall our general scheme for function minimization, Algorithm 9.1. The Newton search direction was defined by $\boldsymbol{H}(\boldsymbol{x}^{(k)})\boldsymbol{p}^{(k)} = -\boldsymbol{g}(\boldsymbol{x}^{(k)})$.

We derived Newton's method by fitting a quadratic function to a function that we were trying to minimize. Equivalently, we can fit a linear function to the gradient $\boldsymbol{g}$:

$$\boldsymbol{g}(\boldsymbol{x}^{(k)} + p) \approx \boldsymbol{g}(\boldsymbol{x}^{(k)}) + \boldsymbol{H}(\boldsymbol{x}^{(k)})\boldsymbol{p} = \boldsymbol{0} \rightarrow \boldsymbol{H}(\boldsymbol{x}^{(k)})\boldsymbol{p} = -\boldsymbol{g}(\boldsymbol{x}^{(k)}).$$

Similarly, for solving nonlinear equations, we derive Newton's method by fitting a linear function to $\boldsymbol{F}$:

$$\boldsymbol{F}(\boldsymbol{x}^{(k)} + p) \approx \boldsymbol{F}(\boldsymbol{x}^{(k)}) + \boldsymbol{J}(\boldsymbol{x}^{(k)})\boldsymbol{p} = \boldsymbol{0} \rightarrow \boldsymbol{J}(\boldsymbol{x}^{(k)})\boldsymbol{p} = -\boldsymbol{F}(\boldsymbol{x}^{(k)}).$$

This gives us the search direction $\boldsymbol{p}^{(k)}$. What about the stepsize α_k? For nonlinear equations there is no notion of downhill and no natural notion of progress because there is no function f. So we just eliminate the linesearch and set $\alpha_k = 1$, or else we use $\|\boldsymbol{F}(\boldsymbol{x})\|$ to measure progress. The resulting algorithm is **Newton's method for nonlinear equations**: Algorithm 24.1. Under some "standard assumptions" (second derivatives exist, the zero is **simple** (i.e., the Jacobian matrix is nonsingular at the zero), etc.), the algorithm is guaranteed to converge to a solution if started close enough to it, and the convergence rate is quadratic.

Algorithm 24.1 Newton's Method for Solving Nonlinear Equations

Given: An initial guess $\boldsymbol{x}^{(0)}$ for the solution.
Set $k = 0$.
while $\boldsymbol{x}^{(k)}$ is not a good enough solution,
  Find a direction $\boldsymbol{p}^{(k)}$ by solving $\boldsymbol{J}(\boldsymbol{x}^{(k)})\boldsymbol{p}^{(k)} = -\boldsymbol{F}(\boldsymbol{x}^{(k)})$.
  Set $\boldsymbol{x}^{(k+1)} = \boldsymbol{x}^{(k)} + \alpha_k \boldsymbol{p}^{(k)}$, where α_k is a scalar.
  Set $k = k + 1$.
end

Alternatively, we can apply a **trust region method** for solving nonlinear equations. In this case we determine the step $\boldsymbol{p}^{(k)}$ by solving the problem

$$\min_{\boldsymbol{p}} \|\boldsymbol{F}(\boldsymbol{x}) + \boldsymbol{J}(\boldsymbol{x})\boldsymbol{p}\|$$

subject to

$$\|\boldsymbol{p}\| \leq h.$$

The objective function is a linear approximation to $\boldsymbol{F}(\boldsymbol{x}+\boldsymbol{p})$, obtained from Taylor series. Solving this problem leads to the **Levenberg–Marquardt** formula

$$(\boldsymbol{J}(\boldsymbol{x})^T \boldsymbol{J}(\boldsymbol{x}) + \lambda \boldsymbol{I})\boldsymbol{p} = -\boldsymbol{J}(\boldsymbol{x})^T \boldsymbol{F}(\boldsymbol{x})$$

for an appropriate choice of the parameter λ. See Section 9.4 for a discussion of the choice of norm and the choice of h.

We experiment with Newton's method in the next challenge.

CHALLENGE 24.1.

Write MATLAB statements to apply 5 steps of Newton's method to the problem

$$x^2 y^3 + xy = 2,$$
$$2xy^2 + x^2 y + xy = 0,$$

starting at the point $x = 5$, $y = 4$.

24.3.2 Alternatives to Newton's Method

We have a variety of methods in the Newton family, just as we did for optimization, and also slower but simpler **fixed-point iterations**.

- **Finite difference Newton method**: If $\boldsymbol{J}$ is not available, we can approximate it using finite differences. Again, as in optimization, this is not recommended; use the next method instead.
- **Inexact Newton method**: Instead of solving the linear system $\boldsymbol{J}(\boldsymbol{x}^{(k)})\boldsymbol{p}^{(k)} = -\boldsymbol{F}(\boldsymbol{x}^{(k)})$ exactly, we can use an iterative method to obtain an approximate solution to this linear system. (See Chapter 28.) The usual choice of iterative method is GMRES, a relative of **conjugate gradients**, and matrix-vector products are evaluated by differencing $\boldsymbol{F}$, just as discussed for the truncated Newton method of Section 9.5.2.
- **Quasi-Newton methods**: If storage is not a limitation, then instead of the inexact Newton method, we can store and update an approximation to $\boldsymbol{J}$. The usual formula is **Broyden's method**:

$$\boldsymbol{B}^{(k+1)} = \boldsymbol{B}^{(k)} + \frac{(\boldsymbol{y} - \boldsymbol{B}^{(k)}\boldsymbol{s})\boldsymbol{s}^T}{\boldsymbol{s}^T \boldsymbol{s}},$$

where $\boldsymbol{y} = \boldsymbol{F}(\boldsymbol{x}^{(k+1)}) - \boldsymbol{F}(\boldsymbol{x}^{(k)})$ and $\boldsymbol{s} = \boldsymbol{x}^{(k+1)} - \boldsymbol{x}^{(k)}$. Then the **secant condition**, discussed in Section 9.5.1, becomes

$$\boldsymbol{B}^{(k+1)}\boldsymbol{s} = \boldsymbol{y},$$

and if $\boldsymbol{s}^T\boldsymbol{v} = 0$, then Broyden's method satisfies

$$\boldsymbol{B}^{(k+1)}\boldsymbol{v} = \boldsymbol{B}^{(k)}\boldsymbol{v}.$$

- **Fixed-point iterations**: These methods are related to stationary iterative methods, used to solve linear systems of equations in Section 28.1. The idea is to reduce the problem to solving a sequence of nonlinear equations in a single variable. As an example, the **nonlinear Gauss-Seidel** iteration is given in Algorithm 24.2. These algorithms are simple to program but very slow, since they make no use of derivatives.

Algorithm 24.2 Gauss–Seidel for Nonlinear Equations

Given: an initial guess $\boldsymbol{x}^{(0)}$ for the system of nonlinear equations $\boldsymbol{F}(\boldsymbol{x}) = \boldsymbol{0}$.
Set $k = 0$.
while the solution is not good enough,
    for $j = 1, \ldots, n$,
        Solve the jth equation $\boldsymbol{F}_j(\boldsymbol{x}) = 0$ for ${x_j}^{(k+1)}$, keeping every variable except x_j at its most recently computed value (${x_\ell}^{(k+1)}$ if $\ell < j$ and $x_\ell^{(k)}$ if $\ell > j$).
    end
    Set $k = k + 1$.
end

In minimization using quasi-Newton methods, we used matrix updating techniques to solve the linear system that determines the step; we do the same thing to solve $\boldsymbol{F}(\boldsymbol{x}) = \boldsymbol{0}$, as discussed in the next two challenges.

CHALLENGE 24.2.

Consider the following function that computes a quasi-Newton direction for solving the system of equations $\boldsymbol{F}(\boldsymbol{x}) = \boldsymbol{0}$:

```
function [p,B] = broyden(s,y,B,F)

% Given a previous Jacobian approximation  B,
% update it using s (the change in x)
% and y (the change in F)
% and then compute a Newton-like direction p.

B = B + (y-B*s)*s'/(s'*s);
p = -B \ F;
```

The function is correct, but it is inefficient. Identify two sources of the inefficiency and propose remedies.

CHALLENGE 24.3.

In Broyden's method for solving nonlinear equations, we need to solve a linear system involving the $n \times n$ matrix

$$\boldsymbol{B}^{(k+1)} = \boldsymbol{B}^{(k)} + \frac{(\boldsymbol{y} - \boldsymbol{B}^{(k)}\boldsymbol{s})\boldsymbol{s}^T}{\boldsymbol{s}^T\boldsymbol{s}}.$$

Recall the Sherman–Morrison–Woodbury formula, discussed in the case study of Chapter 7:

$$(\boldsymbol{A} - \boldsymbol{Z}\boldsymbol{V}^T)^{-1} = \boldsymbol{A}^{-1} + \boldsymbol{A}^{-1}\boldsymbol{Z}(\boldsymbol{I} - \boldsymbol{V}^T\boldsymbol{A}^{-1}\boldsymbol{Z})^{-1}\boldsymbol{V}^T\boldsymbol{A}^{-1}.$$

If we have a way to solve linear systems involving $\boldsymbol{B}^{(k)}$ using p multiplications, how long would it take to solve a linear system involving $\boldsymbol{B}^{(k+1)}$?

As in the minimization case, we get a superlinear convergence rate for the Newton-like methods if we start close enough to the solution and if the direction is (asymptotically) close enough to the Newton direction. The fixed-point iterations are generally easy to implement, since they rely on a method to solve a nonlinear equation in a single variable, but their convergence rate is only linear, not superlinear.

24.4 Continuation Methods

The superlinear convergence result for Newton's method and its relatives is a major strength. It says that once we are close enough to the solution, the method converges quite quickly. But if we do not have a sufficiently good initial guess, then we might fail to converge at all, and the same is true of fixed-point iterations.

Continuation methods resolve this dilemma by always ensuring that we are close enough to the solution to *some* problem, and gradually changing the problem to the one we want to solve. Here is one way to do this:

The basic idea is that we want to solve $\boldsymbol{F}(\boldsymbol{x}) = \boldsymbol{0}$ but we might not have a good enough initial guess to ensure (fast) convergence of Newton's method. But we can choose some easy problem $\boldsymbol{F_a}(\boldsymbol{x}) = \boldsymbol{0}$ for which we know the solution; for example, the solution to $\boldsymbol{F_a}(\boldsymbol{x}) = \boldsymbol{x} - \boldsymbol{a} = \boldsymbol{0}$ is $\boldsymbol{a}$. So we formulate a **homotopy function**

$$\boldsymbol{\rho_a}(\lambda, \boldsymbol{x}) = \lambda \boldsymbol{F}(\boldsymbol{x}) + (1 - \lambda)\boldsymbol{F_a}(\boldsymbol{x}),$$

where λ is a scalar in the interval $[0,1]$. Note that the solution to $\boldsymbol{\rho_a}(0, \boldsymbol{x}) = \boldsymbol{0}$ is known (for our example, it is just $\boldsymbol{a}$), and the solution to $\boldsymbol{\rho_a}(1, \boldsymbol{x}) = \boldsymbol{0}$ is the solution to our desired problem. This is just one way to construct such a $\boldsymbol{\rho_a}(\lambda, \boldsymbol{x})$.

Our solution strategy is to use an approximate solution to $\boldsymbol{\rho_a}(\lambda, \boldsymbol{x}) = \boldsymbol{0}$ for one value of λ as the starting guess for the solution for a somewhat larger value of λ, thus ensuring that the starting guess is good enough that Newton's method performs well. If we iterate this process, we can gradually increase λ to 1 and produce the solution to our original

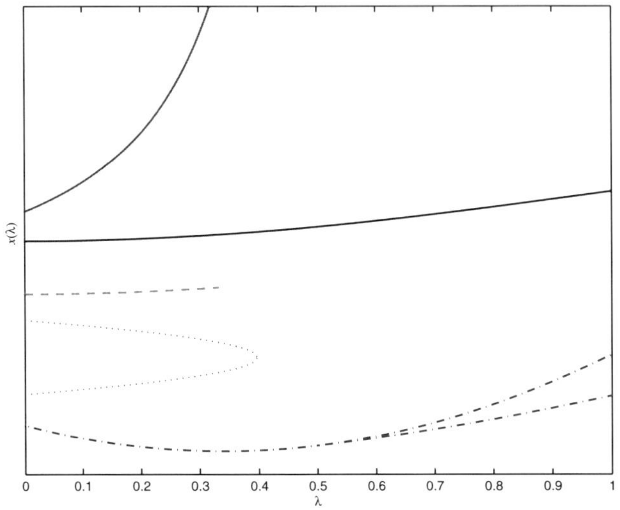

Figure 24.2. *Paths of solutions in a continuation method. Ideally, all paths are simple, as illustrated by the blue solid curve, but troubles can occur. Paths can have turning points (black dotted), bifurcation points (red dot-dashed), or end points (green dashed), or may wander to infinity (magenta solid).*

Algorithm 24.3 Continuation Method for Solving Nonlinear Equations

Given a function ρ_a, we set $\lambda = 0$ and $\hat{x}$ to the solution to $F_a(x) = 0$.
while $\lambda < 1$
  Increase λ a little bit but keep $\lambda \leq 1$.
  Solve $\rho_a(\lambda, x) = 0$ using your favorite algorithm (Newton-like, etc.) with $\hat{x}$ as a starting guess.
  Call the solution $\hat{x}$.
end
Upon termination, we have computed $\hat{x}$ so that $\rho_a(1, \hat{x}) = 0$, so $\hat{x}$ solves the problem $F(x) = 0$.

---

problem. The basic algorithm is Algorithm 24.3. There are some worries in using this method. We are essentially tracing out a path of solutions from $\lambda = 0$ to $\lambda = 1$, but this path may be faulty, as illustrated in Figure 24.2.

- It may have **turning points**, which we can overcome by allowing λ to decrease sometimes.
- It may **bifurcate**.
- The solution might fail to exist for some $\lambda < 1$, and then the path stops.
- The solution path might wander off to infinity.

Therefore, the difficult part of the theory is to construct ρ_a so that we can walk all the way from $\lambda = 0$ to $\lambda = 1$ without failing to find a solution at any intermediate step. We focus on methods that (almost always) construct a successful function ρ_a and that signal failure, if it occurs, so that we can try a different function.

24.4.1 The Theory behind Continuation Methods

The solution path is well-behaved if $\rho(\lambda, x)$ is **transversal to zero**. A function w is transversal to zero on an open domain U if, for any point $\widehat{u} \in U$ such that $w(\widehat{u}) = 0$, the Jacobian matrix at $\widehat{u}$ has full rank. Note that the Jacobian matrix need not be square. Its dimensions are length(w)$\times$ length($\widehat{u}$).

The following challenge gives an example using this jargon.

CHALLENGE 24.4.

Consider using a continuation method to solve the problem

$$F(x) = \begin{bmatrix} x^2y^3 + xy - 2 \\ 2xy^2 + x^2y + xy \end{bmatrix} = \mathbf{0}.$$

Letting $x = [x, y]^T$, our homotopy is

$$\rho_a(\lambda, x) = \lambda F(x) + (1 - \lambda)(x - a).$$

(a) Compute the Jacobian matrix for $\rho_a(\lambda, x)$.

(b) What needs to hold in order that the function ρ_a is transversal to zero on its domain? Why is this likely to be true?

(c) Convince yourself that derivatives of all orders (first, second, ...) exist for the functions in this problem for $[\lambda, x] \in \mathcal{R}^3$, since polynomials have an infinite number of continuous derivatives.

Our continuation method Algorithm 24.3 almost always works, thanks to a **parameterized Sard's theorem** [151, Thm. 2.2]: Let $U = \mathcal{R}^n \times [0, 1) \times \mathcal{R}^n$ and define a function $\rho : U \to \mathcal{R}^n$ that has two continuous derivatives and is transversal to zero on U. (We'll call its variables (a, λ, x), where x is a function of λ.) Choose a point $a \in \mathcal{R}^n$ and define

$$\rho_a(\lambda, x) = \rho(a, \lambda, x).$$

Then the map ρ_a is transversal to zero on $[0, 1) \times \mathcal{R}^n$ for almost all a.

For almost all means except for a set of measure zero; i.e., if we choose a point a at random, it has probability zero of giving a function ρ_a that is not transversal to zero on $[0, 1) \times \mathcal{R}^n$, and therefore probability zero of yielding a path that bifurcates or stops.

The parameterized Sard's theorem gives us a recipe for a successful algorithm: Find a function $\rho_a(a, \lambda, x)$ that satisfies the assumptions of the parameterized Sard's theorem. Then choose a random value of a and follow the path from $\lambda = 0$ to $\lambda = 1$. If we encounter a point that does not have a full-rank Jacobian, then we made an unlucky choice of a, so we try again. Our algorithm traces out a solution path $(\lambda, x(\lambda))$, where $\rho_a(\lambda, x(\lambda)) = \mathbf{0}$.

Our solution path has a variety of other nice properties. If we also assume that the equation $\rho_a(0, x) = \mathbf{0}$ has a unique solution x_0 for any choice of a, then [151, Thm. 2.3] for almost all $a \in \mathcal{R}^n$:

- The Jacobian matrix has (full) rank n at all points $(\lambda, \boldsymbol{x})$.
- The path is smooth, does not intersect itself or any other solution path, and does not bifurcate.
- The path has finite length in any compact subset of $[0,1) \times \mathcal{R}^n$.
- The path reaches the hyperplane $\lambda = 1$ as long as the Jacobians are bounded away from rank-deficient matrices. If the path does not reach $\lambda = 1$, then it goes to infinity.

In summary, the path is very well-behaved! In fact, we can hope to walk along it.

The above result means that we almost always have a solution path that starts at $\lambda = 0$ and goes to our desired solution at $\lambda = 1$. But there may be a countable number of other solution paths, some closed and some with two endpoints at $\lambda = 1$. So we need to be careful not to step onto one of these other paths as we follow our good path.

Notes on the website illustrate how this theory can be used for one class of problems, minimization of a convex function.

24.4.2 Following the Solution Path

To solve the nonlinear system of equations $\boldsymbol{F}(\boldsymbol{x}) = \boldsymbol{0}$, all we need to do is to follow the solution path

$$\boldsymbol{\rho_a}(\lambda, \boldsymbol{x}) = \lambda \boldsymbol{F}(\boldsymbol{x}) + (1-\lambda)\boldsymbol{F_a}(\boldsymbol{x}) = \boldsymbol{0}$$

from $\lambda = 0$ to $\lambda = 1$. In some sense we don't need to be too careful when following the path—we don't care about the values at any points in between $\lambda = 0$ and $\lambda = 1$. Unfortunately, if we are not careful enough, we might jump to other solution paths that make no progress.

In implementing the basic continuation method, Algorithm 24.3, there are two remaining issues: choosing the stepsize in λ and choosing the tolerance in the Newton-like method. In practice, we aim to keep the increase in λ as big as possible (to minimize the number of nonlinear systems we consider) while ensuring that $\widehat{\boldsymbol{x}}$ is a good enough initial guess that the Newton-like method converges quickly. Similarly, we try to keep the tolerance as big as possible, to reduce the work in the Newton-like method, without losing track of our path.

Issues of choosing a stepsize and tolerance are troublesome, and considerations are similar to those in Chapter 20, for solving ordinary differential equations (ODEs). An alternative to Algorithm 24.3 is to use our ODE machinery to solve the homotopy problem. We do this by differentiating our path $\boldsymbol{\rho_a}(\lambda, \boldsymbol{x})$. We could differentiate with respect to λ, but then the derivative would fail to exist at any turning point and we might get in trouble. Instead, we'll introduce a new independent variable and differentiate with respect to it. It is most convenient to define s to be **arc length** and to write

$$\boldsymbol{\rho_a}(\lambda(s), \boldsymbol{x}(s)) = \boldsymbol{0},$$

so

$$\frac{\mathrm{d}}{\mathrm{d}s}\boldsymbol{\rho_a}(\lambda(s), \boldsymbol{x}(s)) = \boldsymbol{0}.$$

We have the initial conditions

$$\lambda(0) = 0, \qquad \boldsymbol{x}(0) = \boldsymbol{a}.$$

When the solution path reaches $\lambda(s) = 1$, we are finished; the resulting $\boldsymbol{x}(s)$ solves our nonlinear system. For uniqueness, we normalize so that

$$\left(\frac{\mathrm{d}\lambda}{\mathrm{d}s}\right)^2 + \left\|\frac{\mathrm{d}\boldsymbol{x}}{\mathrm{d}s}\right\|^2 = 1\,.$$

When $\boldsymbol{F_a}(\boldsymbol{x}) = \boldsymbol{x} - \boldsymbol{a}$, this system takes the form

$$\frac{\mathrm{d}\lambda}{\mathrm{d}s}\boldsymbol{F}(\boldsymbol{x}) + \lambda\boldsymbol{J}(\boldsymbol{x})\frac{\mathrm{d}\boldsymbol{x}}{\mathrm{d}s} - \frac{\mathrm{d}\lambda}{\mathrm{d}s}(\boldsymbol{x}-\boldsymbol{a}) + (1-\lambda)\frac{\mathrm{d}\boldsymbol{x}}{\mathrm{d}s} = \boldsymbol{0}\,,$$

$$\left(\frac{\mathrm{d}\lambda}{\mathrm{d}s}\right)^2 + \sum_{i=1}^{n}\left(\frac{\mathrm{d}x_i}{\mathrm{d}s}\right)^2 = 1\,.$$

We summarize in Algorithm 24.4.

Algorithm 24.4 ODE-Continuation Method for Solving Nonlinear Equations

Solve the ODE

$$\frac{d\lambda}{ds}\boldsymbol{F}(\boldsymbol{x}) + \lambda\boldsymbol{J}(\boldsymbol{x})\frac{d\boldsymbol{x}}{ds} - \frac{d\lambda}{ds}(\boldsymbol{x}-\boldsymbol{a}) + (1-\lambda)\frac{d\boldsymbol{x}}{ds} = \boldsymbol{0}\,,$$

$$\left(\frac{d\lambda}{ds}\right)^2 + \sum_{i=1}^{n}\left(\frac{dx_i}{ds}\right)^2 = 1,$$

with initial conditions $\lambda(0) = 0$ and $\boldsymbol{x}(0) = \boldsymbol{a}$.

When $\lambda = 1$, return the corresponding value $\boldsymbol{x}(1)$.

In principle, we could just plug the system into our favorite ODE solver, and this is possible for prototyping algorithms, but in practice the ODE solver should be tailored to this problem:

- If the arc length parameter s gets too long, restart with a new choice of $\boldsymbol{a}$.
- Take care that we don't wander too far from the solution path but that we don't work too hard in following it exactly.
- When the ODE solver passes the value $\lambda = 1$, use **inverse interpolation**, discussed in the next challenge, to compute the value $\hat{s}$ for which $\lambda(\hat{s}) = 1$.

CHALLENGE 24.5.

Inverse interpolation. Suppose we have a function $f(t)$ and that we are given data points (t_i, f_i), $i = 1, \ldots, n$. Perhaps we choose to approximate $f(t)$ by a function $\hat{f}$ that is a polynomial or a spline or a Fourier series. Then it is easy to compute an approximation to $f(t)$ for any given value of t by evaluating $\hat{f}$.

But suppose we want to find an approximation to the solution to $f(t) = 0$. Our function $\hat{f}$ might be helpful if it is more easy to find a solution to $\hat{f}(t) = 0$ than to find a

POINTER 24.3. Software for Continuation Methods.

Hompack, by Layne Watson and co-workers, is a high-quality system for solving nonlinear equations by continuation methods [153].

POINTER 24.4. Further Reading.

A thorough treatment of Newton's method is given in a book by C. T. Kelley [88]. More variants on Newton's method and fixed-point iterations for solving nonlinear equations can be found in the classic book by Ortega and Rheinboldt [120].

Trust region methods for nonlinear equations are discussed, for example, in [28].

The foundation for continuation methods can be found in books by Allgower and Georg [1] and T. Y. Li [103]. For more information on continuation (homotopy) methods, consult papers by Watson [5, 151, 150, 152].

---

solution to $f(t) = 0$, but a better approach results from turning the graph of the problem on its side.

If each of the values f_i is distinct, then just as f is a function of t, it is also true that t is a function of f. So let's think of the data as (f_i, t_i), $i = 1, \ldots, n$, and let's fit a function $\hat{t}(f)$ to this data. This is called **inverse interpolation**, since we obtain an approximation to the function $f^{-1}(z) = t$, and it is a useful method for solving a single nonlinear equation. Now if we evaluate $t_{comp} = \hat{t}(0)$, then t_{comp} is an approximate solution to $f(t) = 0$.

Write MATLAB statements to approximate the solution to an equation $f(t) = 0$ using inverse interpolation with three function values (t_i, f_i), $i = 1, \ldots, 3$ given as input.

Chapter 25 / Case Study

Variable-Geometry Trusses: What's Your Angle?

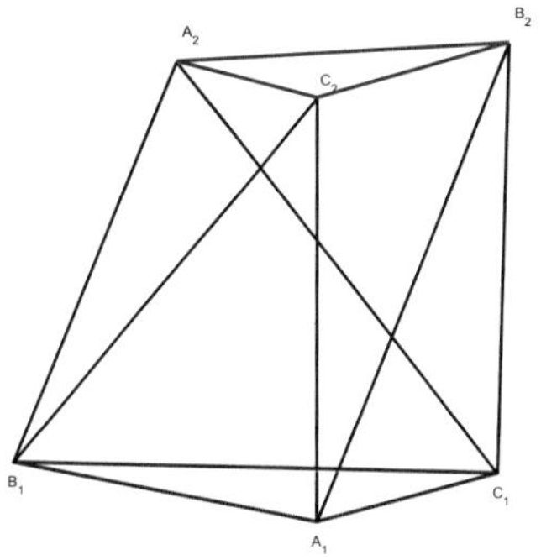

Imagine that we have attached 3 members (beams) together to form a triangle (triangular truss) $A_1B_1C_1$. This is the base of a platform, shown in blue in Figure 25.1. Now take 3 more members and form a triangle $A_2B_2C_2$ from them. This is the (red) "top" of the platform (truss). We use 6 more fixed-length members to attach the top to the bottom:

- Node (vertex) A_1 is attached to B_2 and C_2.
- Node B_1 is attached to A_2 and C_2.
- Node C_1 is attached to A_2 and B_2.

Given the lengths of all of the members, our problem is to determine the (at most 16) possible configurations of the resulting platform.

Computing the coordinates of the vertices is an exercise in geometry. As an example, suppose that the members on the bottom all have length equal to ℓ_b and that the members attaching the top to the bottom all have length ℓ_v. The bottom is an equilateral triangle with each vertex having x, y, and z coordinates; we'll fix it to lie in the plane $z = 0$, with the coordinates of the vertices defined as

$$\begin{bmatrix} A_1 \\ B_1 \\ C_1 \end{bmatrix} = \begin{bmatrix} 0, & 0, & 0 \\ 0, & \ell_b, & 0 \\ \ell_b \cos(\pi/6), & \ell_b \sin(\pi/6), & 0 \end{bmatrix}.$$

We could use the 9 coordinates of the top vertices as our variables, but this would introduce a lot of redundancy into the equations. Instead we choose to work with three angles from which the 9 coordinates can be computed. Let θ_C be the angle that triangle $A_1B_1C_2$ makes with the plane containing the bottom. We can compute the coordinates of C_2 by considering the right triangle with vertices C_2, the vertical projection of C_2 onto the plane of the bottom, and the midpoint $M_1 = (A_1 + B_1)/2$ of the member connecting A_1 and B_1. The length of the hypotenuse is $h_C = \sqrt{\ell_v^2 - \|M_1 - B_1\|^2}$ (using information from the triangle $B_1M_1C_2$). We

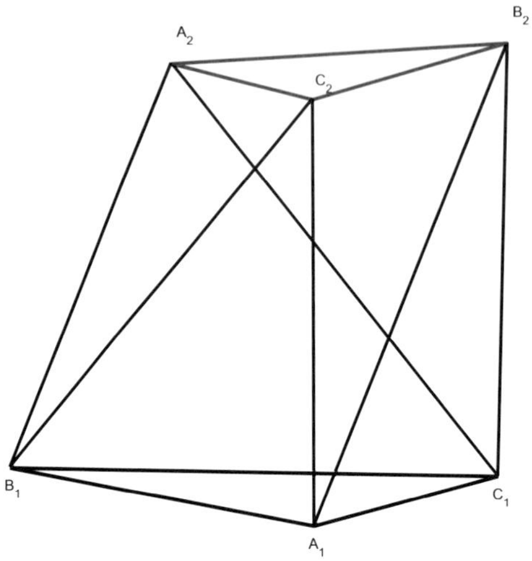

Figure 25.1. *Truss from Watson example. The base, with nodes A_1, B_1, and C_1, is shown in blue and the top, with nodes A_2, B_2, and C_2, in red. See the website for a MATLAB file that generates a rotatable image.*

see that relative to the midpoint M_1, the y-coordinate of C_2 is the same, the x-coordinate is displaced by $h_C \cos(\theta_C)$, and the z-coordinate is displaced by $h_C \sin(\theta_C)$. Similar reasoning leads to formulas for the coordinates of A_2 and B_2.

CHALLENGE 25.1.

Let θ_A be the angle that triangle $A_2B_1C_1$ makes with the plane containing the bottom, and similarly for θ_B and θ_C. Let $d_A = h_A \cos\theta_A$ and similarly for d_B and d_C. Convince yourself that the three nodes of the top of the platform have coordinates

$$\begin{bmatrix} A_2 \\ B_2 \\ C_2 \end{bmatrix} = \begin{bmatrix} (B_1+C_1)/2 \\ (A_1+C_1)/2 \\ (A_1+B_1)/2 \end{bmatrix} + \begin{bmatrix} d_A\cos(\pi/3), & d_A\sin(\pi/3), & h_A\sin(\theta_A) \\ d_B\cos(-\pi/3), & d_B\sin(-\pi/3), & h_B\sin(\theta_B) \\ d_C, & 0, & h_C\sin(\theta_C) \end{bmatrix}.$$

Now that we have the six nodes of the platform, we can verify that a given set of angles $\theta_A, \theta_B, \theta_C$ yields a valid configuration by checking that the resulting lengths of the three members on top match the specified lengths ℓ_{AB}, ℓ_{BC}, and ℓ_{AC}:

$$\boldsymbol{F}(\boldsymbol{\theta}) = \begin{bmatrix} \|A_2 - B_2\|^2 - \ell_{AB}^2 \\ \|B_2 - C_2\|^2 - \ell_{BC}^2 \\ \|A_2 - C_2\|^2 - \ell_{AC}^2 \end{bmatrix} = \begin{bmatrix} 0 \\ 0 \\ 0 \end{bmatrix}. \tag{25.1}$$

This is a system of three nonlinear equations in three variables, and we investigate how the methods of Chapter 24 perform on it.

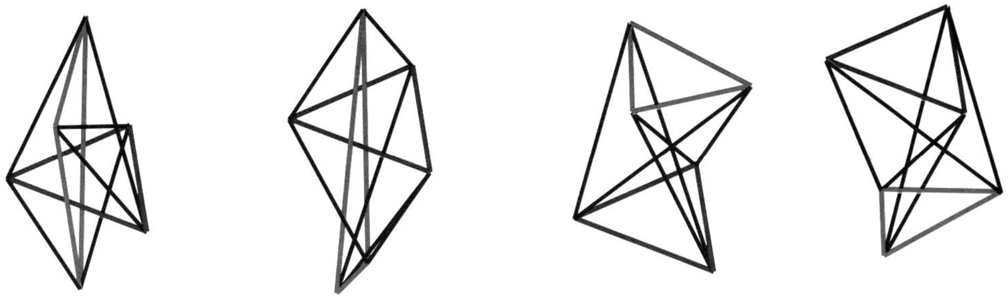

Figure 25.2. *Distinct configurations found for Platform C.*

Our first idea for an algorithm might be to solve the system by choosing some different starting guesses and running our favorite nonlinear equation solver on the problem, starting from each of the guesses. A better idea might be to use a continuation method, but this requires us to construct an easy function. We'll experiment with two homotopies, one derived from geometric reasoning and one derived from algebraic conversion to a polynomial system of equations.

Notice that if we are given three angles, we can easily determine the lengths $\hat{\ell}_{AB}$, $\hat{\ell}_{BC}$, and $\hat{\ell}_{AC}$ of the top members. So we can construct a geometric homotopy that walks from a truss with these lengths to one with the desired lengths. For each value of λ we compute angles corresponding to the lengths

$$\ell_{j,\lambda} = \lambda \ell_j + (1-\lambda)\hat{\ell}_j, \quad j = AB, BC, AC.$$

Let's see how well it works.

CHALLENGE 25.2.

Use the geometric homotopy to solve four problems, with $\ell_v = 34.5$, $\ell_b = 48$, and the lengths of the top members given by

Platform A:	35,	30,	40.
Platform B:	30,	30,	30.
Platform C:	35,	40,	45.
Platform D:	10,	10,	10.

In particular, compute the coordinates of the vertices of the top triangle and see how many unique solutions you can find.

For initial angles, use $\hat{\theta}_A = \pi/6$, $\hat{\theta}_B = 7\pi/16$, $\hat{\theta}_C = \pi/4$, the negatives of these values, these values plus π, and minus these values minus π, giving 64 points in all. Solve each of the four problems, using the continuation algorithm started from the 64 initial angles. Compare with using `fsolve` with these 64 starting points.

To construct a polynomial homotopy, we use a clever variable transformation [5, 4]. Notice that the three equations in (25.1) involve cos, sin, $\cos^2$, and $\sin^2$ of the three

POINTER 25.1. Further Reading.

This case study solves a problem taken from [4, 5]. Platforms such as these are lightweight structures with multiple configurations that require little storage space. They are used, for example, as building blocks for manipulator arms on spacecraft and for other robots. Even more interesting is the analysis of a Stewart platform, in which the vertical members have adjustable lengths. See, for example, `http://en.wikipedia.org/wiki/Stewart_platform`.

---

unknown angles. If

$$t_A = \tan(\theta_A/2),$$

then

$$\cos\theta_A = \frac{1-t_A^2}{1+t_A^2}, \quad \sin\theta_A = \frac{2t_A}{1+t_A^2}.$$

Making this substitution into (25.1) and then multiplying the first and third equations by $(1+t_A^2)^2$ makes the system polynomial in the variable t_A. Applying this same transformation for the other two variables gives us a system of three polynomial equations, each of degree 4. A theorem of Bezoit tells us that there are $4^3 = 64$ solutions, counting multiplicities, although it turns out that in this problem 48 of them are infinite and not of practical use.

The authors [5, 4] suggest using a homotopy with a polynomial with the same structure: we want it to have 16 solutions with the highest-order terms involving the same variables. But we need it to be solved easily; for example:

$$(x_1^2 - \hat{b})(x_2^2 - \hat{c}) = 0,$$
$$(x_2^2 - \hat{a})(x_3^2 - \hat{b}) = 0,$$
$$(x_3^2 - \hat{c})(x_1^2 - \hat{a}) = 0,$$

where $\hat{a}$, $\hat{b}$, and $\hat{c}$ are three numbers. Note that there are 16 real solutions to our easy problem: $(\pm\sqrt{\hat{a}}, \pm\sqrt{\hat{c}}, \pm\sqrt{\hat{b}})$ and $(\pm\sqrt{\hat{b}}, \pm\sqrt{\hat{a}}, \pm\sqrt{\hat{c}})$. If we call our real polynomial problem $\boldsymbol{p}(\boldsymbol{x}) = \boldsymbol{0}$ and our easy problem $\boldsymbol{q}(\boldsymbol{x}) = \boldsymbol{0}$, then our homotopy becomes

$$\boldsymbol{\rho}(\lambda, \boldsymbol{x}) = \lambda \boldsymbol{p}(\boldsymbol{x}) + (1-\lambda)\boldsymbol{q}(\boldsymbol{x}).$$

We use this homotopy, started from each of the 16 solutions to the easy problem.

CHALLENGE 25.3.

Use the polynomial homotopy to compute configurations of the four platforms and compare with using `fsolve` to solve $\boldsymbol{q}(\boldsymbol{x}) = \boldsymbol{0}$ with the same 64 initial guesses from Challenge 25.2, using, for example, $\hat{a} = 1$, $\hat{b} = 4$, and $\hat{c} = 9$.

In Figure 25.2 we display the configurations we found for Platform C. One lesson to take from these experiments is that knowledge of the problem gives great insight into potential solution algorithms, as illustrated by the geometric homotopy.

Chapter 26 / Case Study

Beetles, Cannibalism, and Chaos: Analyzing a Dynamical System Model

The evolution of a system over time can be described by a set of equations called a **dynamical system**. In this case study, we use a dynamical system to model the life cycle of flour beetles in order to estimate key biological parameters that describe their behavior.

If you open a container of flour and see small red rods among the powder, you are probably looking at confused beetles (*Tribolium confusum*) or red beetles (*Tribolium castaneum*), pictured in Figure 26.1. These insects progress through several stages of life, including egg (2–4 days), larva (approximately 14 days), pupa (approximately 14 days), and adult (3 years or more). Their life cycle is complicated by one additional fact: they are cannibalistic. Adults and larvae eat eggs, pupae, and immature adults, and adults also eat larvae.

By using a dynamical system to model the flour beetle's life cycle, we can estimate key parameters of biological significance that describe their behavior. At the same time, we illustrate some mathematical properties of dynamical systems.

The Model

To try to understand the population dynamics, entomologists have developed a model of the beetles with three stages:

- $L(t)$ is the number of (feeding) larvae at time t.
- $P(t)$ is the number of nonfeeding larvae, pupae, and immature adults at time t.
- $A(t)$ is the number of mature adults at time t.

We define five coefficients of interaction, setting all other ones to zero:

- $b > 0$ is the average number of larvae recruited (i.e., tended) by each adult.
- $0 \leq \mu_L \leq 1$ is the probability that a larva dies from a cause other than cannibalism.

Figure 26.1. *Red beetles. Actual length: approximately* 3 *mm. The image* ©*Alex Wild, used with permission; all rights reserved.*

- $0 \le \mu_A \le 1$ is the probability that an adult dies from a cause other than cannibalism.
- $e^{-c_{ea}A(t)}$ is the probability that an egg is not eaten by an adult between time t and time $t+1$. The value c_{ea} (denoting interaction between eggs and adults) is a **rate constant.**
- $e^{-c_{el}L(t)}$ is the probability that an egg is not eaten by a larva between time t and time $t+1$.
- $e^{-c_{pa}A(t)}$ is the probability that a pupa is not eaten by an adult between time t and time $t+1$.

The rate that adults eat larvae is small, and we set it to zero.

The model proposed by Dennis et al. [39] is[10]

$$L(t+1) = bA(t)\exp(-c_{ea}A(t) - c_{el}L(t)), \tag{26.1}$$

$$P(t+1) = L(t)(1-\mu_L), \tag{26.2}$$

$$A(t+1) = P(t)\exp(-c_{pa}A(t)) + A(t)(1-\mu_A). \tag{26.3}$$

Time t is measured in 14-day units. Since L, P, and A can take on noninteger values, we need to interpret their values as the average numbers of individuals over the 14-day period.

[10]The notation $\exp(y)$ means e^y.

CHALLENGE 26.1.

Let $b = 11.6772$, $\mu_L = 0.5129$, $c_{el} = 0.0093$, $c_{ea} = 0.0110$, $c_{pa} = 0.0178$, $L(0) = 70$, $P(0) = 30$, and $A(0) = 70$. To get some experience with this model, plot the populations L, P, and A for 100 time units for three sets of data: $\mu_A = 0.1$, 0.6, and 0.9. Describe the behavior of the populations in these three cases as if you are speaking to someone who is not looking at the graphs.

Equilibria and Stability

It is interesting to determine **equilibria populations**, values for the initial numbers of larvae, pupae, and adults for which the population remains constant. We denote these as A_{fixed}, L_{fixed}, and P_{fixed}. Of course, one such solution is the **extinction solution** of zero larvae, zero pupae, and zero adults. If $c_{el} = 0$, then Dennis et al. provide a nonzero solution, valid when $b > \mu_A/(1-\mu_L)$:

$$A_{fixed} = \log(b(1-\mu_L)/\mu_A)/(c_{ea}+c_{pa}), \tag{26.4}$$

$$L_{fixed} = bA_{fixed}\exp(-c_{ea}A_{fixed}), \tag{26.5}$$

$$P_{fixed} = L_{fixed}(1-\mu_L). \tag{26.6}$$

An equilibrium solution is called **stable** if a colony of beetles with initial population $A(0) \approx A_{fixed}$, $L(0) \approx L_{fixed}$, and $P(0) \approx P_{fixed}$ tends to approach these values as time passes; otherwise the solution is **unstable**. Let $\boldsymbol{x}_t$ be a vector with elements $L(t)$, $P(t)$, and $A(t)$. Then our equations (26.1) – (26.3) are

$$\boldsymbol{x}_{t+1} = \boldsymbol{F}(\boldsymbol{x}_t),$$

where

$$\boldsymbol{F}(\boldsymbol{x}_t) = \begin{bmatrix} bA(t)\exp(-c_{ea}A(t) - c_{el}L(t)), \\ L(t)(1-\mu_L), \\ P(t)\exp(-c_{pa}A(t)) + A(t)(1-\mu_A) \end{bmatrix}.$$

By Taylor series,

$$\boldsymbol{x}_{t+1} = \boldsymbol{F}(\boldsymbol{x}_t) \approx \boldsymbol{x}_{fixed} + \boldsymbol{J}(\boldsymbol{x}_{fixed})(\boldsymbol{x}_t - \boldsymbol{x}_{fixed}),$$

where $\boldsymbol{J}(\boldsymbol{x}_{fixed})$ is the Jacobian of $\boldsymbol{F}$, i.e., the 3×3 matrix of partial derivatives. Therefore,

$$\boldsymbol{x}_{t+1} - \boldsymbol{x}_{fixed} \approx \boldsymbol{J}(\boldsymbol{x}_{fixed})(\boldsymbol{x}_t - \boldsymbol{x}_{fixed}).$$

From Challenge 5.15, we conclude that the new point $\boldsymbol{x}_{t+1}$ tends to be closer to $\boldsymbol{x}_{fixed}$ than $\boldsymbol{x}_t$ is if all eigenvalues of the Jacobian, evaluated at $\boldsymbol{x}_{fixed}$, are inside the unit circle. Therefore, each equilibrium solution for our beetle problem can be labeled as **stable** or **unstable** depending on whether or not the eigenvalues of the Jacobian matrix $\boldsymbol{J}(\boldsymbol{x}_{fixed})$ all lie within the unit circle.

CHALLENGE 26.2.

Let $\mu_L = 0.5$, $\mu_A = 0.5$, $c_{el} = 0.01$, $c_{ea} = 0.01$, and $c_{pa} = 0.01$. Plot A_{fixed}, L_{fixed}, and P_{fixed} for $b = 1.0, 1.5, 2.0, \ldots, 20.0$. To compute these values for each b, use `fsolve`, started from the solution with $c_{el} = 0$, to solve the equations $\widehat{\boldsymbol{F}}(\boldsymbol{x}) = \boldsymbol{F}(\boldsymbol{x}) - \boldsymbol{x} = \boldsymbol{0}$. Provide `fsolve` with the Jacobian matrix for the function $\widehat{\boldsymbol{F}}$, and on your plot, mark the b values for stable equilibria with plus signs.

Stability and Bifurcation

Let's investigate stability a bit more. We know that when $b > \mu_A/(1-\mu_L)$, equations (26.4)–(26.6) give a constant solution to our population model. This means that if we start the model with exactly these numbers of larvae, pupae, and adults, we expect the population at each time to remain constant. Let's see what happens numerically. We'll study the solution as a function of μ_A, with the other parameters set to a particular choice of values. We want to know whether the solution for large values of t is constant, periodic, or **chaotic** with no regular pattern. To decide this, we make a **bifurcation diagram**: we run the LPA iteration (26.1)–(26.3) for various values of μ_A and plot the last 100 values of the LPA iteration as a function of μ_A.

CHALLENGE 26.3.

(a) Let $\mu_L = 0.5128$, $c_{el} = 0.0$, $c_{ea} = 0.01$, and $c_{pa} = 0.09$. For $\mu_A = 0.02, 0.04, \ldots, 1.00$, use the LPA iteration (26.1) – (26.3) to determine the population for 250 time units. On a single graph, plot the last 100 values as a function of μ_A to produce the bifurcation diagram.

(b) Determine the largest of the values $\mu_A = 0.02, 0.04, \ldots, 1$ for which the constant solution is stable (i.e., well-conditioned).

(c) Explain why the bifurcation diagram is not just a plot of L_{fixed} vs. μ_A when the system is unstable.

(d) Give an example of a value of μ_A for which nearby solutions cycle between two fixed values. Give an example of a value of μ_A for which nearby solutions are chaotic (or at least have a long cycle).

Nurturing vs. Cannibalism: Estimating the Parameters

Now that we understand some properties of our dynamical system model $\boldsymbol{x}_{t+1} = \boldsymbol{F}(\boldsymbol{x}_t)$, we use some observed data to try to determine the parameter values. Desharnais and Liu [43] observed four colonies of red beetles for 266 days, making observations every 14 days. With least squares, we can estimate the six parameters in our model using this data. Aside from the initial values $L(0)$, $P(0)$, and $A(0)$, Desharnais and Liu give us three data values

$L_{observed}(t)$, $P_{observed}(t)$, and $A_{observed}(t)$ for each time $t = 1,\ldots,19$. Given values of the six parameters in our model, we can compute predicted values of the populations at each of these times, so we would like to determine parameters that minimize the difference between the predictions and the observations. Since Desharnais and Liu tell us that errors in the logs of the observed values are approximately equal, we minimize the **least squares function**

$$\sum_{t=1}^{19}(\log(L_{observed}(t)) - \log(L_{predicted}(t)))^2$$
$$+\sum_{t=1}^{19}(\log(P_{observed}(t)) - \log(P_{predicted}(t)))^2$$
$$+\sum_{t=1}^{19}(\log(A_{observed}(t)) - \log(A_{predicted}(t)))^2,$$

where $L_{predicted}(t)$, $P_{predicted}(t)$, and $A_{predicted}(t)$ denote the values obtained from (26.1)–(26.3).

CHALLENGE 26.4.

(a) Use `lsqnonlin` to solve the least squares minimization problem, using each of the four sets of data in `beetledata.m`. In each case, determine the six parameters (μ_L, μ_A, c_{el}, c_{ea}, c_{pa}, and b). Set reasonable upper and lower bounds on the parameters, and perhaps start the least squares iteration with the guess $\mu_L = \mu_A = 0.5$, $c_{el} = c_{ea} = c_{pa} = 0.1$, and $b = 10$. Print the solution parameters and the corresponding residual norm.

(b) Compare your results with those computed by Dennis et al., found in `param_dl` in `beetledata.m`. Include a plot that compares the predicted values with the observed values.

When I asked 25 students to solve Challenge 26.4 with the data from the second colony of beetles, they obtained 13 different answers, all of them different from mine! Unfortunately, none of them gave a good fit to the measured data. When solving a nonlinear system of equations by formulating it as a nonlinear least squares problem, it is important to realize that the function may be nonconvex, which means that there might be many local minimizers. This makes it quite difficult for an optimization routine like `lsqnonlin` to find the globally optimal solution. For difficult optimization problems, it sometimes helps to use a **continuation method**, so let's try that.

CHALLENGE 26.5.

Consider the data for the second beetle colony. For each value $b = 0.5, 1.0, \ldots, 50.0$, minimize the least squares function by using `lsqnonlin` to solve for the five remaining parameters. For each value of b after the first, start the minimization from the solution

POINTER 26.1. Further Reading.

The red flour beetle is quite important; its genome was the first beetle genome sequenced [83].

The lifespan and population data are taken from data of Desharnais and Liu [43]. The dynamical systems model that we use was developed by Dennis, Desharnais, Cushing, and Costantino [39]. The developers of the LPA model revisited the problem in [29].

An excellent introduction to dynamical systems, stability, and bifurcation diagrams is given in a textbook by Scheinerman [134].

---

determined for the previous value of b. In what sense is this a continuation method? Plot the square root of the least squares function vs. b, and determine the best set of parameters. How sensitive is the function to small changes in b?

Perform further calculations to estimate the **forward error** (how sensitive the optimal parameters are to small changes in the data) and the **backward error** (how sensitive the model is to small changes in the parameters).

Dennis et al. evaluated the LPA model using a less demanding criterion; they just compared the 1-step predictions of the model with the true values. (This is akin to a local error evaluation for an ordinary differential equation model; we just ask how much error is produced in a single step, assuming that correct values were given at the previous step.) It would be interesting to repeat the sensitivity analysis under this error criterion.

Unit VII

Sparse Matrix Computations, with Application to Partial Differential Equations

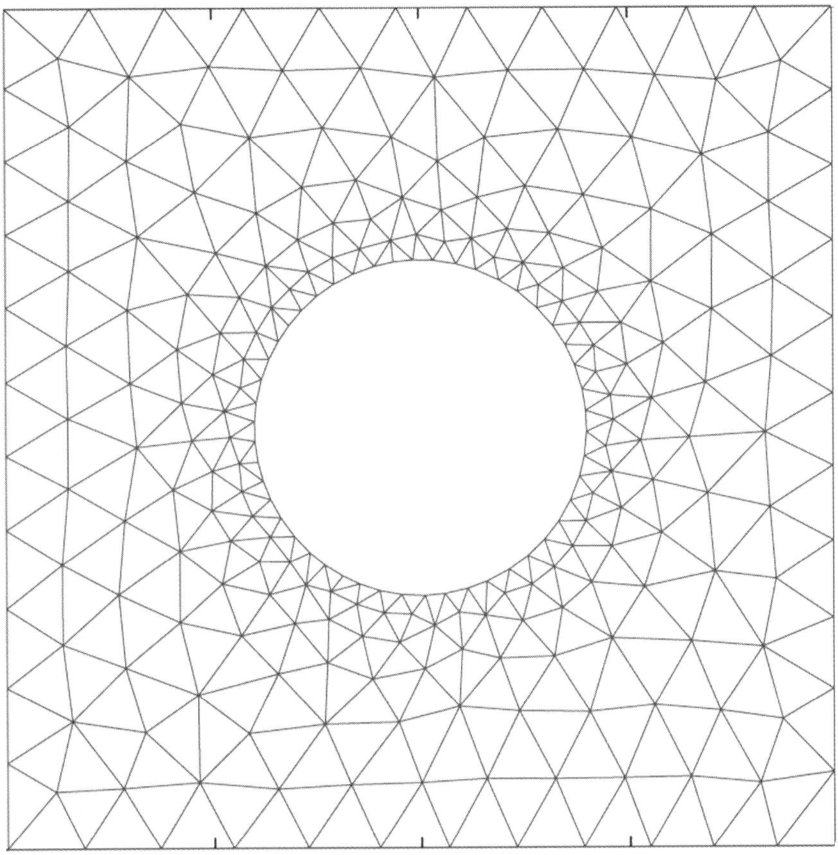

In Unit II, we saw the importance of using the appropriate decomposition to solve matrix problems. As the matrix becomes large, these decompositions become expensive. In this unit, we concentrate on how the cost can be reduced if the matrix is **sparse**; i.e., if there are many zeros. Such matrices arise in many applications but are particularly common in finite difference and finite element models of partial differential equations, as introduced in Chapter 23.

In Chapter 27 we consider reordering the equations and unknowns in a linear system in order to reduce the number of nonzero elements in the matrix factors. Chapter 28 takes a different approach, computing an approximate solution to a linear system or eigenvalue problem without changing the matrix at all.

Four case studies illustrate the use of sparse matrices. In Chapter 29, we solve an optimization problem in order to understand the stresses on a twisted bar. Chapter 30 shows how structure in some problems can be exploited to obtain a fast solver. Eigenvalue problems for differential equations are studied in Chapter 31. Finally, Chapter 32 introduces multigrid methods, a useful technique for solving grid-based problems or for preconditioning Krylov subspace methods.

MASTERY: After you have worked through this unit, you should be able to do the following:

- Convert a matrix from its (usual) dense representation to sparse format, and from sparse format to dense.
- Determine which elements of a matrix fill in when doing Cholesky decomposition.
- Determine which elements are in the profile of a matrix.
- Given a sparse matrix, determine its graph. Given a graph, determine the sparsity of the corresponding matrix.
- Apply our reordering strategies (Cuthill–McKee, minimum degree, and nested dissection) to a given graph.
- Apply the Gauss–Seidel iteration to a given system of linear equations (algebraically or geometrically).
- Construct a basis for a Krylov subspace and, in particular, construct an orthogonal basis.
- Explain why the Krylov methods terminate in at most n iterations with the exact solution.
- Use the CG algorithm and its convergence results.
- Count the multiplications per iteration for the Arnoldi algorithm or the conjugate gradient algorithm.
- Determine the storage requirements for the Arnoldi algorithm or the conjugate gradient algorithm.

- Use the convergence results for GMRES.

  Example: Show that if the preconditioned matrix has only 5 distinct eigenvalues, then GMRES must terminate in at most 5 iterations with the true solution.

  Example: Show that if the preconditioned matrix has 5 small clusters of eigenvalues, then after 5 iterations, GMRES produces a good approximate solution if its eigenvector matrix is well-conditioned.

- Implement a preconditioning algorithm such as Gauss–Seidel.
- Identify iterative algorithms for finding some eigenvalues or singular values of a sparse matrix.
- Solve optimization problems involving sparse matrices.
- Use the Schur decomposition to solve Sylvester equations.
- Write fast algorithms to solve problems whose eigenvectors are related to the Fourier transform.
- Use matrix eigenvalue algorithms to approximate eigenvalues and eigenfunctions of differential operators.
- Form a sequence of nested grids.
- Transfer values between grids in multigrid, given a restriction operator and an interpolation operator.
- Explain and use the V-cycle and nested grids algorithms.

Chapter 27

Solving Sparse Linear Systems: Taking the Direct Approach

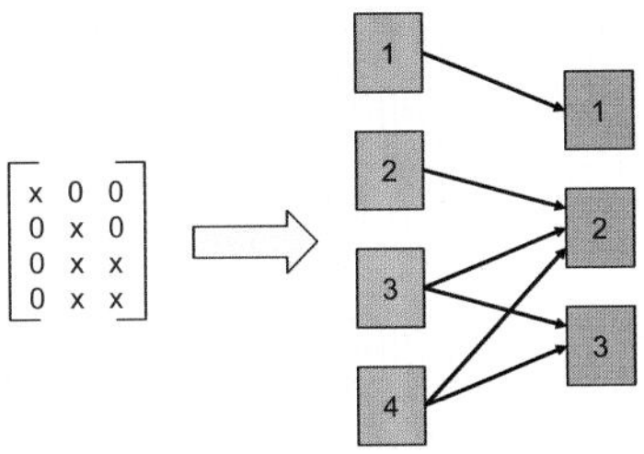

In this chapter, we explore the importance of wise ordering of equations and unknowns when solving large, sparse systems of linear equations. Our examples are drawn from solution of partial differential equations, and such problems are a prime source of such linear systems. Nothing that we do is specific to these problems, though, and you may prefer to work with a matrix from a standard test set (e.g., `wathen` from MATLAB's `gallery` function, or a matrix from the Matrix Market [106].

In addition to **direct methods** such as the Cholesky decomposition that we consider in this chapter, we should consider the **iterative methods** discussed in the next chapter whenever we have a sparse matrix problem.

27.1 Storing and Factoring Sparse Matrices

There are many possible storage schemes for sparse matrices. MATLAB chooses a typical one: store the indices and values of the nonzero elements in column order, so

$$\begin{bmatrix} 2 & 0 & 0 & 0 \\ 0 & 5 & 7 & 0 \\ 1 & 0 & 6 & 0 \\ 0 & 0 & 0 & 8 \end{bmatrix} \quad \text{is stored as} \quad \begin{matrix} (1,1) & 2 \\ (3,1) & 1 \\ (2,2) & 5 \\ (2,3) & 7 \\ (3,3) & 6 \\ (4,4) & 8 \end{matrix} .$$

Storing a sparse matrix in this way takes $3nz$ storage locations, where nz is the number of nonzeros. For small matrices, there is not much difference in storage space between dense and sparse storage schemes, but for problems with thousands or millions of unknowns, it matters a lot! If we have a matrix that has only three nonzeros per row, for example, then we require $9n$ locations to store it in sparse format rather than the n^2 required for dense, and these numbers are quite different for large n.

POINTER 27.1. Existence, Uniqueness, and Stability.

For results on existence, uniqueness, and stability of solutions to linear systems of equations, least squares problems, and eigenvalues, see the pointers in Chapter 5.

If our problem involves a sparse matrix $\boldsymbol{A}$, then we would want its LU or Cholesky factors to be sparse, too, but this is not always the case. Consider, for example, a linear system involving the "arrowhead" matrix

$$\boldsymbol{A}\boldsymbol{x} \equiv \begin{bmatrix} \times & \times & \times & \times & \times & \times \\ \times & \times & 0 & 0 & 0 & 0 \\ \times & 0 & \times & 0 & 0 & 0 \\ \times & 0 & 0 & \times & 0 & 0 \\ \times & 0 & 0 & 0 & \times & 0 \\ \times & 0 & 0 & 0 & 0 & \times \end{bmatrix} \begin{bmatrix} x_1 \\ x_2 \\ x_3 \\ x_4 \\ x_5 \\ x_6 \end{bmatrix} = \begin{bmatrix} b_1 \\ b_2 \\ b_3 \\ b_4 \\ b_5 \\ b_6 \end{bmatrix} \equiv \boldsymbol{b},$$

where $\times$ denotes a nonzero value (we don't care what it is) and 0 denotes a zero. The number of nonzeros is $3n-2$. In the first step of the LU decomposition, we add some multiple of the first row to every other row in order to put zeros in the off-diagonal elements of column 1. Disaster! The matrix is now dense, with $n(n-1)+1$ nonzeros!

There is a simple fix for this problem, though. Let's rewrite our problem by moving the first column and the first row to the end:

$$\begin{bmatrix} \times & 0 & 0 & 0 & 0 & \times \\ 0 & \times & 0 & 0 & 0 & \times \\ 0 & 0 & \times & 0 & 0 & \times \\ 0 & 0 & 0 & \times & 0 & \times \\ 0 & 0 & 0 & 0 & \times & \times \\ \times & \times & \times & \times & \times & \times \end{bmatrix} \begin{bmatrix} x_2 \\ x_3 \\ x_4 \\ x_5 \\ x_6 \\ x_1 \end{bmatrix} = \begin{bmatrix} b_2 \\ b_3 \\ b_4 \\ b_5 \\ b_6 \\ b_1 \end{bmatrix}.$$

We have replaced $\boldsymbol{A}$ by $\boldsymbol{P}\boldsymbol{A}\boldsymbol{P}^T$, where

$$\boldsymbol{P} = \begin{bmatrix} 0 & 1 & 0 & 0 & 0 & 0 \\ 0 & 0 & 1 & 0 & 0 & 0 \\ 0 & 0 & 0 & 1 & 0 & 0 \\ 0 & 0 & 0 & 0 & 1 & 0 \\ 0 & 0 & 0 & 0 & 0 & 1 \\ 1 & 0 & 0 & 0 & 0 & 0 \end{bmatrix}$$

is a **permutation matrix**. A permutation matrix is just an identity matrix with its rows reordered, and this one can be represented by the reordering sequence $\boldsymbol{r} = [2,3,4,5,6,1]$.

In the next challenge, we see the dramatic effect that reordering has on sparsity.

CHALLENGE 27.1.

(a) Verify that the reordered system has the same solution as the original one, and that when we use Gauss elimination (or the Cholesky decomposition) on our reordered system, no new nonzeros are produced. (In particular, the Cholesky factor has at most $2n-1$ nonzeros.) Compare the work in factoring $\boldsymbol{A}$ with that for factoring $\boldsymbol{PAP}^T$.

(b) Show that our reordered system is

$$(\boldsymbol{PAP}^T)(\boldsymbol{Px}) = \boldsymbol{Pb}.$$

Reordering the variables and equations is a powerful tool for maintaining sparsity during factorization, and we investigate some strategies for determining good permutations. Notice that in order to preserve symmetry, we always pair $\boldsymbol{P}$ with $\boldsymbol{P}^T$ in reordering $\boldsymbol{A}$, but for nonsymmetric problems it can be advantageous to choose a different column permutation in place of $\boldsymbol{P}^T$.

27.2 What Matrix Patterns Preserve Sparsity?

Finding the reordering that minimizes the number of nonzeros in $\boldsymbol{L}$ for the Cholesky factorization or $\boldsymbol{U}$ in the LU factorization is generally too expensive. Therefore, we rely on heuristics that give us an inexpensive algorithm to find a reordering but are not guaranteed to produce an optimal ordering. Usually the heuristics do well, but sometimes they produce a very bad reordering. The heuristics all aim to permute the matrix to a form that has little **fill-in** (new nonzeros produced in the factorization), so let's investigate nonzero patterns for which sparsity is preserved.

CHALLENGE 27.2.

(a) A matrix $\boldsymbol{A}$ is a **band matrix** with bandwidth ℓ if $a_{jk}=0$ whenever $|j-k|>\ell$. (An important special case is that of a **tridiagonal matrix**, with $\ell=1$.) Show that the factor $\boldsymbol{L}$ (or $\boldsymbol{U}$) for $\boldsymbol{A}$ also has bandwidth ℓ.

(b) Define the **profile** of a matrix to stretch from the first nonzero in each column to the main diagonal element in the column, and from the first nonzero in each row to the main diagonal element. For example, if

$$\boldsymbol{A} = \begin{bmatrix} \times & 0 & \times & 0 & 0 & 0 \\ 0 & \times & 0 & 0 & \times & 0 \\ 0 & 0 & \times & 0 & \times & 0 \\ \times & 0 & 0 & \times & 0 & 0 \\ 0 & 0 & \times & 0 & \times & 0 \\ 0 & \times & 0 & 0 & 0 & \times \end{bmatrix},$$

then the profile of A contains its nonzeros as well as those zeros marked with $\otimes$:

$$\begin{bmatrix} \times & 0 & \times & 0 & 0 & 0 \\ 0 & \times & \otimes & 0 & \times & 0 \\ 0 & 0 & \times & 0 & \times & 0 \\ \times & \otimes & \otimes & \times & \otimes & 0 \\ 0 & 0 & \times & \otimes & \times & 0 \\ 0 & \times & \otimes & \otimes & \otimes & \times \end{bmatrix},$$

Show that the factor $\boldsymbol{L}$ for a symmetric matrix $\boldsymbol{A}$ has no nonzeros outside the profile of $\boldsymbol{A}$.

From Challenge 27.2 we conclude that a good reordering strategy might try to produce a reordered matrix with a small bandwidth or a small profile.

27.3 Representing Sparsity Structure

The sparsity of a matrix can be encoded in a graph. For example, a symmetric matrix

$$A = \begin{bmatrix} \times & 0 & \times & 0 & 0 & 0 \\ 0 & \times & 0 & 0 & \times & \times \\ \times & 0 & \times & 0 & \times & 0 \\ 0 & 0 & 0 & \times & 0 & 0 \\ 0 & \times & \times & 0 & \times & 0 \\ 0 & \times & 0 & 0 & 0 & \times \end{bmatrix}$$

has upper-triangular nonzero off-diagonal elements $a_{13}, a_{25}, a_{26}, a_{35}$ and corresponds to a graph with 6 **nodes**, one representing each row (or equivalently, each column), and four **edges**, connecting nodes $(1,3)$, $(2,5)$, $(2,6)$, and $(3,5)$. (We omit the edges corresponding to the main diagonal elements, and assume that these elements are nonzero. This is always true for a positive definite matrix.)

In Figure 27.1 we draw the graph for the matrix $\boldsymbol{S}$ that is given in Figure 27.2, corresponding to the finite difference matrix for Poisson's equation on the unit square, discretized with a 5×5 grid of unknowns. Notice that the **degree** of any node (the number of edges it has) is at most 4, and that there are 4 nodes (nodes 1, 5, 21 and 25) of minimum degree, which is 2.

The sparsity of a nonsymmetric matrix can also be represented by a graph. In this case we use two sets of nodes, one for each row and one for each column, and draw an edge from row node i to column node j when a_{ij} is nonzero. This is illustrated in Figure 27.3.

27.4 Some Reordering Strategies for Sparse Symmetric Matrices

We now have the jargon and concepts necessary to discuss some reordering strategies. The $n \times n$ matrices $\boldsymbol{A}$ we consider in this section have three important properties:

- They are **real symmetric** (or complex Hermitian), so that element $a_{jk} = a_{kj}$ for $j,k = 1,\ldots,n$. This forces all of the eigenvalues to be real.

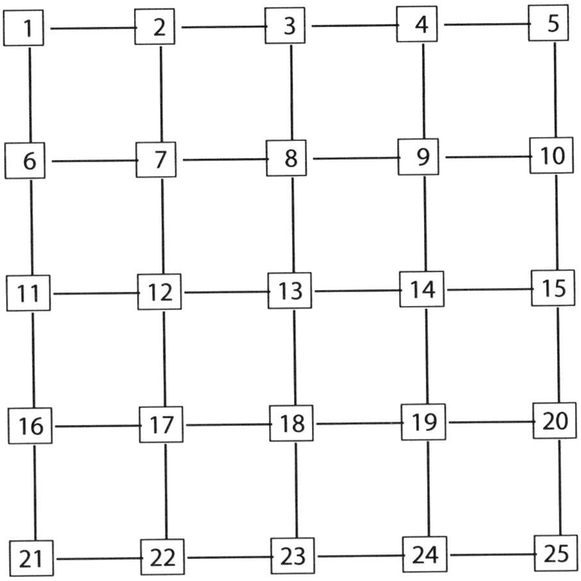

Figure 27.1. *The graph corresponding to the matrix in Figure* 27.2.

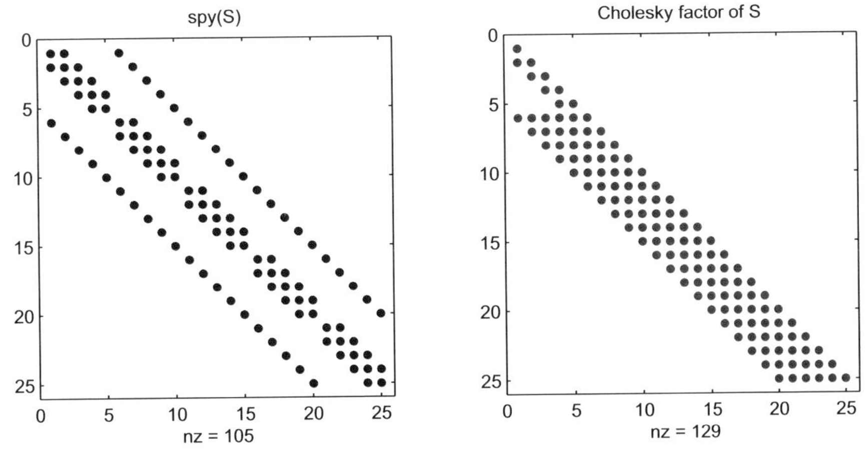

Figure 27.2. *The sparsity pattern in the finite difference matrix* **S** *for Poisson's equation* $-u_{xx} - u_{yy} = f(x, y)$ *on the unit square, discretized with a* 5×5 *grid of unknowns. The display was made with MATLAB's* `spy` *function.*

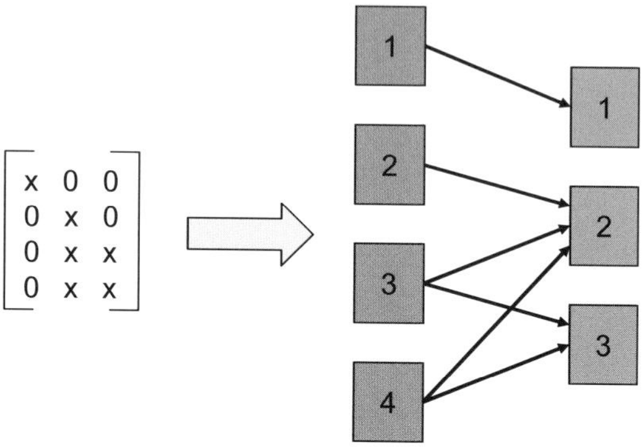

Figure 27.3. *A graph representing the sparsity of a general matrix. There is a blue node for each row and a red node for each column.*

- They are **positive definite**, so that all of the eigenvalues are positive.
- They are **sparse**, meaning that most of the matrix entries are zero, and the number of nonzero elements grows as n rather than as n^2 as the discretization is refined.

The first two properties ensure that if we perform Gauss elimination on the linear systems, we never need to **pivot for stability**, interchanging rows of the matrix in order to put a larger magnitude element on the main diagonal. Also, we can take advantage of the symmetry of $\boldsymbol{A}$ and use the **Cholesky decomposition** of the matrix, factoring $\boldsymbol{A} = \boldsymbol{L}\boldsymbol{L}^*$, where $\boldsymbol{L}$ is a lower-triangular matrix. This requires half the work of Gauss elimination—but it is only stable if $\boldsymbol{A}$ is positive definite.

Since we don't need to pivot for stability, we are completely free to **pivot to preserve sparsity**, and we'll turn our attention to why this is necessary and how to do it effectively.

Strategy 1: Cuthill–McKee. One of the oldest strategies is **Cuthill–McKee**, Algorithm 27.1, which uses the graph to order the rows and columns, working our way out from a node of small degree.

Algorithm 27.1 Cuthill–McKee Reordering

```
Find a node with minimum degree and order it first.
while some node remains unordered,
  for each node that was ordered in the previous step,
    Order all of the unordered nodes that are connected to it, in order of their degree.
  end
end
```

Reverse Cuthill–McKee (doing a final reordering from last to first) often works even better. The result of the ordering on matrix $\boldsymbol{S}$ is shown in Figure 27.4. The ordering tends to give a matrix with small bandwidth.

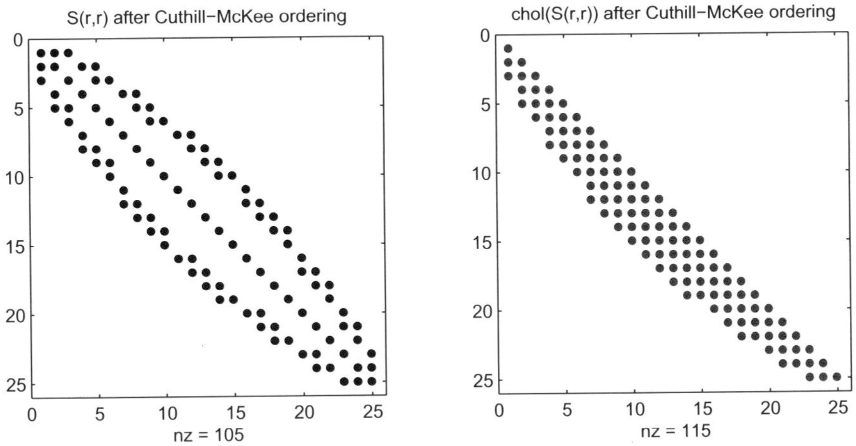

Figure 27.4. *The sparsity pattern after reordering* **S** *using reverse Cuthill–McKee.*

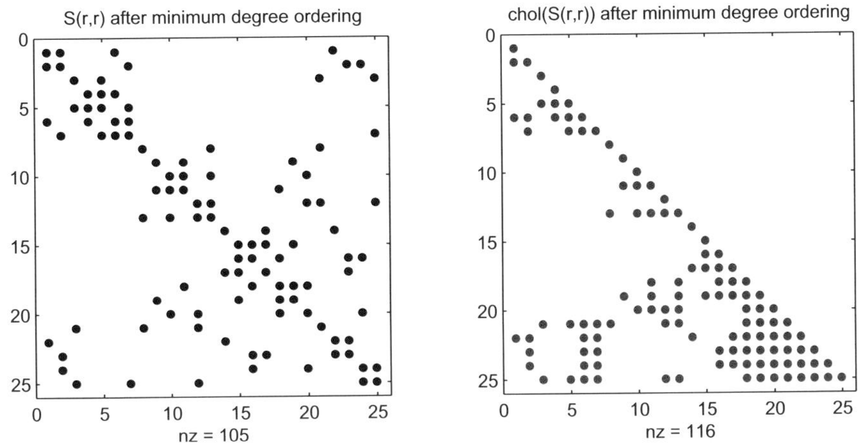

Figure 27.5. *The sparsity pattern after reordering* **S** *using the minimum degree algorithm.*

Strategy 2: Minimum Degree. The minimum degree strategy, Algorithm 27.2, is a greedy strategy that removes nodes from the graph in order of degree. This strategy works rather well in practice but it is relatively expensive, since the degree counts are updated every time a node is deleted. In Figure 27.5, we see the results on our matrix $\boldsymbol{S}$. The reordering gives small profile but not small bandwidth.

Algorithm 27.2 Minimum Degree Reordering

while some node remains unordered,

Choose a node that has minimum degree in the current graph, and order that node next, removing it from the graph. (If there is a tie, choose any of the candidates.)

end

Strategy 3: Nested Dissection. Nested Dissection, Algorithm 27.3, is a recursive algorithm that breaks the graph into pieces. For our example, we bisect the 5×5 grid graph vertically with a red separator and then bisect the remaining two pieces horizontally with blue separators. Nested dissection orders the nodes in the remaining pieces first, followed by the nodes in the separators, and produces the following renumbering:

$$\begin{array}{ccccc} 1 & 2 & 17 & 5 & 6 \\ 3 & 4 & 18 & 7 & 8 \\ 22 & 23 & 19 & 24 & 25 \\ 9 & 10 & 20 & 13 & 14 \\ 11 & 12 & 21 & 15 & 16 \end{array}.$$

The results are shown in Figure 27.6. The matrix looks quite disordered, but the number of nonzeros in the factor is smaller than for our original ordering because the profile is small.

Algorithm 27.3 Nested Dissection Reordering

Start with one "piece" containing the whole graph.
while some piece of the graph has a large number of nodes,
  Consider one such piece.
  Try to break it into two pieces plus a **separator**, with

- approximately the same number of nodes in the two pieces,
- no edges between the two pieces,
- a small number of nodes in the separator.

end
Then order the nodes piece by piece.
Finally, order the nodes in the separators.

In the next challenge, we construct the graph for a sparse matrix and try these three reorderings on it.

CHALLENGE 27.3.

Draw the graph corresponding to the matrix

$$\begin{bmatrix} \times & 0 & 0 & 0 & \times & 0 & 0 & 0 & 0 & 0 \\ 0 & \times & 0 & 0 & 0 & 0 & 0 & \times & 0 & \times \\ 0 & 0 & \times & 0 & \times & 0 & 0 & 0 & \times & 0 \\ 0 & 0 & 0 & \times & 0 & \times & \times & 0 & \times & \times \\ \times & 0 & \times & 0 & \times & 0 & 0 & 0 & 0 & 0 \\ 0 & 0 & 0 & \times & 0 & \times & \times & 0 & 0 & \times \\ 0 & 0 & 0 & \times & 0 & \times & \times & 0 & \times & 0 \\ 0 & \times & 0 & 0 & 0 & 0 & 0 & \times & 0 & 0 \\ 0 & 0 & \times & \times & 0 & 0 & \times & 0 & \times & 0 \\ 0 & \times & 0 & \times & 0 & \times & 0 & 0 & 0 & \times \end{bmatrix}.$$

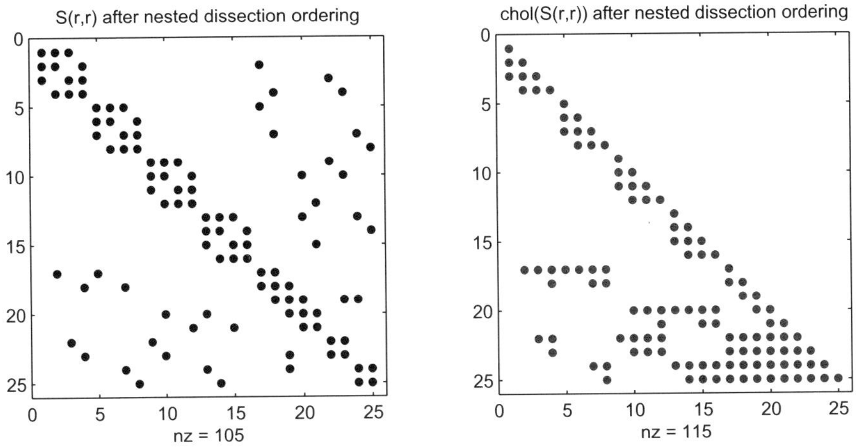

Figure 27.6. *The sparsity pattern after reordering* **S** *using nested dissection.*

Try each of the three reorderings on this matrix. Compare the sparsity of the Cholesky factors of the reordered matrices with the sparsity of the factor corresponding to the original ordering.

Strategy 4: Eigenvector Partitioning. This method, Algorithm 27.4, is less intuitive and much more expensive, but it produces useful orderings. Unlike the previous strategies, we cannot determine the ordering by hand computation, since it involves an eigenvector computation. Because of the expense compared to our other strategies, it should probably only be used on matrices that are to be used multiple times.

Algorithm 27.4 Eigenvector Partitioning

First we form an auxiliary matrix, the **Laplacian** of the graph corresponding to our sparse matrix. This matrix $\boldsymbol{B}$ has the same size and sparsity pattern as $\boldsymbol{A}$. It has -1 in place of each nonzero off-diagonal element of $\boldsymbol{A}$, and the main diagonal elements of $\boldsymbol{B}$ are set so that each of the row sums is zero.

The matrix $\boldsymbol{B}$ is symmetric and (by Gerschgorin's theorem) has no negative eigenvalues. It has a zero eigenvalue (since $\boldsymbol{Be} = \boldsymbol{0}$, where $\boldsymbol{e}$ is the vector of all ones). We compute the eigenvector $\boldsymbol{v}$ corresponding to its next smallest eigenvalue.

Partition the graph into two pieces, one corresponding to nodes with positive entries in $\boldsymbol{v}$, and the other containing the remaining nodes.

If desired, repeat the algorithm recursively on each of the two subgraphs formed by this partition.

Order the nodes piece by piece.

The eigenvector computation is accomplished, for example, by asking the Lanczos algorithm (see Chapter 28) to produce approximations to the two smallest eigenvalues and

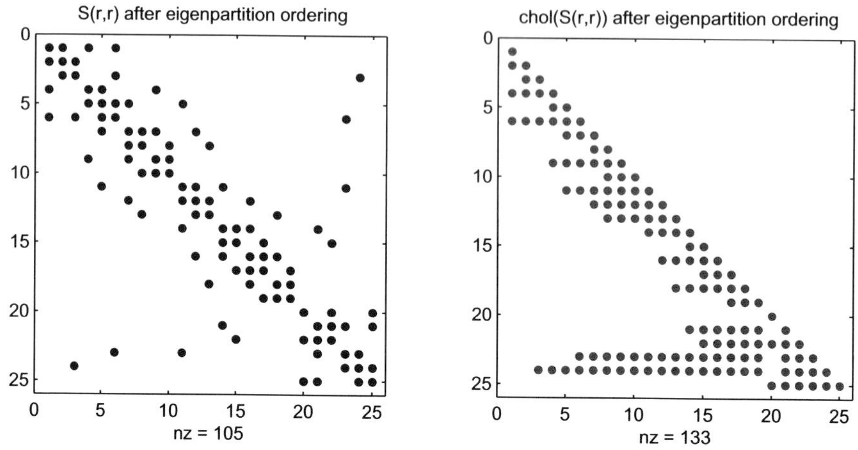

Figure 27.7. *The sparsity pattern after reordering* **S** *using eigenpartitioning.*

POINTER 27.2. Software for Reordering.

MATLAB's `symrcm` implements the reverse Cuthill–McKee ordering. MATLAB's `symamd` is an approximate minimum degree permutation; `symmmd` is exact but more expensive. Nested dissection and the eigenvector orderings are not built-in, but MATLAB's `eigs` can be used for the eigenvector computation. (If `eigs` complains about a singular matrix, send it the Laplacian plus a multiple of the identity; this shifts the eigenvalues but preserves the eigenvectors.) These two orderings are also available in a toolbox written by Gilbert and Teng [57, 56].

their eigenvectors. We are only interested in the signs of the entries in the eigenvectors, so we don't need much accuracy. The results of our example are shown in Figure 27.7. The matrix again looks quite disordered, but the profile remains rather small. This algorithm is not very effective on this matrix.

Bigger problems show trends more clearly. Instead of a 5×5 grid, let's consider a 50×50, giving a matrix of size $n = 2500$. Here is a summary of our results:

Ordering	Nonzeros in $\mathbf{L}$
Original	274689
Rev Cuthill–McKee	189345
Minimum deg	68828
Nested dissection	89733
Eigenpartition	86639

For this very regular graph, minimum degree works the best.

Now that we have our candidate algorithms, we can evaluate them using some test problems.

CHALLENGE 27.4.

You have been hired as a consultant by our start-up company, PoissonIsUs.com, to advise us on solving linear systems of equations. Our business is to solve elliptic partial differential equations in 2 and 3 dimensions. We have limited venture capital funding, so we are starting with a very specific mission: to solve the Poisson equation

$$-u_{xx} - u_{yy} = f(x, y)$$

when $(x, y) \in \Omega \subset \mathcal{R}^2$, or

$$-u_{xx} - u_{yy} - u_{zz} = f(x, y, z)$$

when $(x, y, z) \in \Omega \subset \mathcal{R}^3$. The complete problem specifications also include information about the behavior of the solution u on the boundary of Ω.

The standard method for solving such problems is to **discretize** using either **finite differences** or **finite elements**, and then solve the resulting system of linear equations. In order to get accurate estimates of the solution, this system is made very large, involving thousands or millions of unknowns, so it is important to be very efficient in our solution algorithm. Your job is to evaluate some alternatives.

For testing purposes, we have developed two problems that we believe typical of those that our customers will provide. In the first problem, the domain is a sector of a circle, and the differential equation is discretized using an adaptive finite element grid. In the second, the domain is a three-dimensional box with discretization using finite differences. Use `slit2.m` and `laplace3d.m` (found on the website) with $n = 15$ to generate the linear systems.

Solve the linear systems using as many reorderings as possible: original matrix, reverse Cuthill–McKee ordering, (approximate) minimum degree ordering, nested dissection ordering, eigenvector ordering. (See Pointer 27.2.) Make a table reporting, for each method,

- time to solve the system (include reordering, factorization, forward and back substitution.)
- storage for the matrix factors.
- the final relative residual $\|b - Ax_{computed}\|_2 / \|b\|_2$. (These should all be well below the errors due to discretization, and so they will not be a factor in your recommendation.)

If possible, run larger problems, too. Considering the two-dimensional and three-dimensional problems separately, report to the CEO of PoissonIsUs.com the performance of the various methods and your recommendation for what ordering to use.

27.5 Reordering Strategies for Nonsymmetric Matrices

If your linear system involves a matrix that is nonsymmetric, or symmetric but not positive definite, then reordering for the stability of the factorization must take priority over reordering for sparsity. See Pointer 27.3 for references.

POINTER 27.3. Further Reading.

For more information on sparse matrices, reordering strategies, and graph representation, George and Liu [55] discuss the symmetric positive definite case. Duff, Erisman, and Reid [45] discuss the general case, including the complications added by stability considerations. Timothy Davis's book [35] gives a tutorial tour of a stripped-down but useful solver for sparse linear systems. The eigenvector partitioning method is not considered in these references; it is discussed by Pothen, Simon, and Liou [125], and an intuitive approach to it is given by Demmel [38].

When problems get very large, even a good reordering strategy does not enable us to keep the LU or Cholesky factors in memory, and **iterative methods** must be considered as an alternative. We'll discuss them in Chapter 28.

Chapter 28

Iterative Methods for Linear Systems

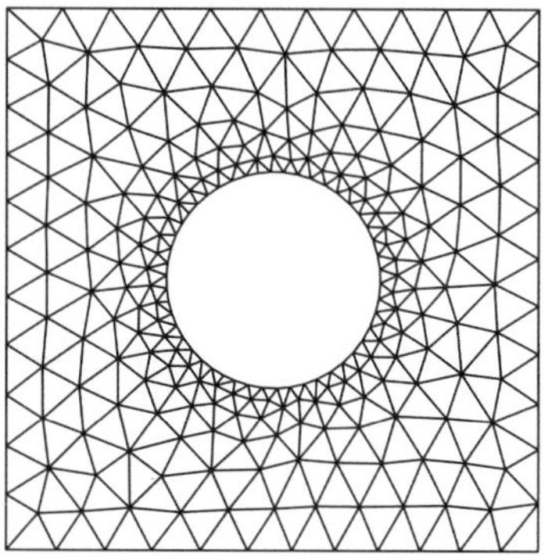

Solving a linear system of equations is one of the easiest computational tasks imaginable. But sometimes when the matrix is very large, the algorithms we learned in linear algebra class or in Chapter 5 take too long, and we need to use a different type of method. In this chapter, we investigate these **iterative methods**.

In **direct methods** such as Gauss elimination, the LU and Cholesky decomposition, or even Cramer's rule, we have a fixed set of operations that yield (using exact arithmetic) the exact solution $\widehat{\boldsymbol{x}}$ to the linear system $\boldsymbol{A}\widehat{\boldsymbol{x}} = \boldsymbol{b}$. The cost is predictable: $O(n!)$ for Cramer's rule and $O(n^3)$ for the other two methods, if the $n \times n$ matrix $\boldsymbol{A}$ is dense (i.e., has only a few zeros). In Chapter 27, we investigated how to make direct methods faster if $\boldsymbol{A}$ is sparse (having many zeros), but sometimes even this is not enough to make direct methods practical for a large problem.

The idea behind iterative methods is to start from an initial guess $\boldsymbol{x}^{(0)}$ for the solution to $\boldsymbol{A}\widehat{\boldsymbol{x}} = \boldsymbol{b}$ and construct a sequence of guesses $\{\boldsymbol{x}^{(0)}, \boldsymbol{x}^{(1)}, \boldsymbol{x}^{(2)}, \ldots\}$ converging to $\widehat{\boldsymbol{x}}$. (Often, $\boldsymbol{x}^{(0)} = \boldsymbol{0}$, but good use can be made of an estimate obtained using other information about the problem or from the known solution to a similar problem.) In contrast to direct methods, iterative methods yield only an approximate solution to the linear system, and although the cost per iteration is fixed, usually proportional to the cost of multiplying a vector by $\boldsymbol{A}$, the number of iterations is not precisely known.

Large sparse systems of linear equations arise in many contexts, but in this chapter we use as an example a finite element model of the steady state heat distribution in two thin plates: a square one and the one shown in Figure 28.1. See the Pointer 28.2 for other examples.

We'll consider two classes of methods, **stationary iterative methods** and **Krylov subspace methods**, and discuss accelerating the Krylov methods using preconditioners.

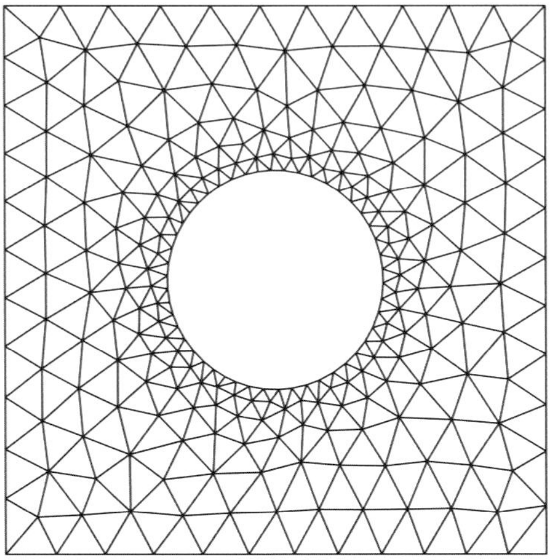

Figure 28.1. *The thin plate for the second example problem is a square with a hole cut out of it.*

28.1 Stationary Iterative Methods (SIMs)

These methods grew up in the engineering literature. They were very popular in the 1960s and are still used sometimes. Today, they are almost never the best algorithms (because they take too many iterations), but we will see in Challenge 28.5 that they are useful **preconditioners** for Krylov subspace methods.

One of the most common SIMs is **Gauss–Seidel**, also called **successive displacements**. We discussed its use for nonlinear systems of equations in Section 24.3.2. For linear systems, it is derived by solving the ith equation of $\boldsymbol{A}\widehat{\boldsymbol{x}} = \boldsymbol{b}$ for $\hat{x}_i$:

$$\hat{x}_i = (b_i - \sum_{j=1}^{i-1} a_{ij}\hat{x}_j - \sum_{j=i+1}^{n} a_{ij}\hat{x}_j)/a_{ii}, \qquad i = 1,\ldots,n. \tag{28.1}$$

Notice that we must require $\boldsymbol{A}$ to have nonzeros on its main diagonal. We don't know $\widehat{\boldsymbol{x}}$, but we can use our latest information on the right-hand side of this equation to get an updated guess for x_i. This yields Algorithm 28.1.

Algorithm 28.1 The Gauss–Seidel Algorithm (One Step)

Given $\boldsymbol{x}^{(k)}$, construct $\boldsymbol{x}^{(k+1)}$ by

$$x_i^{(k+1)} = \left(b_i - \sum_{j=1}^{i-1} a_{ij} x_j^{(k+1)} - \sum_{j=i+1}^{n} a_{ij} x_j^{(k)} \right) / a_{ii}, \qquad i = 1,\ldots,n.$$

We continue the algorithm until either the residual $b - Ax^{(k)}$ is small enough, or the change $x^{(k+1)} - x^{(k)}$ is small enough, or a maximum number of iterations k has been performed.

The algorithm is very easy to program! We only need to store one x vector, and let the most recent guess overwrite the old. We should only use terms in (28.1) corresponding to the nonzeros in A; otherwise the work per iteration is $O(n^2)$ instead of $O(nz)$, where nz is the number of nonzeros in A.

If A is nonsingular and the iteration converges, it converges to $\widehat{x}$, but convergence is not guaranteed. Convergence depends on the properties of A. If we partition A as $L + D + U$, where D contains the diagonal entries, U contains the entries above the diagonal, and L contains the entries below the diagonal, then we can express the iteration as

$$(D + L)x^{(k+1)} = b - Ux^{(k)}.$$

This form of the algorithm is sometimes useful for computation, and it is essential for analyzing convergence, since we can express the iteration as

$$Mx^{(k+1)} = b + Nx^{(k)},$$

splitting the matrix A into $M - N$, with $M = D + L$ and $N = -U$. Equivalently,

$$x^{(k+1)} = Gx^{(k)} + c,$$

where $G = M^{-1}N$ is a matrix that depends on A and $c = M^{-1}b$ is a vector that depends on A and b. Also observe that the true solution $\widehat{x}$ satisfies $\widehat{x} = G\widehat{x} + c$. Subtracting, we see that the error $e^{(k)} = x^{(k)} - \widehat{x}$ satisfies

$$e^{(k+1)} = Ge^{(k)},$$

and it can be shown (See Challenge 5.15) that the error converges to zero for any initial $x^{(0)}$ if and only if all of the eigenvalues of G lie inside the unit circle. Many conditions on A have been found that guarantee convergence of these methods, but it is better to use faster methods. To convince ourselves of this, let's try the method on our sample problems so we can compare with better methods later.

CHALLENGE 28.1.

Use Gauss–Seidel to solve the linear systems of equations generated by the MATLAB function `generateproblem` found on the website. (Set the input parameter `kappa` to zero, and note that you only need a part of the structure generated by this function: the matrices `mesh(k).A`, the right-hand-side vectors `mesh(k).b`, and the convergence tolerance `mesh(k).tol`.) Use an initial guess $x^{(0)} = \mathbf{0}$. The mathematical model of heat distribution is not exact, so we can stop the iteration when the norm of the residual is somewhat smaller than the error in the model. We approximate this by stopping when the norm of the residual has been reduced by a factor of `mesh(k).tol`. As the number of unknowns in the linear system grows, this error decreases. Graph the number of iterations and the solution time as a function of the number of unknowns, choosing a range of problem sizes appropriate for your computer. (Number of iterations is not a very good measure for comparing algorithms, since the work per iteration is so different for the different methods we study.) Compare the time and storage with that for direct solution of the linear systems.

28.2 From SIMs to Krylov Subspace Methods

Stationary iterative methods such as Gauss–Seidel have an interesting property. If $\boldsymbol{x}^{(0)} = \boldsymbol{0}$, then

$$\boldsymbol{x}^{(1)} = \boldsymbol{c}$$

and

$$\begin{aligned}\boldsymbol{x}^{(2)} &= \boldsymbol{G}\boldsymbol{x}^{(1)} + \boldsymbol{c} \\ &= \boldsymbol{G}\boldsymbol{c} + \boldsymbol{c}.\end{aligned}$$

Therefore, $\boldsymbol{x}^{(2)}$ is a linear combination of $\boldsymbol{c}$ and $\boldsymbol{G}\boldsymbol{c}$; in other words,

$$\boldsymbol{x}^{(2)} \in \operatorname{span}\{\boldsymbol{c}, \boldsymbol{G}\boldsymbol{c}\}.$$

It is easy to show that the pattern continues:

$$\boldsymbol{x}^{(k)} \in \operatorname{span}\{\boldsymbol{c}, \boldsymbol{G}\boldsymbol{c}, \boldsymbol{G}^2\boldsymbol{c}, \ldots, \boldsymbol{G}^{k-1}\boldsymbol{c}\} \equiv \mathcal{K}_k(\boldsymbol{G}, \boldsymbol{c}),$$

and we call $\mathcal{K}_k(\boldsymbol{G}, \boldsymbol{c})$ a **Krylov subspace**.

The work per iteration in Gauss–Seidel and other stationary iterative methods is primarily the work involved in multiplying a vector by $\boldsymbol{G}$. So it is natural to ask whether the Gauss-Seidel choice for $\boldsymbol{x}^{(k)}$ is the best choice in the Krylov subspace, or whether we might do better without much extra work or storage.

First we need to define "best." There are two common approaches:

- The **variational approach**: Choose $\boldsymbol{x}^{(k)} \in \mathcal{K}_k(\boldsymbol{G}, \boldsymbol{c})$ to minimize the distance between $\boldsymbol{x}^{(k)}$ and $\widehat{\boldsymbol{x}}$.

- The **projection approach** (also called the **Galerkin approach**): Choose $\boldsymbol{x}^{(k)} \in \mathcal{K}_k(\boldsymbol{G}, \boldsymbol{c})$ to make the residual $\boldsymbol{r}^{(k)} = \boldsymbol{b} - \boldsymbol{A}\boldsymbol{x}^{(k)}$ orthogonal to every vector in $\mathcal{K}_k(\boldsymbol{G}, \boldsymbol{c})$.

Note that each subspace $\mathcal{K}_k$ contains the previous one. If the Krylov subspace does not expand, then the iteration terminates, fortunately with the exact solution. If it does expand, then after k iterations, we have minimized over, or projected against, a subspace of dimension k. Thus, after at most n iterations, Krylov subspace iterations terminate with the true solution! This finite termination property is appealing but less useful than it might seem, since we think of applying these methods when n is so large (tens of thousands, millions, billions) that we can't afford more than a few hundred iterations.

To make the algorithm practical, we need two ingredients. First, a very clever choice of basis vectors for $\mathcal{K}_k$ keeps the work per iteration small and provides enough numerical stability so that we can run for many iterations. The "right" choice is an orthogonal basis in order to preserve numerical stability. Second, a clever choice of $\boldsymbol{G}$ ensures that we converge in a small number of iterations, before the work and the accumulated numerical error overwhelm us.

There are many Krylov subspace methods (See Pointer 28.2) but we focus first on one that is useful when $\boldsymbol{A}$ is symmetric and positive definite. Such matrices arise in many practical situations: for example, in discretizations of self-adjoint elliptic partial differential equations such as the ones in our examples. This Krylov subspace method is called

(preconditioned) conjugate gradients (PCG). It is both a minimization algorithm, in the **energy norm**

$$\|x-\hat{x}\|_A \equiv [(x-\hat{x})^T A(x-\hat{x})]^{1/2},$$

and a projection algorithm, making the residual $b - Ax^{(k)}$ orthogonal to the Krylov subspace. There is a very compact and practical form for the algorithm, in which we keep only the latest x-vector, the latest residual vector r, and two other vectors z and p. We present Algorithm 28.2 using a splitting matrix M as a **preconditioner**. The choice of M determines $G = M^{-1}N$ and can be as simple as taking $M = I$. We discuss other choices in the next section.

Algorithm 28.2 Algorithm PCG for Solving $Ax = b$

Given: symmetric positive definite matrices A and M, a vector b, an initial guess x, and a tolerance tol.
Let $r = b - Ax$, solve $Mz = r$ for z, and let $\gamma = r^T z$, $p = z$, and $\rho = \|r\|$.
for $k = 0, 1, \ldots,$ until $\|r\|/\rho < tol$,
  $\alpha = \gamma/(p^T Ap)$
  $x = x + \alpha p$
  $r = r - \alpha Ap$
  Solve $Mz = r$ for z.
  $\hat{\gamma} = r^T z$
  $\beta = \hat{\gamma}/\gamma$, $\gamma = \hat{\gamma}$
  $p = z + \beta p$
end

The PCG algorithm is quite remarkable: it solves a linear system—exactly—in at most n steps, and the only use of the matrix A is in forming matrix-vector products! The choice of the scalar parameter α at each iteration ensures that the minimization and projection properties hold. The choice of β ensures that the direction vector p is **A-conjugate** to all previous vectors (i.e., $p_i^T Ap_j = 0$ for $i \neq j$). The algorithm derives its name from this conjugacy, plus the choice of the negative gradient of the function $f(x) = \frac{1}{2}x^T Ax - x^T b$ as the initial direction. The vectors p form a basis for the Krylov subspace, as do the (orthogonal) residual vectors r.

Let's compare the performance of this algorithm with that of Gauss–Seidel on our two problems.

CHALLENGE 28.2.

Use PCG to solve the linear systems of equations considered in Challenge 28.1. Use an initial guess $x^{(0)} = \mathbf{0}$ and set $M = I$. Graph the number of iterations and the solution time as a function of the number of unknowns and compare this and the storage with the results of Challenge 28.1.

28.3 Preconditioning CG

The matrix M is called the **preconditioner**, and our next task is to understand what it does and why we might need it. Using the **energy norm**, we have a bound on the error after k iterations of PCG:

$$\|x^{(k)} - \widehat{x}\|_A \leq 2\left(\frac{\sqrt{\kappa(\widehat{G})}-1}{\sqrt{\kappa(\widehat{G})}+1}\right)^k \|x^{(0)} - \widehat{x}\|_A,$$

where $\kappa(\widehat{G})$ is the ratio of the largest and smallest eigenvalues of $\widehat{G} = I - G = I - M^{-1}N = M^{-1}A$.

This gives us three guidelines for choosing a preconditioning matrix M:

- We require that M be symmetric and positive definite.
- For fast iterations, we need to be able to solve linear systems involving M very quickly, since this must be done once per iteration.
- To make the number of iterations small, we want M to be a good approximation to A. For example, it is good if the eigenvalues of $\widehat{G}$ lie in a small interval.

Note that the linear system $Mz = r$ is typically solved using a direct method, so the closer M is to A, the closer we are to solving $Ax = b$ using a direct method. In fact, if $M = A$, then $N = 0$ and the iteration converges in a single iteration; we have created a direct method. So the art of preconditioning is the trade-off between work per iteration (minimized by taking $M = I$) and the number of iterations (minimized by taking $M = A$). Common choices of preconditioning matrices M include the following:

- M = the diagonal of A. This often reduces the effects of poor scaling in the problem formulation.
- M = a banded piece of A.
- M = an incomplete decomposition of A, leaving out inconvenient elements (the **incomplete Cholesky preconditioner**).
- M^{-1} = a sparse approximation to A^{-1} (the **sparse approximate inverse preconditioner** (SAIP)).
- M = a matrix related to A either physically or algebraically. For example, if A is a discretization of a differential operator, M might be a discretization of a related operator that is easier to solve, or M might be the block diagonal piece of the matrix after ordering for nested dissection.
- M might be the matrix from any stationary iterative method (SIM) or from **multigrid** (discussed in the case study of Chapter 32).

Let's try the Incomplete Cholesky preconditioners on our problems.

CHALLENGE 28.3.

Use PCG to solve the linear systems of equations considered in Challenge 28.1. Use an initial guess $\boldsymbol{x}^{(0)} = \boldsymbol{0}$ and set $\boldsymbol{M}$ to be incomplete Cholesky preconditioners generated by `cholinc` with various choices for its parameters `droptol` and `opts`. Compare with the results of Challenges 28.1 and 28.2, but remember that time-per-iteration is very different for the different algorithms.

It takes a bit of thought to decide how to use our SIM preconditioners. The search direction $\boldsymbol{p}$ in PCG is built from the preconditioned residual. We need to express the step taken by a SIM in terms of the residual, and we do this algebra in the next challenge.

CHALLENGE 28.4.

Consider our SIM

$$\boldsymbol{M}\boldsymbol{x}^{(k+1)} = \boldsymbol{N}\boldsymbol{x}^{(k)} + \boldsymbol{b}$$

or

$$\boldsymbol{x}^{(k+1)} = \boldsymbol{M}^{-1}\boldsymbol{N}\boldsymbol{x}^{(k)} + \boldsymbol{M}^{-1}\boldsymbol{b}.$$

Show that if $\boldsymbol{r}^{(k)} = \boldsymbol{b} - \boldsymbol{A}\boldsymbol{x}^{(k)}$, then

$$\boldsymbol{x}^{(k+1)} = \boldsymbol{x}^{(k)} + \boldsymbol{M}^{-1}\boldsymbol{r}^{(k)},$$

and therefore the step taken by the SIM is

$$\boldsymbol{x}^{(k+1)} - \boldsymbol{x}^{(k)} = \boldsymbol{M}^{-1}\boldsymbol{r}^{(k)}.$$

Using Challenge 28.4, we can compute $\boldsymbol{M}^{-1}\boldsymbol{r}$ in the PCG algorithm by taking one step of the SIM starting from the latest PCG iterate, returning the change in $\boldsymbol{x}$ as $\boldsymbol{M}^{-1}\boldsymbol{r}$.

One complication remains: the Gauss Seidel matrix $\boldsymbol{M}$ is not symmetric. In order to use Gauss–Seidel as a preconditioner for PCG, we perform a double iteration, considering the equations in order $1, \ldots, n$ and then again in order $n, \ldots, 1$. This is called the **symmetric Gauss–Seidel** iteration, given in Algorithm 28.3.

Algorithm 28.3 The Symmetric Gauss–Seidel Algorithm (One Step)

Given $\boldsymbol{x}^{(k)}$, construct $\boldsymbol{x}^{(k+1)}$ by

$$x_i^{(k+1/2)} = \left(b_i - \sum_{j=1}^{i-1} a_{ij} x_j^{(k+1/2)} - \sum_{j=i+1}^{n} a_{ij} x_j^{(k)} \right) / a_{ii}, \qquad i = 1, 2, \ldots, n.$$

$$x_i^{(k+1)} = \left(b_i - \sum_{j=1}^{i-1} a_{ij} x_j^{(k+1/2)} - \sum_{j=i+1}^{n} a_{ij} x_j^{(k+1)} \right) / a_{ii}, \qquad i = n, n-1, \ldots, 1.$$

CHALLENGE 28.5.

Use PCG to solve the linear systems of equations considered in Challenge 28.1. Use an initial guess $\boldsymbol{x}^{(0)} = \boldsymbol{0}$ and set $\boldsymbol{M}$ to be the symmetric Gauss–Seidel preconditioner. Compare with the results of Challenges 28.1, 28.2, and 28.3.

The preconditioners that we consider depend on the ordering of the unknowns and equations. We can use the reorderings from Chapter 27 to try to improve our results.

CHALLENGE 28.6.

Repeat your experiments after reordering the matrices using the approximate minimum degree (AMD) reordering. Compare the results of all experiments.

The matrices in the challenges considered in this chapter arise from discretizing an elliptic partial differential equation whose domain is a two-dimensional region. When the region is three-dimensional, the problem sizes grow even faster, and storage rapidly becomes an issue for the Cholesky algorithm.

Our preconditioners did not take advantage of the origin of the problem. Often, the more you know about the underlying problem, the better you can solve it. With a clever choice of preconditioners for this collection of matrices, the number of iterations for PCG is independent of mesh size. We consider such **multigrid methods** in the case study of Chapter 32.

28.4 Krylov Methods for Symmetric Indefinite Matrices and for Normal Equations

If the matrix $\boldsymbol{A}$ is symmetric but indefinite, with both positive and negative eigenvalues, or if we aren't sure whether the symmetric matrix is positive definite or not, then the Krylov variant called SYMMLQ [123] is the algorithm of choice.

Quite frequently, the matrix $\boldsymbol{A}$ for our problem has been created by forming $\boldsymbol{A} = \boldsymbol{B}^T\boldsymbol{B}$ for some matrix $\boldsymbol{B}$. A matrix of this form is guaranteed to be symmetric and positive semidefinite. If $\boldsymbol{B}$ is sparse and has many columns, we surely do not want to form $\boldsymbol{A}$, since it is likely to be much more dense. Luckily, CG does not require the matrix to be formed, only that we can form matrix vector products, and we can do this as $\boldsymbol{B}^T(\boldsymbol{B}\boldsymbol{p})$ for any given vector $\boldsymbol{p}$. Such problems often arise from considering the **normal equations** arising from solving the least squares problem

$$\min_{\boldsymbol{x}} \|\boldsymbol{B}\boldsymbol{x} - \boldsymbol{d}\|_2^2$$

by setting the derivative of the function $\|\boldsymbol{B}\boldsymbol{x} - \boldsymbol{d}\|_2^2$ to zero. Special algorithms have been written for problems like these; in particular, LSQR [122] is an efficient and stable variant.

28.5 Krylov Methods for Nonsymmetric Matrices

Conjugate gradients has many desirable properties. It is both a minimization and a projection method, and the iteration can be performed by storing only a few vectors, whose number does not grow with n. For general matrices, no method has all of these properties. Instead, there are two families of methods.

One family uses the **Arnoldi basis**. This requires storing and computing with k vectors at the kth iteration, so iterations slow down as k increases. The linear system solver based on minimization is called GMRES, while that based on projection is called **Arnoldi** (like the basis). Because of the cost, these methods are normally **restarted** periodically, discarding the basis and using the current residual to form another Krylov subspace.

Construction of the Arnoldi basis requires the definition of a matrix $\widehat{\boldsymbol{G}} = \boldsymbol{M}^{-1}\boldsymbol{A}$ and an initial basis vector which we take to be $\boldsymbol{c} = \boldsymbol{M}^{-1}(\boldsymbol{b} - \boldsymbol{A}\boldsymbol{x})$. At iteration j, the next vector in the orthonormal basis $\boldsymbol{v}_1, \boldsymbol{v}_2, \ldots, \boldsymbol{v}_j$ is the component of $\widehat{\boldsymbol{G}}\boldsymbol{v}_j$ that is orthogonal to $\boldsymbol{v}_1, \ldots, \boldsymbol{v}_j$. We discuss using Gram–Schmidt orthogonalization to construct the basis, but Householder can also be used. Our first basis vector is $\boldsymbol{v}_1 = z/\|z\|$. Now suppose that we have j orthonormal basis vectors $\boldsymbol{v}_1, \ldots, \boldsymbol{v}_j$ for $\mathcal{K}_j(\widehat{\boldsymbol{G}}, \boldsymbol{c})$. Then $z = \widehat{\boldsymbol{G}}\boldsymbol{v}_j \in \mathcal{K}_{j+1}(\widehat{\boldsymbol{G}}, \boldsymbol{c})$. If $z \in \mathcal{K}_j(\widehat{\boldsymbol{G}}, \boldsymbol{c})$, the Arnoldi construction terminates (and we can use our basis vectors to find an exact solution to the linear system). If z is not in this subspace, then we define the next orthonormal basis vector by the process of Gram–Schmidt orthogonalization:

$$\boldsymbol{v}_{j+1} = (z - h_{1,j}\boldsymbol{v}_1 - \cdots - h_{j,j}\boldsymbol{v}_j)/h_{j+1,j},$$

where $h_{i,j} = \boldsymbol{v}_i^* z$ $(i = 1, \ldots, j)$ and $h_{j+1,j}$ is chosen so that $\boldsymbol{v}_{j+1}^* \boldsymbol{v}_{j+1} = 1$. In matrix form, we can express this relation as

$$\widehat{\boldsymbol{G}}\boldsymbol{v}_j = \begin{bmatrix} \boldsymbol{v}_1 & \boldsymbol{v}_2 & \ldots & \boldsymbol{v}_{j+1} \end{bmatrix} \begin{bmatrix} h_{1,j} \\ h_{2,j} \\ \vdots \\ h_{j+1,j} \end{bmatrix},$$

so after k steps we have

$$\widehat{\boldsymbol{G}}\boldsymbol{V}_k = \boldsymbol{V}_{k+1}\boldsymbol{H}_k, \tag{28.2}$$

where $\boldsymbol{H}_k$ is a $(k+1) \times k$ matrix with entries h_{ij} (zero if $i > j+1$) and $\boldsymbol{V}_k$ is $n \times k$ and contains the first k basis vectors as its columns. Algorithm 28.4 constructs the Arnoldi basis. Note that we compute $\widehat{\boldsymbol{G}}\boldsymbol{v}$ by first computing $\boldsymbol{A}\boldsymbol{v}$ and then by solving a linear system involving the matrix $\boldsymbol{M}$, since forming $\widehat{\boldsymbol{G}}$ is generally inefficient in time and storage.

Equation (28.2) is very important: Since the algorithm ordinarily terminates with a zero vector after n vectors have been formed,[11] we have actually factored our matrix

$$\widehat{\boldsymbol{G}} = \boldsymbol{V}_n \boldsymbol{H}_n \boldsymbol{V}_n{}^{-1}.$$

Therefore, the matrix $\boldsymbol{H}_n$ is closely related to $\widehat{\boldsymbol{G}}$: it is formed by a similarity transform and therefore has the same eigenvalues. In fact, the leading $(k+1) \times k$ or $k \times k$ piece of H_n (available after k steps) is in some sense a good approximation to $\widehat{\boldsymbol{G}}$, and we exploit this fact in the GMRES algorithm and when we need approximate eigenvalues of $\widehat{\boldsymbol{G}}$.

[11] The set of vectors for which the algorithm terminates earlier than this has measure zero.

Algorithm 28.4 Constructing the Arnoldi Basis

$[\boldsymbol{V}_{m+1}, \boldsymbol{H}_m]$ = Arnoldi$(m, \widehat{\boldsymbol{G}}, \boldsymbol{c})$
Given: a positive integer m, a matrix $\widehat{\boldsymbol{G}}$, and a vector $\boldsymbol{c}$.
Use m steps of Gram–Schmidt orthogonalization (Algorithm 5.2), with $\boldsymbol{a}_1 = \boldsymbol{c}$ and $\boldsymbol{a}_j = \widehat{\boldsymbol{G}}\boldsymbol{q}_{j-1}$, to compute a $n \times (m+1)$ matrix $\boldsymbol{Q}$ and an $(m+1) \times (m+1)$ matrix $\boldsymbol{R}$.
Let $\boldsymbol{H}_m$ be columns 2 through $m+1$ of $\boldsymbol{R}$ and let $\boldsymbol{V}_{m+1} = \boldsymbol{Q}$.

Algorithm 28.5 Algorithm RESTARTED GMRES **for Solving $\boldsymbol{Ax} = \boldsymbol{b}$**

Given: nonsingular matrices $\boldsymbol{A}$ and $\boldsymbol{M}$, a vector $\boldsymbol{b}$, an initial guess $\boldsymbol{x}$, a restart parameter m, and a tolerance tol.
Let $\boldsymbol{r} = \boldsymbol{b} - \boldsymbol{Ax}$ and $\rho = \|\boldsymbol{r}\|$.
while $\|\boldsymbol{r}\| > tol * \rho$,
  Determine z by solving $\boldsymbol{M}z = \boldsymbol{r}$.
  $[\boldsymbol{V}_{m+1}, \boldsymbol{H}_m]$ = Arnoldi$(m, \widehat{\boldsymbol{G}}, z)$, where $\widehat{\boldsymbol{G}} = \boldsymbol{M}^{-1}\boldsymbol{A}$ (not explicitly formed).
  Solve $\boldsymbol{H}_m^* \boldsymbol{H}_m \boldsymbol{y} = \boldsymbol{H}_m^* \boldsymbol{V}_{m+1}^* z$.
  Set $\boldsymbol{x} = \boldsymbol{x} + \boldsymbol{V}_m \boldsymbol{y}$ and $\boldsymbol{r} = \boldsymbol{r} - \boldsymbol{A}(\boldsymbol{V}_m \boldsymbol{y})$.
end

The RESTARTED GMRES algorithm is given in Algorithm 28.5. It generates Arnoldi basis vectors from the starting vector z that satisfies $\boldsymbol{M}z = \boldsymbol{b} - \boldsymbol{Ax}$. Then it solves the least squares problem

$$\min_{\boldsymbol{y}} \|z - \widehat{\boldsymbol{G}}\boldsymbol{V}_m \boldsymbol{y}\|$$

and adds $\boldsymbol{V}_m \boldsymbol{y}$ on to the current iterate, thus reducing the residual norm $\|\boldsymbol{r}\|$. Making use of the facts that $\widehat{\boldsymbol{G}}\boldsymbol{V}_m = \boldsymbol{V}_{m+1}\boldsymbol{H}_m$ and $\boldsymbol{V}_{m+1}^T \boldsymbol{V}_{m+1} = \boldsymbol{I}$, the normal equations defining the solution to the least squares problem can be written as

$$\boldsymbol{H}_m^* \boldsymbol{H}_m \boldsymbol{y} = \boldsymbol{H}_m^* \boldsymbol{V}_{m+1}^* z,$$

and these equations can be solved efficiently using a Givens version of the QR decomposition of $\boldsymbol{H}_m$, since $\boldsymbol{H}_m$ is already upper Hessenberg.

The convergence of GMRES depends on the eigenvalues λ_j of $\widehat{\boldsymbol{G}}$. Among all polynomials of degree m with constant coefficient equal to 1, choose the one with the minimum maximum value at the eigenvalues and call that value ϵ:

$$\epsilon = \min_{\substack{p(0)=1 \\ \text{degree}(p) \le m}} \max_{j=1,\dots,n} |p(\lambda_j)|.$$

After one cycle of GMRES, updating $\boldsymbol{x}_{new} = \boldsymbol{x}_{old} + \boldsymbol{V}_m \boldsymbol{y}$, we have the residual bound

$$\|\boldsymbol{r}_{new}\| \le \epsilon\kappa(\boldsymbol{U})\|\boldsymbol{r}_{old}\|,$$

where $\boldsymbol{U}$ is the matrix of eigenvectors of $\widehat{\boldsymbol{G}}$ and $\kappa(\boldsymbol{U}) = \|\boldsymbol{U}\|_2 \|\boldsymbol{U}^{-1}\|_2$ is the ratio of largest to smallest singular values of $\boldsymbol{U}$. Unfortunately if the matrix of eigenvectors is very ill-conditioned, then $\epsilon\kappa(\boldsymbol{U})$ may be greater than 1, making the bound useless.

POINTER 28.1. Solving Large Eigenvalue Problems.

Problems such as stress analysis of large structures and energy analysis of molecular configurations lead to very large eigenvalue problems. Surprisingly, the largest eigenvalue computations are performed for information retrieval. Google, for example, uses a matrix of size equal to the number of webpages on the Internet (in the billions) to compute **PageRanks®** of each page. These PageRanks are used to determine which websites are returned in response to a query [15].

Restarted GMRES (with $m < n$) can **stagnate** on some problems, repeatedly computing the same iterate $\boldsymbol{x}$. (Taking $m = n$, cannot fail but is generally too expensive.) This is unfortunate, but in such a case, the algorithm can be run again with a different starting guess or a different preconditioner $\boldsymbol{M}$.

A second family of Krylov methods for solving nonsymmetric problems uses the **Lanczos basis**. Like CG, the computations can be done by storing only a few vectors, but in the nonsymmetric case, occasionally the iteration breaks down, producing a zero vector before the solution is found. (GMRES does not break down.) The linear system solver based on minimization is called MR (or QMR), while that based on projection is called BI-CG.

28.6 Computing Eigendecompositions and SVDs with Krylov Methods

The residuals in the CG method are orthogonal, as are the vectors in the Arnoldi basis. Suppose that we take n steps of our iteration (impractical, but interesting conceptually). Then (in exact arithmetic) the CG or the Arnoldi-based methods terminate with the exact solution to our linear system. But, rather remarkably, as we saw in Section 28.5, we have also secretly computed a matrix decomposition. If we let $\boldsymbol{V}$ be the matrix whose columns are the basis vectors normalized to length 1 and let $\boldsymbol{M} = \boldsymbol{I}$, then $\boldsymbol{V}^*\boldsymbol{A}\boldsymbol{V}$ is tridiagonal for CG and upper-Hessenberg (zeros below the first subdiagonal) for Arnoldi, and, because we have computed a similarity transform, the eigenvalues are the same as those of $\boldsymbol{A}$. The entries in $\boldsymbol{V}^*\boldsymbol{A}\boldsymbol{V}$ are derived from the scalar parameters computed in CG or Arnoldi.

Now we are never going to take n steps of CG or Arnoldi, but after k steps, we have computed $\boldsymbol{V}_k$ (the first k columns of $\boldsymbol{V}$) and the upper $k \times k$ block of $\boldsymbol{V}^*\boldsymbol{A}\boldsymbol{V}$. We can obtain good approximations to some of the eigenvalues of $\boldsymbol{A}$ by using the QR algorithm from Section 5.5 to find the eigenvalues and eigenvectors of this rather small $k \times k$ matrix. If λ is an eigenvalue and $\boldsymbol{w}$ is an eigenvector of the small matrix, then λ is an approximate eigenvalue of $\boldsymbol{A}$ with approximate eigenvector $\boldsymbol{V}_k\boldsymbol{w}$.

If this sounds complicated, it is! It is necessary to incorporate many tricks and safeguards, including selective reorthogonalization against eigenvectors that have already converged and implicit restarting to keep k small. As usual, it is important to choose high-quality software like `eigs` to do this computation.

The Lanczos basis yields approximations to singular values rather than eigenvalues, and high-quality sofware like MATLAB's `svds` is available.

POINTER 28.2. Further Reading.

Ortega [119] gives a good introduction to SIMs such as Jacobi, Gauss-Seidel, and successive over-relaxation (SOR).

Saad's book [132] on Krylov subspace methods is an excellent reference for algorithms such as PCG, GMRES, MR, QMR, Arnoldi, and BI-CG. MATLAB has implementations of several of these algorithms, and implementations of SYMMLQ and LSQR are also available [122, 123]. In some situations, it is a good idea to let the preconditioner $\boldsymbol{M}$ change at each iteration, resulting in an algorithm called **flexible-**GMRES [132].

MATLAB functions `eigs` and `svds` use Krylov subspace methods to compute eigenvalue-eigenvector pairs and singular value-vector triplets [100, 140].

The mesh depicted in Figure 28.1 was produced by the `distmesh` system of Persson and Strang [124].

Interesting sparse matrices for further experimentation can be found in the `gallery` function in MATLAB, the Matrix Market [106], or Davis's collection [36].

Chapter 29 / Case Study

Elastoplastic Torsion: Twist and Stress

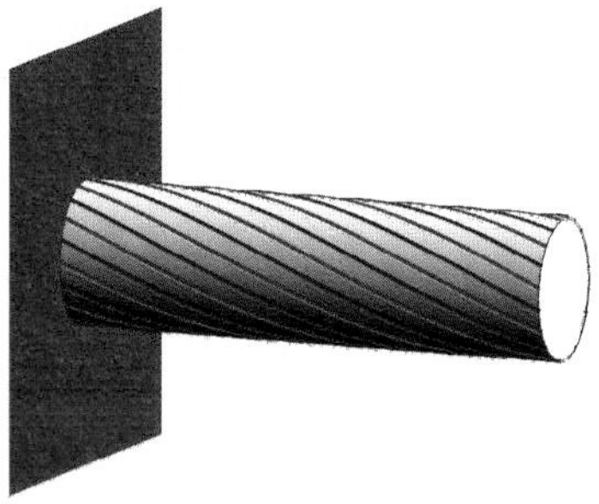

This case study focuses on the stress induced in a rod by twisting it. We'll investigate two situations: first, when the stress is small enough that the rod behaves elastically, and second, when we pass the elastic-plastic boundary. The solution to our problem involves repeated solution of sparse systems of linear equations.

Consider a long rod made of metal, plastic, rubber, or some other homogeneous material mounted on a wall, as shown in Figure 29.1 (left). Hold the rod at the end and twist counterclockwise, as shown in the figure (right). This **torsion** (twisting) causes stresses in the rod. If the force we apply is small enough, then the rod behaves as an **elastic body**, and when we release it, it returns to its original state. But if we apply a lot of twisting force, we eventually change the structure of the rod; some portion of it behaves **plastically** and is permanently changed. If the whole rod behaves elastically, or if it all behaves plastically, then modeling is rather easy. More difficult cases occur when there is a mixture of elastic and plastic behavior, and in this problem we investigate the behavior of the rod over a full range of torsion.

The Elastic Model

As usual in mathematical modeling, we make simplifying assumptions to make the computation tractable. We assume that the torsional force is evenly distributed throughout the rod, and that the rod has uniform cross sections. Under these circumstances, we can understand the system by modeling the stress in any single cross section. We'll call the interior of the two-dimensional cross section Ω and its boundary Γ.

The standard elastic model involves the **stress function** $u(x,y)$ on Ω, where the quantities $-\partial u(x,y)/\partial x$ and $\partial u(x,y)/\partial y$ are the stress components. If we set the net force to zero at each point in the cross-section, we obtain

$$\nabla^2 u \equiv \frac{\partial^2 u}{\partial x^2} + \frac{\partial^2 u}{\partial y^2} \equiv u_{xx} + u_{yy} = -2G\theta \quad \text{in} \quad \Omega,$$

$$u = 0 \quad \text{on} \quad \Gamma,$$

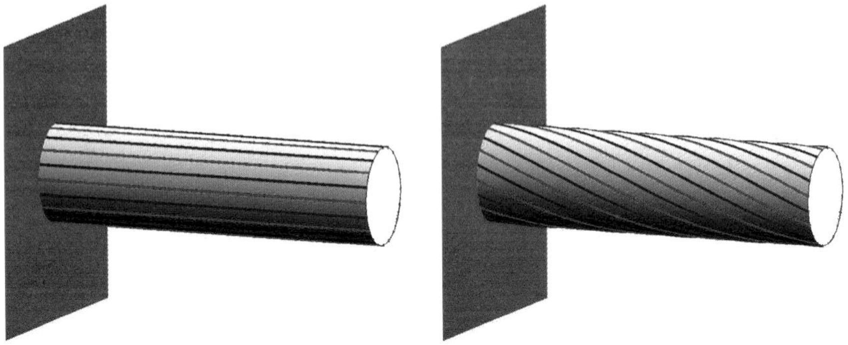

Figure 29.1. *A rod before torsion is applied (left) and after (right).*

where G is the **shear modulus** of the material and θ (radians) is the angle of twist per unit length. In order to guarantee existence of a smooth solution to our problem, we'll assume that the boundary Γ is smooth; in fact, in our experiments, Γ is an ellipse.

An alternate equivalent formulation is derived from minimizing an energy function

$$E(u) = \frac{1}{2}\int\int_{\Omega} \|\nabla u(x,y)\| \mathrm{d}x\mathrm{d}y - 2G\theta \int\int_{\Omega} u(x,y)\mathrm{d}x\mathrm{d}y.$$

The magnitude of the gradient $\|\nabla u(x,y)\| = \sqrt{(\partial u(x,y)/\partial x)^2 + (\partial u(x,y)/\partial y)^2}$ is the **shear stress** at the point (x,y), an important physical quantity. At any point where the shear stress exceeds the **yield stress** σ_0, the material becomes plastic, and our standard model is no longer valid.

For simple geometries (e.g., a circle), we can solve this problem analytically. But, for the sake of generality and in preparation for the more difficult elastoplastic problem, we consider numerical methods. Discretization by **finite differences** would be a possibility, but the geometry makes the flexibility of **finite elements** attractive. The case study of Chapter 23 gave a simple example of their use. We can use a finite element package to formulate the matrix $\boldsymbol{K}$ that approximates the operator $-\nabla^2 u$ on Ω, and also assemble the right-hand side $\boldsymbol{b}$ so that the solution to the linear system $\boldsymbol{Ku} = \boldsymbol{b}$ is the approximation to $u(x,y)$ at the nodes (x_i, y_i) of the finite element mesh. Since the boundary Γ and the forcing function $-2G\theta$ are smooth, we expect **optimal order approximation** of the finite element solution to the true solution as the mesh is refined: for piecewise linear elements on triangles, for example, this means that the error is $O(h^2)$, where h is a measure of the size of the triangles.

In Challenge 29.1, we see what this model predicts for the sheer stress on our rod.

CHALLENGE 29.1.

Suppose that the cross-section Ω of the rod is the interior of a circle of radius one, and let $G = 5$ and $\theta = 1$. Use a finite element package to approximate the stress function. Plot the approximate solution and describe what it says about the stress. Solve again using a finer mesh and estimate the error in your approximation $1/2\boldsymbol{u}^T K\boldsymbol{u} - \boldsymbol{b}^T\boldsymbol{u}$ to $E(u)$.

Note that by symmetry, we could reduce our computational domain in Challenge 29.1 to a quarter circle, setting the normal derivative of u along the two straight edges to zero.

The Elastoplastic Model

As the value of θ is increased, the maximum value of the shear stress $\|\nabla u(x,y)\|$ increases, eventually exceeding the yield stress of the rod, and then our model breaks down because the rod is no longer behaving elastically. We can extend our model to this case by adding constraints: we still minimize the energy function, but we don't allow stresses larger than the yield stress:

$$\begin{aligned} &\min_u \quad E(u) \\ \|\nabla u(x,y)\| &\leq \sigma_0, \quad (x,y) \in \Omega \\ u &= 0 \text{ on } \Gamma \end{aligned}$$

The new constraints $\|\nabla u(x,y)\| \leq \sigma_0$ are nonlinear, but we can reduce them to linear by a simple observation: if we start at the boundary and work our way in, we see that the constraint is equivalent to saying that $|u(x,y)|$ is bounded by σ_0 times the (shortest) distance from (x,y) to the boundary.

So the next (and most challenging) ingredient in solving our problem is an algorithm for determining these distances. In the next two challenges, we develop and implement such an algorithm.

CHALLENGE 29.2.

Derive an algorithm for finding the distance $d(z)$ between a given point $z = [z_1, z_2]^T$ and a ellipse. In other words, solve the problem

$$\min_{x,y} (x - z_1)^2 + (y - z_2)^2$$

subject to

$$\left(\frac{x}{\alpha}\right)^2 + \left(\frac{y}{\beta}\right)^2 = 1$$

for given parameters α and β. Note that the distance is the square root of the optimal value of the objective function $(x - z_1)^2 + (y - z_2)^2$. The problem can be solved using Lagrange multipliers, as a calculus student would. You need only consider points z on or inside the ellipse, but handle all of the special cases: $\alpha = \beta$, z has a zero coordinate, etc.

In Challenge 29.2, we see that a rather simple sounding mathematical problem becomes complicated when we handle the special cases properly. When we consider the fact that computers do their arithmetic inexactly, we see that an algorithm for computing distances to an ellipse must also account for difficulties encountered, for example, when a component of z is near zero, and we face the difficulties of this algorithm in Challenge 29.3. Implementing reliable software requires a great deal of attention to details.

CHALLENGE 29.3.

Program your distance algorithm, document it, and produce a convincing validation of the implementation by designing a suitable set of tests and discussing the results. Use a reliable rootfinder such as `fzero` to solve the nonlinear equation.

Now we have the elements in place to solve our elastoplastic torsion problem. We discretize $E(u)$ using finite elements, and we use our distance function to form the constraints, resulting in the problem

$$\min_{\boldsymbol{u}} 1/2\boldsymbol{u}^T \boldsymbol{K}\boldsymbol{u} - \boldsymbol{b}^T \boldsymbol{u}$$
$$-\sigma_0 \boldsymbol{d} \le \boldsymbol{u} \le \sigma_0 \boldsymbol{d},$$

where $d_i = d(x_i, y_i)$ and the ith component of $\boldsymbol{u}$ approximates the solution at (x_i, y_i). Because the matrix $\boldsymbol{K}$ is symmetric positive definite (due to the elliptic nature of the differential equation), the solution to the problem exists and is unique. This is a **quadratic programming** problem. Algorithms for solving it include **active set strategies** and the newer **interior-point methods**; see Chapter 10.

CHALLENGE 29.4.

Solve the elastoplastic problem on a mesh that you estimate gives an error of less than 0.1 in the function $E(u)$. Use the parameters $G = 1$, $\sigma_0 = 1$, and $\beta = 1$. Let $\alpha\theta = 0$, 0.25, 0.50,...,5 and $\beta/\alpha = 1$, 0.8, 0.65, 0.5, 0.2. Plot a few representative solutions. On a separate graph, for each value of β/α, plot a curve $T/(\sigma_0\alpha^3)$ vs $G\alpha\theta/\sigma_0$, where T is the estimate of the torque, the integral of u over the domain Ω. (This gives you 5 curves, one for each value of β/α.) On the same plot, separate the elastic solutions (those for which no variable is at its bound) from the elastoplastic ones. Estimate the errors in the data points of your plot.

We solved this problem on a rod with a simple cross-section. Think about how you could extend our methods to more complicated shapes!

POINTER 29.1. Further Reading.

This case study comes from a paper coauthored by Wei H. Yang [117] in 1978. At that time, we worked very hard to develop memory- and time-efficient algorithms to solve the elastoplastic problem so that we wouldn't need a supercomputer. Now, sufficient computational resources are available in laptops.

Selvadurai [136, Sec. 9.9] gives an excellent derivation of the elastic model equation. He also discusses the history of the model, noting that there were several incorrect models before Barre de Saint-Venant proposed a correct one.

The solution to Challenge 29.1 requires access to a package to generate finite element meshes and stiffness matrices. A stand-alone package such as PLTMG [6] or MATLAB's PDE Toolbox routines (`initmesh, refinemesh, assempde, pdeplot`) can be used. An introduction to finite element formulations can be found, for example, in books by Bathe [8] and Brenner and Scott [18].

For Challenge 29.4, you need a quadratic programming algorithm, such as MATLAB's `quadprog` from the Optimization Toolbox. Quadratic programming is discussed in textbooks such as that by Bazaraa, Sherali, and Shetty [9].

Chapter 30 / Case Study

Fast Solvers and Sylvester Equations: Both Sides Now

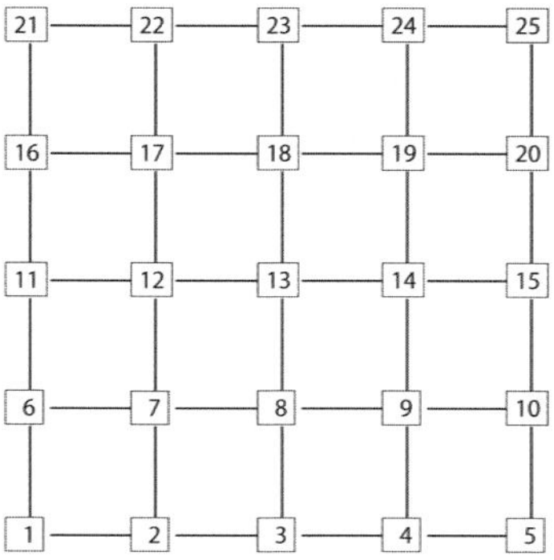

In Chapter 27, we considered large sparse systems of linear equations. One conclusion we can draw from that study is that it is very important to exploit structure in our matrix in order to keep storage and computational costs low. In that chapter we exploited sparsity; now we consider a set of problems that have an additional type of structure.

Among the problems considered in Chapter 27 was the Poisson equation

$$-u_{xx} - u_{yy} = f(x, y)$$

with $(x, y) \in \Omega \subset \mathcal{R}^2$, and with appropriate boundary conditions specified. One strategy was to discretize the equation by choosing mesh points. We write an equation for each mesh point by approximating the derivatives u_{xx} and u_{yy} by finite differences. This gives a system of linear equations $\boldsymbol{Au} = \boldsymbol{f}$ to solve for estimates of the value of u each of the mesh points.

In one of our examples, Ω was a unit square with the mesh points chosen so that they were equally spaced, and we were given zero boundary conditions. If we have a 5×5 grid of mesh points, for example, we might order them as in Figure 30.1. If we let $x_j = jh$ and $y_k = kh$, with $h = 1/6 = 1/(n+1)$, then we create two vectors

$$\boldsymbol{u} \approx [u(x_1, y_1), \ldots, u(x_n, y_1), u(x_1, y_2), \ldots, u(x_n, y_2), \ldots, u(x_1, y_n), \ldots, u(x_n, y_n)]^T,$$
$$\boldsymbol{f} = [f(x_1, y_1), \ldots, f(x_n, y_1), f(x_1, y_2), \ldots, f(x_n, y_2), \ldots, f(x_1, y_n), \ldots, f(x_n, y_n)]^T,$$

and write our approximation to $-u_{xx}$ at all of the mesh points as

$$\boldsymbol{A}_x \boldsymbol{u} \equiv \frac{1}{h^2} \begin{bmatrix} \boldsymbol{T} & \boldsymbol{0} & \boldsymbol{0} & \boldsymbol{0} & \boldsymbol{0} \\ \boldsymbol{0} & \boldsymbol{T} & \boldsymbol{0} & \boldsymbol{0} & \boldsymbol{0} \\ \boldsymbol{0} & \boldsymbol{0} & \boldsymbol{T} & \boldsymbol{0} & \boldsymbol{0} \\ \boldsymbol{0} & \boldsymbol{0} & \boldsymbol{0} & \boldsymbol{T} & \boldsymbol{0} \\ \boldsymbol{0} & \boldsymbol{0} & \boldsymbol{0} & \boldsymbol{0} & \boldsymbol{T} \end{bmatrix} \boldsymbol{u}.$$

POINTER 30.1. Exploiting Problem Structure.

In this case study, we use the structure of a problem with n^2 unknowns to reduce the amount of computation from $O(n^6)$ (using the Cholesky decomposition) to $O(n^4)$ (exploiting sparsity) and then to $O(n^3)$ (using the Sylvester structure), a substantial savings when n is large. Then, knowing just a bit more about the structure of the problem allows further reduction, to $O(n^2 \log_2 n)$when n is a power of 2. Since we are computing n^2 answers, this is close to optimal, and it illustrates the value of exploiting every possible bit of structure in our problems.

In this equation, the ith component of the 25×1 vector $\boldsymbol{u}$ is our approximation to u at the ith mesh point. The matrix $\mathbf{0}$ is a 5×5 matrix of zeros, and the matrix $\boldsymbol{T}$ is defined by

$$\boldsymbol{T} = \begin{bmatrix} 2 & -1 & 0 & 0 & 0 \\ -1 & 2 & -1 & 0 & 0 \\ 0 & -1 & 2 & -1 & 0 \\ 0 & 0 & -1 & 2 & -1 \\ 0 & 0 & 0 & -1 & 2 \end{bmatrix}.$$

Similarly, our approximation to $-u_{yy}$ at all of the mesh points is

$$\boldsymbol{A}_y \boldsymbol{u} \equiv \frac{1}{h^2} \begin{bmatrix} 2\boldsymbol{I} & -\boldsymbol{I} & \mathbf{0} & \mathbf{0} & \mathbf{0} \\ -\boldsymbol{I} & 2\boldsymbol{I} & -\boldsymbol{I} & \mathbf{0} & \mathbf{0} \\ \mathbf{0} & -\boldsymbol{I} & 2\boldsymbol{I} & -\boldsymbol{I} & \mathbf{0} \\ \mathbf{0} & \mathbf{0} & -\boldsymbol{I} & 2\boldsymbol{I} & -\boldsymbol{I} \\ \mathbf{0} & \mathbf{0} & \mathbf{0} & -\boldsymbol{I} & 2\boldsymbol{I} \end{bmatrix} \boldsymbol{u},$$

where $\boldsymbol{I}$ is the identity matrix of dimension 5×5. So, we want to solve the linear system

$$(\boldsymbol{A}_x + \boldsymbol{A}_y)\boldsymbol{u} = \boldsymbol{f}. \tag{30.1}$$

This gives us a problem with a sparse matrix, and since our usual grids are $n \times n$ where n is much bigger than 5, it is important to exploit the sparsity.

But in this particular problem—an equally-spaced grid over a square (or a rectangle)—there is even more structure that we can exploit, and in the next challenge we see how to write our problem in a more compact form.

CHALLENGE 30.1.

Show that equation (30.1) can be written as

$$(\boldsymbol{B}_y \boldsymbol{U} + \boldsymbol{U}\boldsymbol{B}_x) = \boldsymbol{F}, \tag{30.2}$$

where the matrix entry u_{jk} is our approximation to $u(x_j, y_k)$, $f_{jk} = f(x_j, y_k)$, and $\boldsymbol{B}_y = \boldsymbol{B}_x = (1/h^2)\boldsymbol{T}$.

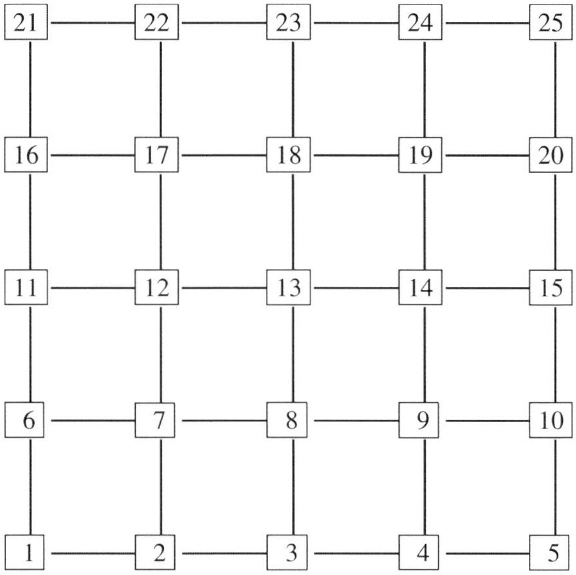

Figure 30.1. *The mesh points for our problem.*

Equation (30.2) is called a **Sylvester equation**. (It is also a **Lyapunov equation** since $\boldsymbol{B}_y = \boldsymbol{B}_x^T$.) It looks daunting because the unknowns $\boldsymbol{U}$ appear on both the left and the right sides of other matrices. But, as we just showed, the problem is equivalent to a system of linear equations. The advantage of the Sylvester equation formulation is that it gives us a compact representation of our problem in terms of the data matrix $\boldsymbol{F}$ and two tridiagonal matrices of dimension n. (In fact, since $\boldsymbol{B}_x = \boldsymbol{B}_y$, we really have only one tridiagonal matrix.) One way to solve our problem efficiently is by using the **Schur decomposition** of a matrix. We'll consider this algorithm in the next challenge, developing both a row-oriented and a column-oriented algorithm. As we saw in Chapter 3, the choice between these algorithms depends on how the elements of the matrix are arranged in the computer memory.

CHALLENGE 30.2.

(a) Consider the Sylvester equation $\boldsymbol{LU} + \boldsymbol{UR} = \boldsymbol{C}$, where $\boldsymbol{L}$ is lower triangular and $\boldsymbol{R}$ is upper triangular. Show that we can easily determine the elements of the matrix $\boldsymbol{U}$ either row by row or column by column. How many multiplications does this algorithm require?

(b) By examining your algorithm, determine conditions on the main diagonal elements of $\boldsymbol{L}$ and $\boldsymbol{R}$ (i.e., the eigenvalues) that are necessary and sufficient to ensure that a solution to the Sylvester equation exists.

(c) Now suppose that we want to solve the Sylvester equation $\boldsymbol{AU} + \boldsymbol{UB} = \boldsymbol{C}$, where $\boldsymbol{A}$, $\boldsymbol{B}$, and $\boldsymbol{C}$ of dimension $n \times n$ are given. ($\boldsymbol{A}$ and $\boldsymbol{B}$ are unrelated to the previously described matrices.) Let $A = WLW^*$ and $B = YRY^*$, where $WW^* = W^*W = I$, $YY^* = Y^*Y = I$, L is lower triangular, and R is upper triangular. (This is called a **Schur decomposition** of the two matrices.) Show that we can solve the Sylvester equation by applying the algorithm derived in part (a) to the equation $L\widehat{U} + \widehat{U}R = \widehat{C}$, where $\widehat{U} = W^*UY$ and $\widehat{C} = W^*CY$.

The reason for using the Schur decomposition in the previous problem is that the most compact form we can achieve using a unitary matrix transformation is triangular form. We consider only unitary transforms in order to preserve stability. One disadvantage of the Schur algorithm applied to real nonsymmetric matrices is that we need to do complex arithmetic, and the resulting computed matrix U may have small imaginary part due to rounding error, even though the true matrix is guaranteed to be real.

We could solve (30.2) using the eigendecomposition in place of the Schur decomposition. This is less efficient for general matrices but more efficient when B_x and B_y have eigenvectors related to the vectors of the discrete Fourier transform, as in our case, and it is stable for symmetric matrices, since in that case the eigenvector matrix is real orthogonal. In fact, the Schur decomposition reduces to the eigendecomposition when A and B are real symmetric, since the matrices L and R are then also symmetric and therefore diagonal. Let's see how we can sometimes efficiently solve our problem using an eigendecomposition.

Suppose that B_x and B_y are any two real symmetric matrices that have the same eigenvectors, so that

$$B_x = V\Lambda_x V^T \quad \text{and} \quad B_y = V\Lambda_y V^T,$$

where the columns of V are the eigenvectors (normalized to length 1) and the entries of the diagonal matrices Λ_x and Λ_y are the eigenvalues. (Note that since B_x and B_y are symmetric, the columns of V are orthogonal so that $V^TV = VV^T = I$.) Substituting our eigendecompositions, equation (30.2) becomes

$$V\Lambda_y V^T U + UV\Lambda_x V^T = F,$$

and multiplying this equation by V^T on the left and by V on the right, we obtain

$$\Lambda_y V^T UV + V^T UV\Lambda_x = V^T FV.$$

Letting $Y = V^T UV$, we have an algorithm:

- Form the matrix $\widehat{F} = V^T FV$.
- Solve the equation $\Lambda_y Y + Y\Lambda_x = \widehat{F}$, where Λ_x and Λ_y are diagonal.
- Form the matrix $U = VYV^T$.

CHALLENGE 30.3.

Determine a way to implement the second step of the algorithm using only $O(n^2)$ arithmetic operations.

Since we have n^2 entries of U to compute in solving our problem, the second step is optimal order, so the efficiency of the algorithm depends on the implementation of the first and third steps. In general, each matrix-matrix product of $n \times n$ matrices takes $O(n^3)$ operations, so our complete algorithm would also take $O(n^3)$. In some special cases, though, the matrix products can be computed more quickly, and this is the case for our Poisson problem, as we see in the next challenge.

CHALLENGE 30.4.

(a) The eigenvalues and eigenvectors of $\boldsymbol{B}_x$ are known. Denote the elements of the vector $\boldsymbol{v}_j$ by

$$v_{kj} = \alpha_j \sin \frac{kj\pi}{n+1},$$

where α_j is chosen so that $\|\boldsymbol{v}_j\| = 1$. Show that $\boldsymbol{B}_x \boldsymbol{v}_j = \lambda_j \boldsymbol{v}_j$, where $\lambda_j = (2 - 2\cos\frac{j\pi}{n+1})/h^2$, for $j = 1, 2, \ldots, n$.

(b) Show that multiplication by the matrix $\boldsymbol{V}$ or $\boldsymbol{V}^T$ can be accomplished by a **discrete Fourier (sine) transform** or inverse Fourier (sine) transform of length n, where the discrete sine transform of a vector $\boldsymbol{x}$ is defined by

$$y_k = \sum_{j=1}^{n} x_j \sin(jk\pi/(n+1)).$$

This can be accomplished in $O(n \log_2 n)$ operations if n is a power of 2 and in some larger but still modest number of operations if n is a composite number with many factors.

Using the multiplication algorithm from Challenge 30.4, we can solve equation (30.2) in $O(n^2 \log_2 n)$ operations when n is a power of 2, considerably less than the $O(n^3)$ operations generally required, or the $O(n^4)$ required for the sparse Cholesky decomposition applied to our original matrix problem! (The reordering strategies discussed in Chapter 27 would reduce the factorization complexity somewhat but would not achieve $O(n^2 \log_2 n)$.)

CHALLENGE 30.5.

Write a well-documented program to solve the discretization of the differential equation using the Schur-based algorithm of Challenge 30.2 and using the algorithm developed in Challenge 30.4. (Debug the Schur-based algorithm using randomly generated real nonsymmetric matrices.) Test your algorithms for $n = 2^p$, with $p = 2, \ldots, 9$ choosing the true solution matrix $\boldsymbol{U}$ randomly. Compare the results of the two algorithms with backslash for accuracy and time.

In MATLAB, the functions `dst` and `idst` from the PDE Toolbox are useful in solving this problem. If the PDE Toolbox is not available, the results of a fast Fourier transform can be manipulated to obtain the desired result. Also of use is `schur`, which has an option to return either an upper-triangular (and possibly complex) factor or a real block upper-triangular factor with 1×1 or 2×2 blocks on the main diagonal.

POINTER 30.2. Further Reading.

The method used in Challenge 30.4 can be extended. There are fast solvers for solving 3-d Poisson problems, for problems on rectangles, circles, and other simple domains, and for problems with different boundary conditions [144].

The Schur algorithm for the Sylvester equation is due to Bartels and Stewart [7].

When we discuss number of operations, we consider the traditional algorithms for matrix product. There are faster versions (e.g., that of Strassen) but Miller showed that the stability is not as good [107]; see [79, Sec. 23.2.2].

The Sylvester equation also arises in state space design in control theory [33] and in image processing [69]. In fact, our Kronecker product problem in Chapter 6 can be written as a Sylvester equation [69, Sec. 4.4.2].

Chapter 31 / Case Study

Eigenvalues: Valuable Principles

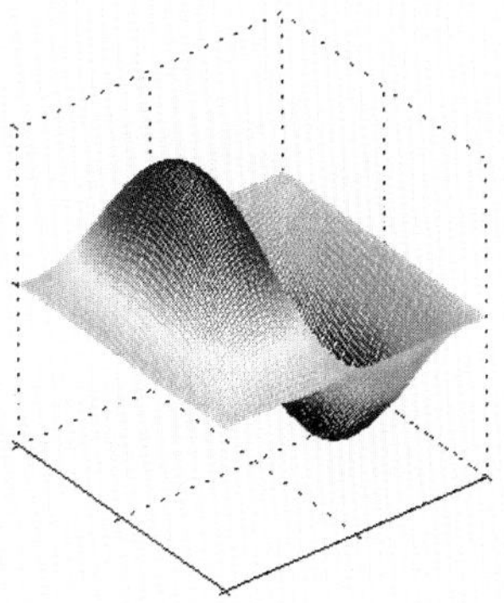

In this case study, we study eigenvalue problems arising from partial differential equations. Eigenvalues can help us solve a differential equation analytically, and they can also provide valuable information about the behavior of a physical system. We'll study properties of eigenvalues and use these properties to design a drum with a particular fundamental frequency of vibration.

What Is an Eigenvalue of an Operator?

To begin, recall that an **eigenvector** of a matrix is a vector $\boldsymbol{w}$ with the property that multiplication of the vector by the matrix simply scales the vector. The scale factor λ is called an **eigenvalue**, or **principal value** of the matrix.

The eigensystem (eigenvalues and eigenvectors) of $\boldsymbol{A}$ has several nice properties, summarized in Pointers 5.1 and 5.6. When the eigenvalues are distinct, the eigenvectors are unique, except that they can be multiplied by any nonzero number. The eigenvectors are **linearly independent**, so they form a basis for $\mathcal{R}^n$. In fact, if $\boldsymbol{A}$ is real symmetric or complex Hermitian, then eigenvectors corresponding to distinct eigenvalues are orthogonal. In this case, the smallest eigenvalue is the value of the function

$$\min_{\boldsymbol{w} \neq \boldsymbol{0}} \frac{\boldsymbol{w}^T \boldsymbol{A} \boldsymbol{w}}{\boldsymbol{w}^T \boldsymbol{w}}.$$

The other eigenvalues can also be characterized as solutions to minimization problems (or maximization problems).

Now, instead of a matrix, let's consider a differential operator. As an example, define

$$\mathcal{A}u = -u''$$

for $x \in \Omega = (0,1)$, and require that u satisfy the boundary conditions $u(0) = u(1) = 0$. Notice that for $j = 1,2,\ldots,$

$$\mathcal{A} \sin(j\pi x) = (j\pi)^2 \sin(j\pi x).$$

In other words, we have found functions $w_j(x) = \sin(j\pi x)$, called **eigenfunctions** of $\mathcal{A}$, that satisfy the zero boundary conditions and have the special property that applying $\mathcal{A}$ just scales the function. We call the scale factor the **eigenvalue** of $\mathcal{A}$, and we can abbreviate the relation as

$$\mathcal{A} w_j = \lambda_j w_j,$$

where $\lambda_j = (j\pi)^2$.

Let $\mathcal{H}$ be the set of functions w that are zero on the boundary of Ω, and for which the integrals over Ω of $|w|^2$ and $\|\nabla w\|^2$ exist. All of the properties that we listed for eigenvectors also hold for eigenfunctions:

- When the eigenvalues are distinct, the eigenfunctions are unique, except that they can be multiplied by any nonzero number.

- Define an inner product

  $$(u, v) = \int_\Omega u(x)v(x)\mathrm{d}x.$$

  We say that $\mathcal{A}$ is **self-adjoint** (i.e., symmetric) if $(u, \mathcal{A}v) = (\mathcal{A}u, v)$ for all choices of functions u and v in $\mathcal{H}$. In such a case, eigenfunctions w_m and w_ℓ corresponding to distinct eigenvalues are orthogonal, meaning that $(w_m, w_\ell) = 0$.

- The eigenfunctions are linearly independent, so they form a basis for functions defined on Ω that satisfy the zero boundary conditions.

- For self-adjoint problems, the smallest eigenvalue solves the minimization problem

  $$\min_{w \in \mathcal{H}, w \neq 0} \frac{(w, \mathcal{A}w)}{(w, w)}. \tag{31.1}$$

  Let's find the eigenvalues for a particular two-dimensional problem.

CHALLENGE 31.1.

Define the domain $\Omega = (0, b) \times (0, b)$. Consider the elliptic partial differential equation

$$-u_{xx}(x, y) - u_{yy}(x, y) = \lambda u(x, y)$$

for $(x, y) \in \Omega$, with $u(x, y) = 0$ on the boundary of Ω.

Show that the function

$$w_{m\ell}(x, y) = \sin(m\pi x/b)\sin(\ell\pi y/b),$$

where m and ℓ are positive integers, satisfies the differential equation and the boundary conditions. Determine the corresponding eigenvalue $\lambda_{m\ell}$.

How Can We Compute Approximations to the Eigenvalues?

Suppose we want to compute approximations to the eigenvalues and eigenfunctions of

$$\mathcal{A}u = -\nabla \cdot (a\nabla u)$$

on the domain Ω, with $u = 0$ on the boundary of Ω. Assume that $a > 0$ is a smooth function. In Challenge 27.4, we computed an approximate solution to a differential equation by replacing it by a matrix problem. Here we do the same thing:

- Replace $\mathcal{A}$ by $\boldsymbol{A}_h$, where $\boldsymbol{A}_h$ is the **finite difference** or **finite element** approximation to $-\nabla \cdot (a\nabla u)$. The parameter h describes the mesh size for the finite difference approximation or the triangle diameter for the finite element approximation.

- Use the eigenvalues $\lambda_{k,h}$ of $\boldsymbol{A}_h$ as approximations to the eigenvalues of $\mathcal{A}$. Since $\mathcal{A}$ has an infinite number of eigenvalues and $\boldsymbol{A}_h$ has finitely many, we can't hope to get good approximations to all eigenvalues of $\mathcal{A}$, but the smallest ones should be well approximated.

- For finite differences, the eigenvectors of $\mathcal{A}_h$ contain approximate values of the eigenfunctions at the mesh points.

- For finite elements, the eigenvectors of $\mathcal{A}_h$ contain coefficients in an expansion of an approximation to the eigenfunction in the finite element basis.

Suppose Ω is a convex polygon and we use a piecewise linear finite element approximation. Let $\lambda_{k,h}$ be the kth eigenvalue of $\boldsymbol{A}_h$ and let λ_k be the kth eigenvalue of $\mathcal{A}$ (ascending order). Then there exist constants C and h_0, depending on k, such that when h is small enough,

$$\lambda_k \leq \lambda_{k,h} \leq \lambda_k + Ch^2.$$

CHALLENGE 31.2.

In this challenge, we study the elliptic eigenvalue problem $-\nabla \cdot (\nabla u) = \lambda u$ on the square $(-1,1) \times (-1,1)$ with zero boundary conditions. We know the true eigenvalues from Challenge 31.1, so we can determine how well the discrete approximation performs.

(a) Form a finite difference or finite element approximation to the problem and find the eigenfunctions corresponding to the 5 smallest eigenvalues. A few of them are pictured in Figure 31.1.

- Describe in words the shape of each of these eigenfunctions. How does the shape change as the eigenvalue increases?

- Theory tells us that we have good approximations with a coarse grid only for the eigenfunctions corresponding to the smallest eigenvalues. How does the shape of the eigenfunctions make this result easier to understand?

(b) Create 5 plots, for eigenvalues 1, 6, 11, 16, and 21, of the error in the approximate eigenvalue vs. $1/h^2$. (Use at least 4 different matrix sizes, with the finest $h < 1/50$.)

- What convergence rate do you observe for each eigenvalue?

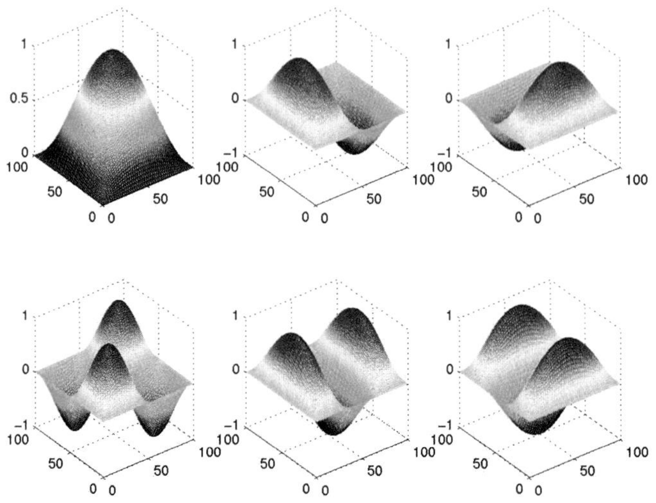

Figure 31.1. *Eigenfunctions corresponding to the eigenvalues* $\lambda = 4.9348$, 12.3370, 12.3370 *(top row) and* $\lambda = 19.7392$, 24.6740, 24.6740 *(bottom row). For multiple eigenvalues, the choice of basis functions for the eigenfunctions is not unique.*

- How does it compare with the theoretical convergence rate?
- Explain any discrepancy.
- Are all of the eigenvalues well-approximated by coarse meshes?

Some Useful Properties of Eigenvalues

Eigenvalues of elliptic operators have many useful properties. We'll consider two of them in the next challenge.

CHALLENGE 31.3.

(a) Suppose Ω is a domain in $\mathcal{R}^n$. Suppose $\mathcal{A}w = \lambda w$ in Ω with $w = 0$ on the boundary, where $\mathcal{A}$ is the self-adjoint operator

$$\mathcal{A}w = -\nabla \cdot (a \nabla w)$$

and $\boldsymbol{a} = [a_1, a_2] > [0,0]$. Prove that $\lambda_1 \geq 0$. Hint: Use integration by parts to replace $(w, \mathcal{A}w)$ by

$$\int_\Omega \boldsymbol{a}(\boldsymbol{x}) \nabla w(\boldsymbol{x}) \cdot \nabla w(\boldsymbol{x}) \mathrm{d}\boldsymbol{x}.$$

(b) Suppose we have two domains $\Omega \subseteq \tilde{\Omega}$. Prove that $\lambda_1(\Omega) \geq \lambda_1(\tilde{\Omega})$. Hint: Notice that the eigenfunction for Ω can be extended to be a candidate for the minimization problem for $\tilde{\Omega}$.

POINTER 31.1. Software.

MATLAB's `pdetool` provides finite element algorithms to solve Challenges 31.2 and 31.4. You may want to make use of `initmesh`, `refinemesh`, `pdeeig`, `squareg`, and `squareb1`.

In Challenge 31.4, you are solving a nonlinear equation: find a value of α so that the smallest eigenvalue (which is a function of α) equals a given value. The eigenvalue is a monotonic function of α, increasing as α increases, and a root finder like MATLAB's `fzero`, discussed in Chapter 24, can be used.

POINTER 31.2. Further Reading.

A good introduction to the eigenvalues of differential operators and theory of finite difference and finite element methods is given by Gockenbach [59]; for a more advanced treatment, see, for example, Larsson and Thomée [99].

How Are Eigenvalues and Eigenfunctions Used?

The eigenvalues and eigenfunctions are useful mathematical quantities. If the eigensystem for a differential operator on a domain Ω can be computed analytically, then the solution to the differential equation involving that operator and that domain can be expressed as a linear combination of the eigenfunctions, and it is then relatively simple to determine the coefficients.

The eigenvalues and eigenfunctions are also useful physical quantities. Suppose we model the vibration of a drum with surface Ω through the problem

$$-u_{tt} - c^2 \nabla \cdot (\nabla u) = 0 \text{ in } \Omega.$$

We impose the boundary conditions $u(x,t) = 0$ for x on the boundary of Ω for all $t > 0$, holding the edge of the drum fixed against its rim. The eigenvalues λ_j of $\nabla \cdot (\nabla u)$ determine the **characteristic frequencies** of vibration of the drum, and $c\sqrt{\lambda_1}/(2\pi)$ is sometimes called the **fundamental frequency**. If we excite the drum so that it vibrates according to the corresponding eigenfunction, then the vibration persists.

By Challenge 31.3, we know, for example, that the fundamental frequency of a square drum $\tilde{\Omega}$ of size $a \times a$ is no higher than that of a circular drum Ω of diameter a since $\Omega \subset \tilde{\Omega}$.

CHALLENGE 31.4.

Determine the dimension of a square drum that has fundamental frequency equal to 1 when $c = 1$. Use numerical methods to find an elliptical domain $\alpha x^2 + 2\alpha y^2 < 1$ with the same fundamental frequency.

You might repeat Challenge 31.4 for domains of different shapes, or for different differential operators.

Chapter 32

Multigrid Methods: Managing Massive Meshes

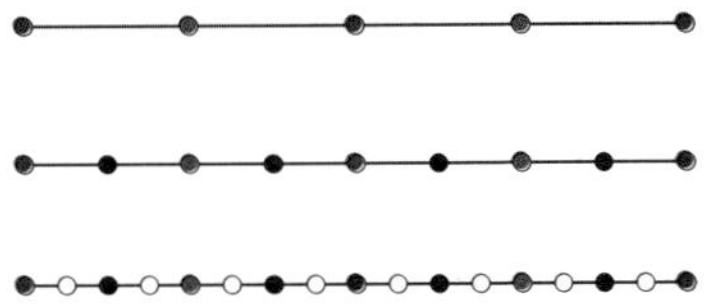

In Chapter 28, we investigated iterative methods for solving large, sparse linear systems of equations. We saw that the Gauss–Seidel method was intolerably slow, but various forms of preconditioned conjugate gradient algorithms (PCG) gave us reasonable results.

The test problems we used were discretizations of elliptic partial differential equations, and for these problems, there is a faster class of methods, called **multigrid** algorithms. Surprisingly, the Gauss–Seidel method (or some variant) is one of the two main ingredients in these algorithms!

To introduce the ideas, let's drop back to a somewhat simpler problem, a special case of one that we considered in Chapter 23.

A Simple Example

Suppose we want to solve the differential equation

$$-u_{xx}(x) = f(x)$$

on the domain $x \in [0,1]$, with $u(0) = u(1) = 0$. We know from Chapter 23 that we can approximate the solution by defining a **grid** or **mesh** $x_j = jh$, where $h = 1/(n+1)$ for some integer n. Then we can determine approximate values $u_j \approx u(x_j)$, $j = 1,\dots,n$, using **finite difference** or **finite element** approximations. If we choose finite differences, then we have

$$-u_{xx}(x_j) \approx \frac{-u_{j-1} + 2u_j - u_{j+1}}{h^2},$$

so we obtain a system of equations $\boldsymbol{Au} = \boldsymbol{f}$ with $\boldsymbol{f} = [f(x_1),\dots,f(x_n)]^T$, $\boldsymbol{u} = [u_1,\dots,u_n]^T$, and $\boldsymbol{A}$ equal to the $n \times n$ tridiagonal matrix

$$\boldsymbol{A} = \frac{1}{h^2}\begin{bmatrix} 2 & -1 & & & \\ -1 & 2 & -1 & & \\ & \ddots & \ddots & \ddots & \\ & & -1 & 2 & -1 \\ & & & -1 & 2 \end{bmatrix}.$$

Recall that in the Gauss–Seidel method of (28.1), we take an initial guess $\boldsymbol{u}^{(0)}$ for the solution and then update the guess by cycling through the equations, solving equation i for the ith variable u_i, so that given $\boldsymbol{u}^{(k)}$, our next guess $\boldsymbol{u}^{(k+1)}$ becomes

$$u_i^{(k+1)} = \left(f_i - \sum_{j=1}^{i-1} a_{ij} u_j^{(k+1)} - \sum_{j=i+1}^{n} a_{ij} u_j^{(k)} \right) / a_{ii}, \qquad i = 1, \ldots, n.$$

In our case, this reduces to

$$u_i^{(k+1)} = h^2 (f_i + u_{i-1}^{(k+1)} + u_{i+1}^{(k)})/2, \qquad i = 1, \ldots, n,$$

where we define $u_0^{(k)} = u_{n+1}^{(k)} = 0$ for all k.

It is easy to see how Gauss–Seidel can be very slow on a problem like this. Suppose, for example, that we take $\boldsymbol{u}^{(0)} = \boldsymbol{0}$ and that $\boldsymbol{f}$ is zero except for a 1 in its last position. Then $\boldsymbol{u}^{(1)}$ is zero except for its last entry, $\boldsymbol{u}^{(2)}$ is zero except for its last two entries, and it takes n iterations to get a guess that has a nonzero first entry. Since the true solution has nonzeros everywhere, this is not good!

The problem is that although Gauss–Seidel is good at fixing the solution locally, the information is propagated much too slowly globally, across the entire solution vector.

So if we are going to use Gauss–Seidel effectively, we need to couple it with a method that has good global properties.

A Multigrid Algorithm

When we set up our problem, we chose a value of n, probably guided by the knowledge that the error in the finite-difference approximation is proportional to h^2. There is a whole family of approximations, defined by different choices of h, and we denote the system of equations obtained using a grid length $h = 1/(n+1)$ by

$$\boldsymbol{A}_h \boldsymbol{u}_h = \boldsymbol{f}_h.$$

The grids on the interval $[0,1]$ corresponding to $h = 1/2$, $1/4$, $1/8$, and $1/16$ are shown in Figure 32.1.

A large value of h gives a **coarse grid**. The dimension n of the resulting linear system of equations is very small, though, so we can solve it fast using either a direct or an iterative method. Our computed solution $\boldsymbol{u}_h$ has the same overall shape as the true solution u but loses a lot of local detail. In contrast, if we use a very **fine grid** with a small value of h, then the linear system of equations is very large and much more expensive to solve, but our computed solution $\boldsymbol{u}_h$ is very close to u.

In order to get the best of both worlds, we might use a coarse-grid solution as an initial guess for the Gauss–Seidel iteration on a finer grid. To do this, we must set values for points in the finer grid that are not in the coarse grid. If someone gave us a solution to the system corresponding to h, then we could obtain an approximate solution for the system corresponding to $h/2$ by **interpolating** those values:

- For points in the finer grid that are common to the coarser grid, we just take their values.

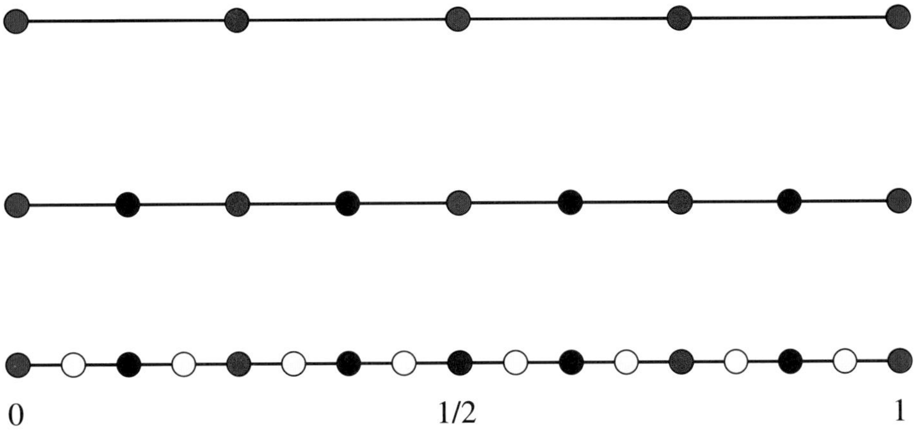

Figure 32.1. *Four levels of nested grids on the interval* $[0, 1]$. *The coarsest grid, with* $h = 1/2$, *consists of the blue points. Adding the red points gives* $h = 1/4$. *Including the black points gives* $h = 1/8$, *and including all of the points gives the finest grid, with* $h = 1/16$.

- For points in the finer grid that are midpoints of two points in the coarser grid, we take the average of these two values.

This defines an **interpolation operator** $\boldsymbol{P}_h$ that takes values in a grid with parameter h and produces values in the grid with parameter $h/2$. For example, because our boundary conditions are zero,

$$P_{1/8} = \begin{bmatrix} 1/2 & 0 & 0 & 0 & 0 & 0 & 0 \\ 1 & 0 & 0 & 0 & 0 & 0 & 0 \\ 1/2 & 1/2 & 0 & 0 & 0 & 0 & 0 \\ 0 & 1 & 0 & 0 & 0 & 0 & 0 \\ 0 & 1/2 & 1/2 & 0 & 0 & 0 & 0 \\ 0 & 0 & 1 & 0 & 0 & 0 & 0 \\ 0 & 0 & 1/2 & 1/2 & 0 & 0 & 0 \\ 0 & 0 & 0 & 1 & 0 & 0 & 0 \\ 0 & 0 & 0 & 1/2 & 1/2 & 0 & 0 \\ 0 & 0 & 0 & 0 & 1 & 0 & 0 \\ 0 & 0 & 0 & 0 & 1/2 & 1/2 & 0 \\ 0 & 0 & 0 & 0 & 0 & 1 & 0 \\ 0 & 0 & 0 & 0 & 0 & 1/2 & 1/2 \\ 0 & 0 & 0 & 0 & 0 & 0 & 1 \\ 0 & 0 & 0 & 0 & 0 & 0 & 1/2 \end{bmatrix}.$$

The process of solving the problems on the sequence of nested grids gives us a **nested iteration** algorithm, Algorithm 32.1, for our sample problem. The termination tolerance for the ∞-norm of the residual $\boldsymbol{f}_h - \boldsymbol{A}_h\tilde{\boldsymbol{u}}_h$ on grid h should be proportional to h^2, since that matches the size of the truncation error. This algorithm runs from coarse grid to finest and is useful (although rather silly for one-dimensional problems). But there is a better way.

Algorithm 32.1 Nested Iteration

Set $k = 1$, $h = 1/2$, and $\tilde{\boldsymbol{u}}_h = 0$.
while the approximation is not good enough,
  Set $k = k+1$, $n = 2^k - 1$, and $h = 1/(n+1)$.
  Form the matrix $\boldsymbol{A}_h$ and the right-hand side $\boldsymbol{f}_h$, and use the Gauss–Seidel iteration, with the initial guess $\boldsymbol{P}_{2h}\tilde{\boldsymbol{u}}_{2h}$, to compute an approximate solution $\tilde{\boldsymbol{u}}_h$ to $\boldsymbol{A}_h\boldsymbol{u}_h = \boldsymbol{f}_h$.
end

The V-Cycle

We can do better if we run from finest grid to coarsest grid and then back to finest. This algorithm has 3 ingredients:

- An iterative method that converges quickly if most of the error is **high frequency**—oscillating rapidly—which happens when the overall shape of the solution is already identified. Gauss–Seidel generally works well.

- A way to transfer values from a coarse grid to a fine one—**interpolation** or **prolongation**.

- A way to transfer values from a fine grid to a coarse one—**restriction**.
  We let $\boldsymbol{R}_h$ be the operator takes values on grid $h/2$ and produces values on grid h.

We already have matrices $\boldsymbol{P}_h$ for interpolation, and (for technical reasons related to preserving the self-adjointness of the problems considered here) we choose $\boldsymbol{R}_h = \boldsymbol{P}_h^T$.

We define the V-cycle idea recursively in Algorithm 32.2. In using this algorithm, we can define $\boldsymbol{A}_{2h} = \boldsymbol{R}_{2h}\boldsymbol{A}_h\boldsymbol{P}_{2h}$. This definition is key to extending the multigrid algorithm beyond problems that have a geometric grid; see Pointer 32.1 for a reference on these **algebraic multigrid** methods. But for now, let's see how it works on our original problem.

Algorithm 32.2 V-Cycle

$\boldsymbol{v}_h$ = V-cycle($\boldsymbol{v}_h, \boldsymbol{A}_h, \boldsymbol{f}_h, \eta_1, \eta_2$)
if h is the coarsest grid parameter **then**
  Compute $\boldsymbol{v}_h$ to solve $\boldsymbol{A}_h\boldsymbol{v}_h = \boldsymbol{f}_h$ and return.
end
Perform η_1 Gauss–Seidel iterations on $\boldsymbol{A}_h\boldsymbol{u}_h = \boldsymbol{f}_h$ using $\boldsymbol{v}_h$ as the initial guess, obtaining an approximate solution that we still call $\boldsymbol{v}_h$.
Let $\boldsymbol{v}_{2h}$ = V-cycle($\boldsymbol{0}, \boldsymbol{A}_{2h}, \boldsymbol{R}_{2h}(\boldsymbol{f}_h - \boldsymbol{A}_h\boldsymbol{v}_h), \eta_1, \eta_2$).
Set $\boldsymbol{v}_h = \boldsymbol{v}_h + \boldsymbol{P}_{2h}\boldsymbol{v}_{2h}$.
Perform η_2 Gauss–Seidel iterations on $\boldsymbol{A}_h\boldsymbol{u}_h = \boldsymbol{f}_h$ using $\boldsymbol{v}_h$ as the initial guess, obtaining an approximate solution that we still call $\boldsymbol{v}_h$.

CHALLENGE 32.1.

Work through the V-cycle algorithm to see exactly what computations it performs on our simple example for the sequence of grids defined in Figure 32.1. Estimate the amount of work, measured by the number of floating-point multiplications performed.

The standard multigrid algorithm repeats the V-cycle until convergence and is given in Algorithm 32.3.

Algorithm 32.3 Standard Multigrid Algorithm for Solving $A_h \boldsymbol{u}_h = \boldsymbol{f}_h$

Initialize $\boldsymbol{u}_h = \mathbf{0}$.
while the termination criteria are not satisfied,
  Compute $\boldsymbol{\Delta u}_h$ = V-cycle($\mathbf{0}, \boldsymbol{A}_h, \boldsymbol{r}_h, \eta_1, \eta_2$), where $\boldsymbol{r}_h = \boldsymbol{f}_h - \boldsymbol{A}_h \boldsymbol{u}_h$.
  Update $\boldsymbol{u}_h = \boldsymbol{u}_h + \boldsymbol{\Delta u}_h$.
end

Cost of Multigrid

Guided by Challenge 32.1, we can estimate the work for multigrid applied to a more general problem. One step of the Gauss–Seidel iteration on a grid of size h costs about $nz(h)$ multiplications, where $nz(h)$ is the number of nonzeros in A_h. We'll call $nz(h)$ multiplications a **work unit**.

Note that $nz(h) \approx 2nz(2h)$ since A_{2h} has about half as many rows as A_h. So performing one Gauss–Seidel step on each grid $h, h/2, \ldots, 1$ costs less than $nz(h)(1 + 1/2 + 1/4 + \cdots) = 2nz(h)$ multiplications $\equiv$ 2 work units.

So the cost of a V-Cycle is at most 2 times the cost of $(\eta_1 + \eta_2)$ Gauss–Seidel iterations on the finest grid plus a modest amount of additional computational overhead. Is it just as efficient in storage?

CHALLENGE 32.2.

Convince yourself that the storage necessary for all of the matrices and vectors is also a modest multiple of the storage necessary for the finest grid.

We know that stationary iterative methods like Gauss–Seidel are usually very slow (take many iterations), so the success of multigrid relies on the fact that we need only a few iterations on each grid, because the error is mostly local. Thus the total amount of work to solve the full problem to a residual of size $O(h^2)$ is a small number of work units.

Since it is rather silly to use anything other than sparse Gauss elimination to solve a system involving a tridiagonal matrix, we won't implement the algorithm for our one-dimensional problem. Note, though, that our algorithm readily extends to higher dimensions; we just need to define $\boldsymbol{A}_h$ and $\boldsymbol{P}_h$ for a nested set of grids in order to use the multigrid V-cycle algorithm.

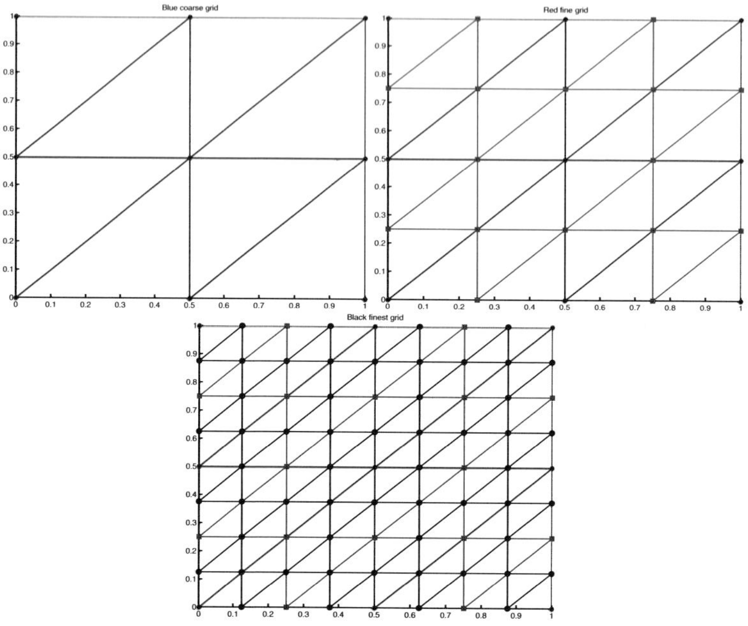

Figure 32.2. *The blue gridpoints (top left) define the coarse grid. The blue and the red gridpoints (top right) define the finer grid. The blue, red, and black gridpoints (bottom) define the finest grid.*

Multigrid for Two-Dimensional Problems

Our first challenge in applying multigrid to two-dimensional problems is to develop a sequence of **nested grids**. Since we discussed finite difference methods for the one-dimensional problem, let's focus on finite element methods for the two-dimensional problem, using a triangular grid and piecewise-linear basis functions.

It is most convenient to start from a **coarse grid** and obtain our **finest grid** through successive refinements. Consider the grid in Figure 32.2 that divides the unit square into 8 triangles with height $h = 1/2$. The grid points are marked in blue. Consider taking the midpoints of each side of one of triangles and drawing the triangle with those points as vertices. If we do this for each triangle, we obtain the red grid points in Figure 32.2 and the red triangles. Each of the original blue triangles has been replaced by 4 triangles, each having 1 or 3 red sides, and each triangle has height $h = 1/4$. If we repeat this process, we obtain the black grid points of Figure 32.2 and a grid length $h = 1/8$.

Writing a program for grid refinement on a general triangular grid takes a bit of effort; see `refine.m` on the website.

Interpolating from one grid to the next finer one is easy. For example, given the blue grid values, we obtain values for the blue and red grid by following two rules: blue gridpoints retain their values, and red grid values are defined as the average of the nearest two values on the blue edge containing it. As before, we take the restriction operator to be the transpose of the interpolation operator.

So we have all the machinery necessary to apply multigrid to two-dimensional problems, and we experiment with it in the next challenge.

CHALLENGE 32.3.

Write a program that applies the multigrid V-cycle iteration to the two-dimensional problems used in Chapter 28. The MATLAB program `generateproblem.m` produces a structure called `mesh` which contains, in addition to the matrices and right-hand side information, the operators `P` and the coordinates `p` of the grid points. The differential equation is

$$-u_{xx}(x,y) - u_{yy}(x,y) + \kappa u(x,y) = f(x,y)$$

for $(x,y) \in [-1,1] \times [-1,1]$ (`myproblem=1`) or for this domain with a hole cut out (`myproblem=2`). The boundary conditions are that u is zero on the boundary of the square, and (for the second case) the normal derivative is zero at the boundary of the hole. Set $\kappa = 0$ and compare the time for solving the problem using multigrid to the methods defined in Chapter 28.

If the partial differential equation is elliptic, as it is for $\kappa \geq 0$, it is not too hard to achieve convergence in a small number of work units. In fact, multigrid experts would say that if we don't achieve it, then we have chosen either the iteration or the interpolation/restriction pair "incorrectly." For problems that are not elliptic, though, things get a bit more complicated, as we see in the next challenge.

CHALLENGE 32.4.

Repeat your experiment from Challenge 32.3, but use $\kappa = 10$ and 100 and then $\kappa = -10$ and -100. (When $\kappa \neq 0$, the differential equation is called the **Helmholtz equation**.) The differential equation remains elliptic for positive κ but not for negative. How was convergence of multigrid affected?

We see that the problem is much harder to solve for negative values of κ. There are two reasons for this: first, the matrix A_h is no longer positive definite, so we lose a lot of nice structure, and second, finer grids are necessary to represent the solution accurately. In order to restore convergence in a small number of work units for the nonelliptic problem, we must make the algorithm more complicated; for example, we might use multigrid as a preconditioner for a Krylov subspace method, bringing us back to the methods used in the Chapter 28. More information about this can be found in the references in Pointer 32.1.

POINTER 32.1. Further Reading.

The multigrid idea dates back to R. P. Fedorenko in 1964. A good introduction is given in a tutorial by Briggs, Henson, and McCormick [20]. "Multigrid" ideas are useful even when there is no natural geometric "grid" underlying the problem; the resulting method is called **algebraic multigrid** and is briefly discussed in [20].

It is also useful to use multigrid if only a portion of the grid is refined from one level to the next; for example, we might want to refine only in regions in which the solution is rapidly changing, where the current grid does not capture its behavior accurately enough. These adaptive methods are also discussed in [20].

One multigrid approach to solving the Helmholtz equation with negative κ is given in [46].

Bibliography

[1] E.L. Allgower and K. Georg. *Introduction to Numerical Continuation Methods*. SIAM, Philadelphia, PA, 2003. (Cited on p. 296.)

[2] E. Anderson, Z. Bai, C. Bischof, S. Blackford, J. Demmel, J. Dongarra, J. Du Croz, A. Greenbaum, S. Hammarling, A. McKenney, and D. Sorensen. LAPACK *Users' Guide, Third Edition*. SIAM, Philadelphia, 1999. (Cited on pp. 38, 78, 79.)

[3] Douglas Arnold. Some disasters attributable to bad numerical computing. `http://www.ima.umn.edu/~arnold/disasters/disasters.html`. Retrieved May 2008. (Cited on p. 14.)

[4] V. Arun. *The Solution of Variable-Geometry Truss Problems Using New Homotopy Continuation Methods*. PhD thesis, Mechanical Engineering Department, Virginia Polytechnic Institute and State University, Blacksburg, Virginia, September 1990. (Cited on pp. 299, 300, 300.)

[5] V. Arun, C. F. Reinholtz, and L. T. Watson. Application of new homotopy continuation techniques to variable geometry trusses. *J. Mech. Des.*, 114:422–428, 1992. (Cited on pp. 296, 299, 300, 300.)

[6] Randolph E. Bank. Software: Pltmg 10.0. `http://scicomp.ucsd.edu/~reb/software.html`), 2007. Retrieved May 2008. (Cited on p. 339.)

[7] R. Bartels and G. W. Stewart. Algorithm 432: The solution of the matrix equation $AX - XB = C$. *Communications of the ACM*, 15(9):820–826, 1972. (Cited on p. 346.)

[8] K.-J. Bathe. *Finite Element Procedures*. Prentice Hall, Englewood Cliffs, NJ, 1996. (Cited on p. 339.)

[9] Mokhtar S. Bazaraa, Hanif D. Sherali, and C. M. Shetty. *Nonlinear Programming: Theory and Algorithms, Third Edition*. Wiley Interscience, New York, 2006. (Cited on p. 339.)

[10] Isabel Beichl, Dianne P. O'Leary, and Francis Sullivan. Approximating the number of monomer-dimer coverings in periodic lattices. *Physical Review E*, 64:016701.1–6, 2001. (Cited on p. 202.)

[11] Isabel Beichl and Francis Sullivan. The importance of importance sampling. *Computing in Science and Engineering*, 1(2):71–73, March-April 1999. (Cited on pp. 202, 212.)

[12] A. Bellen and M. Zennaro. *Numerical Methods for Delay Differential Equations*. Oxford Science Pub., 2003. (Cited on p. 264.)

[13] Edward Beltrami. *Mathematics for Dynamic Modeling, Second Edition*. Academic Press, Boston, MA, 1998. (Cited on p. 267.)

[14] Jon Bentley. *Programming Pearls, Second Edition*. Addison-Wesley, Reading, MA, 2000. (Cited on p. 43.)

[15] Michael W. Berry and Murray Browne. *Understanding Search Engines: Mathematical Modeling and Text Retrieval, Second Edition*. SIAM, Philadelphia, PA, 2005. (Cited on p. 333.)

[16] Åke Björck. *Numerical Methods for Least Squares Problems*. SIAM Press, Philadelphia, PA, 1996. (Cited on p. 66.)

[17] K. E. Brenan, S. L. Campbell, and L. R. Petzold. *Numerical Solution of Initial-Value Problems in Differential-Algebraic Equations*, volume 14 of Classics in Applied Mathematics. SIAM, Philadelphia, PA, 1995. Revised and corrected reprint of the 1989 original. (Cited on p. 257.)

[18] Susanne C. Brenner and L. Ridgeway Scott. *The Mathematical Theory of Finite Element Methods, Third Edition*. Springer, New York, 2008. (Cited on p. 339.)

[19] R. P. Brent. *Algorithms for Minimization without Derivatives*. Prentice-Hall, Englewood Cliffs, NJ, 1973. Reprinted by Dover, 2002. (Cited on p. 286.)

[20] William L. Briggs, Van Emden Henson, and Steve F. McCormick. *A Multigrid Tutorial, Second Edition*. SIAM, Philadelphia, PA, 2000. (Cited on p. 360.)

[21] Nicholas F. Britton. *Essential Mathematical Biology*. Springer Undergraduate Mathematics Series. Springer-Verlag London Ltd., London, 2003. (Cited on pp. 219, 264.)

[22] C. G. Broyden. Quasi-Newton methods. In W. Murray, editor, *Numerical Methods for Unconstrained Optimization*, pages 87–106. Academic Press, New York, 1972. (Cited on p. 133.)

[23] Russel E. Caflisch. Monte Carlo and quasi-Monte Carlo methods. *Acta Numerica*, 7:1–49, 1998. (Cited on pp. 202, 212, 212.)

[24] James Callahan. The spread of a contagious illness. `http://www.math.smith.edu/~callahan/ili/pde.html`, 1996. Retrieved May 2008. (Cited on p. 264.)

[25] A. H. A. Clayton and W. H. Sawyer. Site-specific tryptophan dynamics in class A amphipathic helical peptides at a phospholipid bilayer interface. *Biophysical Journal*, 79:1066–1073, 2000. (Cited on p. 168.)

[26] Earl A. Coddington. *An Introduction to Ordinary Differential Equations*. Prentice-Hall, Englewood Cliffs, NJ, 1961. (Cited on p. 226.)

[27] Thomas F. Coleman and Arun Verma. Admit-1: automatic differentiation and MATLAB interface toolbox. *ACM Trans. Math. Softw.*, 26(1):150–175, 2000. (Cited on p. 122.)

[28] A.R. Conn, N.I.M. Gould, and P.L. Toint. *Trust-Region Methods*. SIAM, Philadelphia, PA, 2000. (Cited on pp. 133, 296.)

[29] R. F. Costantino, Robert A. Desharnais, J. M. Cushing, Brian Dennis, Shandelle M. Henson, and Aaron A. King. Nonlinear stochastic population dynamics: The flour beetle Tribolium as an effective tool of discovery. *Advances in Ecological Research*, 37:101–141, 2005. (Cited on p. 306.)

[30] D. Coppersmith and S. Winograd. Matrix multiplication via arithmetic progressions. *J. Symbolic Comput.*, 9:251–280, 1990. (Cited on p. 37.)

[31] Patrick Cousot. Methods and logics for proving programs. In J. van Leeuwen, editor, *Handbook of Theoretical Computer Science, Vol B: Formal Models and Semantics*, pages 843–993. North Holland, Amsterdam, The Netherlands, 1990. Chapter 15. (Cited on p. 43.)

[32] Germund Dahlquist and Åke Björck. *Numerical Methods in Scientific Computing, Volume 1*. SIAM, Philadelphia, PA, 2008. (Cited on pp. 107, 212, 283.)

[33] Biswa Datta. *Numerical Methods for Linear Control Systems*. Elsevier Academic Press, Boston, MA, 2004. (Cited on p. 346.)

[34] Ian Davidson. Understanding k-means non-hierarchical clustering. Technical report, Computer Science Department, State University of New York, Albany, Technical Report 02-2, `http://www.cs.albany.edu/~davidson/courses/CSI635/UnderstandingK-MeansClustering.pdf`, 2002. Retrieved August 2007. (Cited on p. 155.)

[35] Timothy A. Davis. *Direct Methods for Sparse Linear Systems*. SIAM, Philadelphia, PA, 2006. (Cited on p. 322.)

[36] Timothy A. Davis. University of Florida sparse matrix collection. `http://www.cise.ufl.edu/research/sparse/matrices`, 2007. Retrieved May 2008. (Cited on p. 334.)

[37] James W. Demmel. *Applied Numerical Linear Algebra*. SIAM, Philadelphia, PA, 1997. (Cited on p. 70.)

[38] James W. Demmel. Lecture notes on spectral partitioning. `http://www.cs.berkeley.edu/~demmel/cs267/lecture20/lecture20.html`, 1999. Retrieved August 2007. (Cited on p. 322.)

[39] Brian Dennis, Robert A. Desharnais, J. M. Cushing, and R. F. Costantino. Nonlinear demographic dynamics: Mathematical models, statistical methods, and biological experiments. *Ecological Monographs*, 65:261–281, 1995. (Cited on pp. 302, 306.)

[40] J. E. Dennis, Jr. and R. B. Schnabel. Least change secant updates for quasi-Newton methods. *SIAM Rev.*, 21(4):443–459, 1979. (Cited on p. 133.)

[41] J. E. Dennis, Jr. and Robert B. Schnabel. *Numerical Methods for Unconstrained Optimization and Nonlinear Equations*, volume 16 of Classics in Applied Mathematics. SIAM, Philadelphia, PA, 1996. Corrected reprint of the 1983 original. (Cited on p. 133.)

[42] S. E. Derenzo, M. J. Weber, W. W. Moses, and C. Dujardin. Measurements of the intrinsic rise times of common inorganic scintillators. *IEEE Trans. Nucl. Sci.*, NS-47:860–864, 2000. (Cited on p. 168.)

[43] Robert A. Desharnais and Laifu Liu. Stable demographic limit cycles in laboratory populations of Tribolium castaneum. *Journal of Animal Ecology*, 56(3):885–906, 1987. (Cited on pp. 304, 306.)

[44] I. S. Dhillon, Y. Guan, and J. Kogan. Iterative clustering of high dimensional text data augmented by local search. In *Proceedings of the 2002 IEEE International Conference on Data Mining*, `http://www.cs.utexas.edu/users/dml/Software/gmeans.html`, 2002. Retrieved August 2007. (Cited on p. 155.)

[45] I. S. Duff, A. M. Erisman, and J. K. Reid. *Direct Methods for Sparse Matrices*. Oxford Press, 1986. (Cited on p. 322.)

[46] Howard C. Elman, Oliver G. Ernst, and Dianne P. O'Leary. A multigrid method enhanced by Krylov subspace iteration for discrete Helmholtz equations. *SIAM J. Sci. Comput.*, 23:1291–1315, 2001. (Cited on p. 360.)

[47] Haw-ren Fang and Dianne P. O'Leary. Stable factorizations of symmetric tridiagonal and triadic matrices. *SIAM J. Matrix Anal. Appl.*, 28:576–595, 2006. (Cited on p. 133.)

[48] Ricardo Fierro, Gene H. Golub, Per Christian Hansen, and Dianne P. O'Leary. Regularization by truncated total least squares. *SIAM J. Sci. Comput.*, 18:1223–1241, 1997. (Cited on p. 174.)

[49] George S. Fishman. *Monte Carlo: Concepts, Algorithms and Applications*. Springer, London, 2003. (Cited on p. 30.)

[50] R. Fletcher. *Practical Methods of Optimization, Second Edition*. Wiley-Interscience, New York, 2001. (Cited on pp. 116, 121, 122, 133.)

[51] Geoffrey C. Fox, Roy D. Williams, and Paul C. Messina. *Parallel Computing Works*. Morgan Kaufmann, San Francisco, CA, 1994. (Cited on p. 212.)

[52] Keinosuke Fukunaga. *Introduction to Statistical Pattern Recognition, Second Edition*. Academic Press, Boston, MA, 1990. (Cited on p. 155.)

[53] Walter Gander. Heisenberg effects in computer arithmetic. `http://www.inf.ethz.ch/news/focus/res_focus/april_2005`. Retrieved May 2008. (Cited on p. 14.)

[54] C. W. Gear. *Numerical Initial Value Problems in Ordinary Differential Equations*. Prentice-Hall, Englewood Cliffs, NJ, 1971. (Cited on p. 257.)

[55] Alan George and Joseph W. Liu. *Computer Solution of Large Sparse Positive Definite Systems*. Prentice-Hall, Englewood Cliffs, NJ, 1981. (Cited on p. 322.)

[56] John R. Gilbert, Gary L. Miller, and Shang-Hua Teng. Geometric mesh partitioning: Implementation and experiments. *SIAM J. Sci. Comput.*, 19:2091–2110, 1998. (Cited on p. 320.)

[57] John R. Gilbert and Shang-Hua Teng. Mesh partitioning and graph separator toolbox. `http://www.cerfacs.fr/algor/Softs/MESHPART/`, 2002. Retrieved May 2008. (Cited on p. 320.)

[58] P. E. Gill, G. H. Golub, W. Murray, and M. A. Saunders. Methods for modifying matrix factorizations. *Math. Comp.*, 28:505–535, 1974. (Cited on pp. 95, 126.)

[59] Mark S. Gockenbach. *Partial Differential Equations*. SIAM, Philadelphia, PA, 2002. (Cited on pp. 280, 351.)

[60] Gene H. Golub. Some modified matrix eigenvalue problems. *SIAM Rev.*, 15:318–334, 1973. (Cited on p. 95.)

[61] Gene H. Golub, Per Christian Hansen, and Dianne P. O'Leary. Tikhonov regularization and total least squares. *SIAM J. Matrix Anal. Appl.*, 21:185–194, 1999. (Cited on p. 174.)

[62] G.H. Golub and W. Kahan. Calculating the singular values and pseudo-inverse of a matrix. *SIAM J. Numer. Anal*, 2(3):205–224, 1965. (Cited on p. 85.)

[63] Gene H. Golub and Victor Pereyra. Separable nonlinear least squares: the variable projection method and its applications. *Inverse Problems*, 19:R1–R26, 2003. (Cited on p. 168.)

[64] Gene H. Golub and Charles F. Van Loan. *Matrix Computations, Third Edition*. Johns Hopkins Studies in the Mathematical Sciences. Johns Hopkins University Press, Baltimore, MD, 1996. (Cited on pp. 53, 56, 69, 79, 95, 103, 162.)

[65] Yuqian Guan. Gmeans—a clustering tool in ping-pong style. `http://www.cs.utexas.edu/users/dml/Software/gmeans.html`, 2002. Retrieved May 2008. (Cited on p. 155.)

[66] Mads Haahr. Random.org. `http://www.random.org/`, 1998. Retrieved May 2008. (Cited on p. 194.)

[67] Jack K. Hale and Sjoerd M. V. Lunel. *Introduction to Functional-Differential Equations*, volume 99 of Applied Mathematical Sciences. Springer-Verlag, New York, 1993. (Cited on p. 264.)

[68] Per Christian Hansen. *Rank-Deficient and Discrete Ill-Posed Problems*. SIAM, Philadelphia, PA, 1998. (Cited on pp. 86, 174.)

[69] Per Christian Hansen, James M. Nagy, and Dianne P. O'Leary. *Deblurring Images: Matrices, Spectra, and Filtering*. SIAM, Philadelphia, PA, 2006. (Cited on pp. 84, 86, 346, 346.)

[70] Simon Haykin. *Adaptive Filter Theory, Fourth Edition*. Prentice-Hall, Englewood Cliffs, NJ, 1996. (Cited on p. 103.)

[71] Michael Heath. *Scientific Computing: An Introductory Survey, Second Edition*. McGraw Hill, Boston, MA, 2002. (Cited on pp. xiii, xiv, 185.)

[72] Peter Hellekalek et al. Random number generators. `http://random.mat.sbg.ac.at/`, 2006. Retrieved May 2008. (Cited on p. 194.)

[73] Richard W. Hendler, Lel A. Drachev, Salil Bose, and Manjoj K. Joshi. On the kinetics of voltage formation in purple membranes of Halobacterium salinarium. *European Journal of Biochemistry*, 267:5879–5890, 2000. (Cited on p. 168.)

[74] John L. Hennessy and David A. Patterson. *Computer Architecture: A Quantitative Approach, Second Edition*. Morgan Kaufmann, San Francisco, CA, 1996. (Cited on p. 38.)

[75] Peter Henrici. *Discrete Variable Methods in Ordinary Differential Equations*. John Wiley & Sons, New York, 1962. (Cited on p. 223.)

[76] Magnus R. Hestenes and Eduard Stiefel. Methods of conjugate gradients for solving linear systems. *J. Research Nat. Bur. Standards*, 49:409–436, 1952. (Cited on p. 127.)

[77] Desmond J. Higham. An algorithmic introduction to numerical simulation of stochastic differential equations. *SIAM Rev.*, 43(3):525–546, 2001. (Cited on p. 264.)

[78] Desmond J. Higham and Nicholas J. Higham. *MATLAB Guide, Second Edition*. SIAM, Philadelphia, PA, 2005. (Cited on pp. 3, 43.)

[79] Nicholas J. Higham. *Accuracy and Stability of Numerical Algorithms, Second Edition*. SIAM Press, Philadelphia, PA, 2002. (Cited on pp. 22, 30, 37, 346.)

[80] Frank C. Hoppensteadt and Charles S. Peskin. *Mathematics in Medicine and the Life Sciences*, volume 10 of Texts in Applied Mathematics. Springer-Verlag, New York, 1992. (Cited on p. 219.)

[81] Roger A. Horn and Charles R. Johnson. *Topics in Matrix Analysis*. Cambridge University Press, Cambridge, 1994. Corrected reprint of the 1991 original. (Cited on p. 86.)

[82] Thomas Huckle. Collection of software bugs. `http://www5.in.tum.de/~huckle/bugse.html`. Retrieved May 2008. (Cited on p. 14.)

[83] Human Genome Sequencing Center. Tribolium castaneum genome project. `http://www.hgsc.bcm.tmc.edu/projects/tribolium/`, 2007. Retrieved May 2008. (Cited on p. 306.)

[84] Anil K. Jain and Richard C. Dubes. *Algorithms for Clustering Data*. Prentice Hall Advanced Reference Series. Prentice-Hall, Englewood Cliffs, NJ, 1988. (Cited on p. 155.)

[85] A. K. Jain, M. N. Murty, and P. J. Flynn. Data clustering: A review. *ACM Comput. Surv.*, 31(3):264–323, 1999. (Cited on p. 155.)

[86] M. A. Jenkins and J. F. Traub. Algorithm 419: zeros of a complex polynomial. *Communications of the ACM*, 15(2):97–99, 1972. (Cited on p. 286.)

[87] D. S. Jones and B. D. Sleeman. *Differential Equations and Mathematical Biology*. Chapman & Hall/CRC Mathematical Biology and Medicine Series. Chapman & Hall/CRC, Boca Raton, FL, 2003. (Cited on p. 264.)

[88] C. T. Kelley. *Solving Nonlinear Equations with Newton's Method*. SIAM, Philadelphia, PA, 2003. (Cited on p. 296.)

[89] Claire Kenyon, Dana Randall, and Alistair Sinclair. Approximating the number of monomer-dimer coverings of a lattice. *J. Statist. Phys.*, 83(3-4):637–659, 1996. (Cited on p. 202.)

[90] S. Kirkpatrick, C. D. Gelatt, and M. P. Vecchi. Optimization by simulated annealing. *Science*, 220(4598):671–680, 1983. (Cited on p. 202.)

[91] Donald E. Knuth. *The Art of Computer Programming. Vol. 1: Fundamental Algorithms, Third Edition*. Addison-Wesley Publishing Co., Reading, MA, 1997. (Cited on p. 194.)

[92] Donald E. Knuth. *The Art of Computer Programming. Vol. 2: Seminumerical Algorithms, Third Edition*. Addison-Wesley Publishing Co., Reading, MA, 1997. (Cited on p. 219.)

[93] Donald E. Knuth. *Literate Programming*. Lecture Notes 27, Center for the Study of Language and Information, Stanford, CA, 1992. (Cited on p. 43.)

[94] Gina Bari Kolata. *Flu: The Story of the Great Influenza Pandemic of 1918 and the Search for the Virus That Caused It*. Farrar, Straus, and Giroux, New York, 1999. (Cited on p. 219.)

[95] T.G. Kolda, R.M. Lewis, and V. Torczon. Optimization by direct search: New perspectives on some classical and modern methods. *SIAM Rev.*, 45(3):385–482, 2003. (Cited on p. 133.)

[96] Jack B. Kuipers. *Quaternions and Rotation Sequences: A Primer with Applications to Orbits, Aerospace and Virtual Reality*. Princeton University Press, Princeton, NJ, 2002. (Cited on p. 162.)

[97] Peter Kunkel and V. Mehrmann. *Differential-Algebraic Equations. Analysis and Numerical Solution*. European Mathematical Society. EMS Textbooks, 2006. (Cited on pp. 248, 250, 257.)

[98] David Lane, Joan Lu, Camille Peres, and Emily Zitek. Online statistics: An interactive multimedia course of study. `http://onlinestatbook.com`. Retrieved May 2008. (Cited on pp. 185, 189.)

[99] Stig Larsson and Vidar Thomée. *Partial Differential Equations with Numerical Methods*, volume 45 of *Texts in Applied Mathematics*. Springer-Verlag, Berlin, 2003. (Cited on pp. 223, 251, 252, 280, 351.)

[100] R.B. Lehoucq, D.C. Sorensen, and C. Yang. *ARPACK Users' Guide: Solution of Large-Scale Eigenvalue Problems with Implicitly Restarted Arnoldi Methods*. SIAM, Philadelphia, PA, 1998. (Cited on p. 334.)

[101] Walter Leighton. *Ordinary Differential Equations*. Wadsworth Publishing Co., Belmont, CA, 1966. (Cited on p. 271.)

[102] Steven J. Leon. *Linear Algebra with Applications, Sixth Edition*. Prentice-Hall, Englewood Cliffs, NJ, 2002. (Cited on p. 271.)

[103] T. Y. Li. Numerical Solution of Polynomial Systems by Homotopy Continuation Methods. *Handbook of Numerical Analysis*, 11:209–304, 2003. (Cited on p. 296.)

[104] K. J. R. Liu, D. P. O'Leary, G. W. Stewart, and Y-J. J. Wu. URV ESPRIT for tracking time-varying signals. *IEEE Trans. Signal Proc.*, 42:3441–3448, 1994. (Cited on p. 103.)

[105] Ivo Marek. Iterative aggregation/disaggregation methods for computing some characteristics of Markov chains. II. Fast convergence. *Appl. Numer. Math.*, 45(1):11–28, 2003. (Cited on p. 219.)

[106] Mathematical and Computational Sciences Division. Matrix market. `http://math.nist.gov/MatrixMarket/`. Retrieved May 2008. (Cited on pp. 311, 334.)

[107] W. Miller. Computational complexity and numerical stability. *SIAM J. Comput.*, 4(2):97–107, 1975. (Cited on p. 346.)

[108] Cleve B. Moler. *Numerical Computing with MATLAB*. SIAM, Philadelphia, PA, 2004. (Cited on pp. xiii, xiv.)

[109] A. M. Mood and F. A. Graybill. *Introduction to the Theory of Statistics*. McGraw-Hill, New York, 1963. (Cited on p. 202.)

[110] R. E. Moore. *Interval Analysis*. Prentice-Hall, Englewood Cliffs, NJ, 1966. (Cited on p. 30.)

[111] Stephen G. Nash and Ariela Sofer. *Linear and Nonlinear Programming*. McGraw-Hill, Boston, MA, 1996. (Cited on pp. 30, 117, 125, 127, 133, 148.)

[112] M. Nauenberg, F. Kuttner, and M. Furman. Method for evaluating one-dimensional path integrals. *Physical Review A*, 13(3):1185–1189, March 1976. (Cited on p. 212.)

[113] C. R. Nave. The energy distribution function. http://hyperphysics.phy-astr.gsu.edu/hbase/quantum/disfcn.html, 2005. Retrieved May 2008. (Cited on p. 189.)

[114] Arkadi Nemirovski. Lecture notes on interior point polynomial time methods in convex programming. Technical Report ISYE 8813, Georgia Institute of Technology School of Industrial and Systems Engineering, Spring 2004. (Cited on p. 148.)

[115] Jorge Nocedal and Stephen J. Wright. *Numerical Optimization, Second Edition.* Springer, New York, 2006. (Cited on p. 148.)

[116] K. Ogata. *Modern Control Engineering, Third Edition.* Prentice-Hall, Englewood Cliffs, NJ, 1996. (Cited on p. 271.)

[117] Dianne P. O'Leary and Wei H. Yang. Elasto-plastic torsion by quadratic programming. *Computer Methods in Applied Mechanics and Engineering*, 16:361–368, 1978. (Cited on p. 339.)

[118] Suely Oliveira and David Stewart. *Writing Scientific Software: A Guide to Good Style.* Cambridge University Press, Cambridge, England, 2006. (Cited on p. 43.)

[119] James M. Ortega. *Numerical Analysis: A Second Course.* Academic Press, New York, 1972. (Cited on pp. 53, 181, 334.)

[120] J. M. Ortega and W. C. Rheinboldt. *Iterative Solution of Nonlinear Equations in Several Variables.* Academic Press, New York, 1970. (Cited on p. 296.)

[121] Michael L. Overton. *Numerical Computing with IEEE Floating Point Arithmetic.* SIAM, Philadelphia, PA, 2001. (Cited on pp. 9, 10, 22.)

[122] C. C. Paige and M. A. Saunders. LSQR: An algorithm for sparse linear equations and sparse least squares. *ACM Trans. Math. Softw.*, 8:43–71, 1982. http://www.stanford.edu/group/SOL/software/lsqr.html. (Cited on pp. 330, 334.)

[123] C. C. Paige and M. A. Saunders. Solution of sparse indefinite systems of linear equations. *SIAM J. Numer. Anal.*, 12:617–629, 1975. http://www.stanford.edu/group/SOL/software/symmlq.html. (Cited on pp. 330, 334.)

[124] Per-Olof Persson and Gilbert Strang. A simple mesh generator in MATLAB. *SIAM Review*, 46:329–345, 2004. (Cited on p. 334.)

[125] A. Pothen, H. Simon, and K.-P. Liou. Partitioning sparse matrices with eigenvectors of graphs. *SIAM J. Matrix Anal. Appl.*, 11:430–452, 1990. (Cited on p. 322.)

[126] Armin Pruessner and Dianne P. O'Leary. Blind deconvolution using a regularized structured total least norm approach. *SIAM J. Matrix Anal. Appl.*, 24:1018–1037, 2003. (Cited on p. 181.)

[127] R. Roy and T. Kailath. ESPRIT—estimation of signal parameters via rotational invariance techniques. In *IEEE Trans. ASSP*, 34:984–995, 1989. (Cited on p. 103.)

[128] W.J. Rugh. *Linear System Theory, Second Edition*. Prentice-Hall, Englewood Cliffs, NJ, 1996. (Cited on p. 271.)

[129] Bert W. Rust. Fitting nature's basic functions. *Computing in Science and Engineering*. 3(5):84-89, 2001; 3(6):60–64, 2001; 4(4):72-77, 2002; 5(2):74-79, 2003. (Cited on p. 168.)

[130] Bert W. Rust and Dianne P. O'Leary. Confidence intervals for discrete approximations to ill-posed problems. *Journal of Computational and Graphical Statistics*, 3:67–96, 1994. (Cited on p. 30.)

[131] Bert W. Rust and Dianne P. O'Leary. Residual periodograms for choosing regularization parameters for ill-posed problems. *Inverse Problems*, 24:034005, 2008. (Cited on p. 84.)

[132] Yousef Saad. *Iterative Methods for Sparse Linear Systems, Second Edition*. SIAM, Philadelphia, PA, 2003. (Cited on p. 334.)

[133] Hanan Samet. *Foundations of Multidimensional and Metric Data Structures*. Morgan Kaufmann, San Francisco, CA, 2005. (Cited on p. 38.)

[134] Edward R. Scheinerman. *Invitation to Dynamical Systems*. Prentice-Hall, Englewood Cliffs, NJ, 1996. `http://www.ams.jhu.edu/~ers/invite.html`. Retrieved October 2008. (Cited on p. 306.)

[135] Gerhart I. Schuëller, Helmut J. Pradlwarter, and P. S. Koutsourelakis. A comparative study of reliability estimation procedures for high dimensions. In *16th ASCE Engineering Mechanics Conference*. Seattle, Washington, 2003. `http://www.ce.washington.edu/em2003/proceedings/papers/777.pdf`. Retrieved October 2008. (Cited on p. 212.)

[136] A. P. S. Selvadurai. *Partial Differential Equations in Mechanics 2*. Springer, New York, 2000. (Cited on p. 339.)

[137] Richard I. Shrager and Richard W. Hendler. Some pitfalls in curve-fitting and how to avoid them: A case in point. *J. Biochem. Biophys. Methods*, 36:157–173, 1998. (Cited on p. 168.)

[138] Smithsonian National Air and Space Museum. How things fly: Roll, pitch, and yaw. `http://www.nasm.si.edu/exhibitions/gal109/NEWHTF/HTF541B.HTM`, 1996. Retrieved May 2008. (Cited on p. 162.)

[139] G. W. Stewart. *Matrix Algorithms, Volume 1: Basic Decompositions*. SIAM, Philadelphia, PA, 1998. (Cited on pp. 53, 69, 78, 79.)

[140] G. W. Stewart. *Matrix Algorithms, Volume 2: Eigensystems*. SIAM, Philadelphia, PA, 2001. (Cited on p. 334.)

[141] G. W. Stewart. An updating algorithm for subspace tracking. *IEEE Trans. Signal Proc.*, 40:1535–1541, 1992. (Cited on p. 103.)

[142] William J. Stewart. *Introduction to the Numerical Solution of Markov Chains.* Princeton University Press, 1994. (Cited on p. 219.)

[143] Volker Strassen. Gaussian elimination is not optimal. *Numer. Math.*, 13:354–356, 1969. (Cited on p. 37.)

[144] Paul N. Swarztrauber and Roland A. Sweet. Algorithm 541: Efficient Fortran subprograms for the solution of separable elliptic partial differential equations. *ACM Trans. Math. Softw.*, 5(3):352–364, 1979. (Cited on p. 346.)

[145] Julius T. Tou and Rafael C. Gonzales. *Pattern Recognition Principles.* Addison-Wesley Publishing Co., 1974. (Cited on p. 155.)

[146] Emanuele Trucco and Alessandro Verri. *Introductory Techniques for 3-D Computer Vision.* Prentice-Hall, Englewood Cliffs, NJ, 1998. (Cited on p. 162.)

[147] Sabine Van Huffel and Joos Vanderwalle. *The Total Least Squares Problem.* SIAM, Philadelphia, PA, 1991. (Cited on p. 174.)

[148] Charles F. Van Loan. *Introduction to Scientific Computing, Second Edition.* Prentice-Hall, Englewood Cliffs, NJ, 2000. (Cited on pp. xiii, xiv, 3, 47, 107, 107, 212, 257, 257, 283.)

[149] Kees Vuik. Some disasters caused by numerical errors. `http://ta.twi.tudelft.nl/nw/users/vuik/wi211/disasters.html`. Retrieved May 2008. (Cited on p. 14.)

[150] Layne T. Watson. Numerical linear algebra aspects of globally convergent homotopy methods. *SIAM Rev.*, 28(4):529–545, 1986. (Cited on p. 296.)

[151] Layne T. Watson. Probability-one homotropies in computational science, *J. Comp. Appl. Math.*, 140:785–807, 2002. (Cited on pp. 293, 296.)

[152] Layne T. Watson. Theory of globally convergent probability-one homotopies for nonlinear programming. *SIAM J. Optim.*, 11(3):761–780, 2000. (Cited on p. 296.)

[153] L. T. Watson, M. Sosonkina, R. C. Melville, A. P. Morgan, and H. F. Walker. Algorithm 777: Hompack90: A suite of Fortran 90 codes for globally convergent homotopy algorithms. *ACM Trans. Math. Softw.*, 23:514–549, 1997. (Cited on pp. 286, 296.)

[154] Eric Weisstein. Euler angles. `http://mathworld.wolfram.com/EulerAngles.html`. Retrieved May 2008. (Cited on p. 162.)

[155] Hai Zhuge. Exploring an epidemic in an e-science environment. *Communications of the ACM*, 48(9):109–114, 2005. (Cited on p. 219.)

Index